Spationomy

Vít Pászto • Carsten Jürgens
Polona Tominc • Jaroslav Burian
Editors

Spationomy

Spatial Exploration of Economic Data and Methods of Interdisciplinary Analytics

Springer Open

Editors
Vít Pászto
Department of Informatics and Applied Mathematics
Moravian Business College Olomouc
Olomouc, Czech Republic

Department of Geoinformatics
Palacký University Olomouc
Olomouc, Czech Republic

Carsten Jürgens
Geography, Geomatics Group
Ruhr-University Bochum
Bochum, Germany

Jaroslav Burian
Department of Geoinformatics
Palacký University Olomouc
Olomouc, Czech Republic

Polona Tominc
Faculty of Economics and Business
University of Maribor
Maribor, Slovenia

ISBN 978-3-030-26625-7 ISBN 978-3-030-26626-4 (eBook)
https://doi.org/10.1007/978-3-030-26626-4

This Springer imprint is published by the registered company Springer Nature Switzerland AG.
The registered company address is: Gewerbestrasse 11, 6330 Cham, Switzerland

Preface

In 2015, I was asked to prepare a project proposal for a new Erasmus+ KA2 Strategic Partnership call. The official documents describe the programme to fund “transnational projects designed to develop and share innovative practices and promote cooperation, peer learning, and exchanges of experiences in the fields of education, training, and youth”. Due to my professional training in geoinformatics (as we call “GIScience” in Czechia) from Palacký University Olomouc and my position at the time a Moravian Business College Olomouc, it was the first thing that crossed my mind to merge the two main fields taught at both institutions, i.e. geoinformatics and economy. Later on, the whole idea was growing, and the project proposal was summoning. The strategic partnership also included Ruhr-Universität Bochum (Germany) and the University of Maribor (Slovenia) as partners, and the project activities were designed to help meet the main goals – to share innovative practices and promote cooperation in education. A project titled “Spatial exploration of economic data – methods of interdisciplinary analytics” with the acronym “Spationomy” was submitted. Unsuccessfully. But only for the first time.

In 2016, we went through the whole cycle of a project preparation again, updating the document with fresh ideas and incorporating all the reviewers' suggestions. The project “Spationomy” was submitted again. Successfully. Starting with the kick-off meeting in October 2016, the story begins. The story full of project activities planning and designing, international and interdisciplinary co-operation, lecture preparations, scientific papers writing, simulation game framework elaboration, organisation of excursions and a lot of negotiations with local restaurants, pubs, accommodation facilities and so on. Simply said, the 3-year story of the hard work. A lot of efforts were sacrificed to enrol students into the project (28 students each year) – not because of the project (non)attractiveness but nobody knew nothing about “Spationomy”. The greatest challenge was, therefore, to convince economy students why they should learn something about geoinformatics; and vice versa. We, as project team staff members, were aware of the tremendous potential of the fusion of two seemingly distinct disciplines. But the project was mainly about the education of young people, so we needed to find ways to approach them with “Spationomy”. We succeeded, and every year, the number of students' applications exceeded the available places for them. On behalf of the team, I have to say that we were lucky of the students who participated –

great young spirits eager to learn something new. I want to express my gratitude and thank all the students for their involvement. Thank you.

Why was there such a buzz around the project? What was so special? Well, it must be judged by someone else. But if I may add my perspective, the combination of all activities covering a very appealing mix of lectures, workshops, events, physical gatherings and virtual meetings that set up a unique educational environment and knowledge sharing platform. It was appreciated by all involved parties – students, staff/academics, practitioners and also DG Education and Culture of the European Commission and Czech Erasmus+ National Agency, which selected "Spationomy" to be further evaluated "from outside" in a case study on the impact of Erasmus+ Strategic Partnerships. The case study results were very positive. More information about the project activities and outcomes are on the project website (www.spationomy.mvso.cz) or the official Erasmsus+ websites.

As it is usual in this kind of projects, there must be measures to evaluate the project quantitatively. In the case of "Spationomy", these major achievements were labelled as intellectual outputs. One of the intellectual outputs was called Spationomy methodology and should represent the main pedagogical/curricular material of the project. It was planned to be in the form of a textbook to guide a reader from data sources, through basic principles of all involved disciplines, to their practical applications and simulation game. In the project team, we had a long discussion about the form of the textbook. We finally decided to write "a proper" book under the Springer publishing house. That is the reason why you hold this book and can read all the topics that we promised to write about. During my studies, I was told that the introduction part or a preface of a longer text should be composed at the very end of the writing process. I never followed this rule. Until now. That is why, after all, I feel an urgent need to stress out how much demanding and time-consuming was the book preparation. It is the result of all editors' and authors' hard work besides their ordinary duties at their institutions. Most of the authors are members (I use present tense by intention) of the "Spationomy" team, and it has been my pleasure to work with them for 3 years. Therefore, I take this opportunity to thank all of them for their contributions to this book and all the hard work they gave to the "Spationomy" project. Thank you so much.

It is also my pleasant duty to gratefully acknowledge the support by the Erasmus+ project "Spationomy" (no. 2016-1-CZ01-KA203-024040) funded by the European Union. Without this support, this book would never be alive.

Finally and most importantly, I want to thank my family and closest friends for all the patience they had during my work on this book and the whole project. Thank you.

Enjoy reading!

Olomouc, Czech Republic Vít Pászto

Contents

Part I

Methodological Overview

Data Sources

1

Vít Pászto, Andreas Redecker, Karel Macků, Carsten Jürgens, and Nicolai Moos

Abstract

This chapter is devoted to the overview of the data fundamentals as regards data models and sources accompanied by geomatics, remote sensing, and economy. Description of such data sources is complemented with the basics from respective disciplines to provide a thematic context to the reader. The chapter starts with a summary of the most commonly used data models, starting with tabular and attribute formats. It is then followed by the spatial data models, including vector and raster data core principles. Since the geospatial domain is heterogeneous in terms of different data formats, the list of interoperability data sources and services is provided. Emphasis is also given to the international and selected national data sources, both non-spatial and spatial. This part is mainly covering the economic (socio-demographic) topics. At last, a remote sensing perspective on data sources is introduced, pointing out the most important Earth observation data. The whole chapter focuses on the major data models and sources, so it serves as a gateway to further exploration of existing data storages.

Keywords

Data models · Formats · Data sources · Data portals · Satellite archives

V. Pászto (✉)
Department of Informatics and Applied Mathematics, Moravian Business College Olomouc, Olomouc, Czech Republic

Department of Geoinformatics, Palacký University Olomouc, Olomouc, Czech Republic
e-mail: vit.paszto@gmail.com

A. Redecker · C. Jürgens · N. Moos
Geography, Geomatics Group, Ruhr-University Bochum, Bochum, Germany
e-mail: andreas.redecker@rub.de; carsten.juergens@rub.de; nicolai.moos@rub.de

K. Macků
Department of Geoinformatics, Palacký University Olomouc, Olomouc, Czech Republic
e-mail: karel.macku@upol.cz

1.1 Data Models

1.1.1 Basic Tabular and Attribute Data Formats (by Vít Pászto)

In this section, the most used data formats will be briefly introduced. Some of the data providers offer several options regarding data formats. Therefore, it is liable to mention the main characteristics of such formats.

1.1.1.1 TXT

This is the most common data format using plain text. The text could be supplemented by the special symbols for row endings, blank spaces, and tabulators. The suffix for this data format is .txt. Since the format is mainly plain text (with very

V. Pászto et al. (eds.), *Spationomy*, https://doi.org/10.1007/978-3-030-26626-4_1

limited options for formatting), it is possible to open the .txt file in most of the software and even with the simple text editors (like Wordpad). Thus, the greatest advantage of this format is its interoperability.

1.1.1.2 CSV

Comma Separated Value (CSV) is a simple and standardised format for data storage. Individual records are separated by comma (in some case by a semicolon, blank space, or another tabulator) and the format is classified as a delimiter-separated format family. Most of the tabular software is capable of working with CSV. The format is interoperable, interchangeable and in most cases in the form of a plain text (storing both text and numbers).

1.1.1.3 XLS/XLSX

Files with .XLS/XLSX extension are formats of Microsoft Office package, namely with Excel, and is one of the most used and widespread format. The data is stored in tables, which are organized in spreadsheet and sheets. Label XLS/XLSX is basically "only" a suffix for an Open Office XML scheme (OOXML). The format is binary (i.e. needs specialised software/plugins to be opened), while XLSX represents zipped XML file, and was introduced by Microsoft in 2007. Data stored in XLS could be still opened in the newer Excel version. Thus backward compatibility is secured.

1.1.1.4 XML

This abbreviation stands for eXtensible Markup Language (XML) using structured markup language with constructs such as tags, elements, and attributes. Since its introduction in 1996, XML has become a basis for many other formats (e.g. XHTML, SVG, KML, Microsoft Office, OpenOffice and others). XML files are used mainly for data exchange due to its simplicity, openness, and platform independence. Moreover, the format is machine-readable, easily and quickly searchable and convertible to other formats. Most of the metadata is stored in XML files.

1.1.2 Spatial Data Models (by Andreas Redecker)

Performing scientific analysis implies the use of data and systems that can process this data to gain new insights into the characteristics and interdependencies of research objects. Taking advantage of the information on where an object is located and how it is delimited leads to the field of spatial analysis. It implies the use of a Geographic Information System (GIS) that can process a special type of data referred to as spatial data, geospatial data, geographic data or just geodata.

The understanding of the meaning of GIS varies from "just an application" to "a system of hardware, software and geodata". The latter refers to the fact that besides a particular program, also the data used has to be suitable for spatial analysis. This can also apply to the requirements for the hardware, depending on the kind and size of the geodata utilised. Considering the available data and the aims of a spatial analysis, it can be necessary to use different software products to apply the appropriate methods to the geodata.

Depending on the source and purpose of the geodata there are two completely different models to represent objects from the real world: Raster data and vector data.

1.1.2.1 Raster Data

Raster geodata represents an area in the real world by an array of square cells with a certain edge length referred to as resolution in ground units (mostly meters). It is spatially referenced to the real-world space by the coordinate of the centre of the upper-left cell and – if necessary – rotation angles for orientation (Fig. 1.1).

For each raster cell value can be stored that represents the characteristic of the represented object within the area of the cell. These values can be of different numeric types like integer or float to represent the desired properties of the object such as height, temperature, brightness etc. or to document codes for classes of land use as a result from a classification process. Different rasters with the same geometric properties can be superimposed to constitute a layer stack like a common colour image that consists of three layers, each of them representing one colour of red, green and blue.

To save raster models digitally many different data formats are available as with imagery from any kind of digital camera. These can be complemented with geospatial parameters in additional files of the same name but with a different suffix.

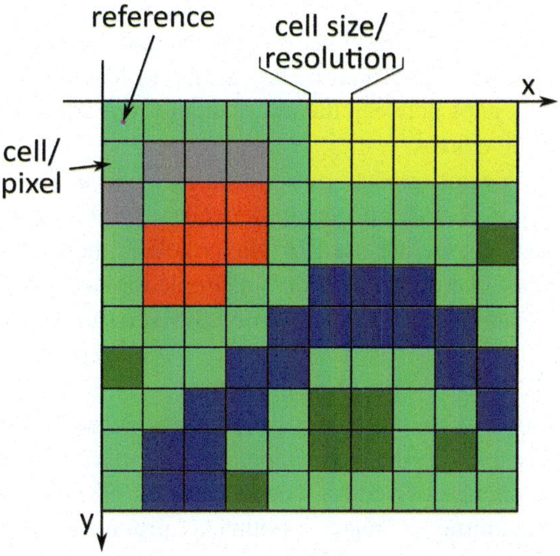

Fig. 1.1 Schematic example of a raster geometry in an image coordinate system. (Source: Author)

Special geodata related file types for raster models are holding the geospatially relevant information within the header of the file (Table 1.1).

All of these file types support compression techniques to reduce the amount of data that has to be stored on storage systems like SSDs or HDDs. Some of the codecs (coder/decoder) that are used to compress and uncompress the data are not able to completely recover the original condition of a raster and are referred to as lossy codecs. For imagery that only has to be viewed visually, these might suffice. But for most geospatial analyses of raster data, the use of lossless compression codecs is vital.

1.1.2.2 Vector Data

Vector Data – also referred to as feature data – represent individual objects (features) of the real world. These are modelled as geometries at a certain location holding attributes about their specific properties. A collection of similar features with a same set of properties forms a feature class.

Depending on the geometric dimension of the objects modelled, a feature class consists of points, lines or polygons.

- Points are defined by x-, y- and – if desired – z-coordinates describing the location of an object. They do not have an extent.

Table 1.1 Common raster-formats, extensions and codecs for geodata (Source: author)

	File extension	Spatial reference in an external file	Spatial reference stored internally	Available codecs for compression	quality
Tagged Image File Format (TIFF)	.tif	.tfw		LZW	lossless
Geo-TIFF	.tif		x	LZW	lossless
Graphics Interchange Format (GIF)	.gif	.gfw		LZW	lossless
Portable Network Graphics (PNG)	.png	.pgw		DEFLATE	lossless
Joint Photographic Experts Group (JPEG File Interchange Format) (JPEG/JFIF)	.jpg	.jgw		JPEG	lossy
JPG2000	.jp2		x	JPEG 2000 (wavelet)	lossless/ lossy
Image (IMG)	.img		x	DR-RLE	lossless
Enhanced Compressed Wavelet (ECW)	.ecw		x	wavelet	lossy
MrSID	.sid		x	wavelet	lossless

- Lines are represented by a series of connected points (vertices). They describe a pathway with a direction and a certain length, but also no expanse.
- Polygons are described by a line with a joint start- and endpoint. They describe an area with a certain acreage and a perimeter accordingly.

Every feature class consists of the mentioned geometries and an attribute table connected to these. Each row (a record) in the attribute table – together with the corresponding geometry – represents one object or feature respectively. With so-called multipart features, several geometries can make up one object connected to one record in the attribute table (Fig. 1.2).

1.1.2.3 Tabular Data

In addition to geodata with a direct spatial relation (expressed by the coordinates of points or vertices), other data with only indirect spatial relation can easily be incorporated in GIS-analyses. Here indirect spatial relation refers to an attribute that can be linked to a feature class holding the same information in its attribute table. Indirect spatial relation can be realised by administrative codes, ids, addresses etc.

The simplest file-type to hold this kind of data is a text file with separated values. Here a special character is used to delimit the columns within each line of the file.

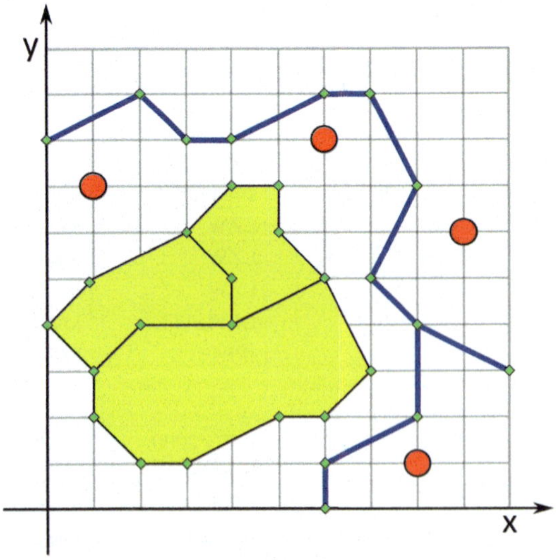

Fig. 1.2 Schematic example of the three different feature geometries (point, line, polygon). (Source: Author)

If tabular data directly contains columns holding coordinates of a known spatial reference system, in many GIS, it can be directly transformed to vector-geodata (point features).

1.1.2.4 Topology

A special characteristic of some vector-geodata models is the ability to deal with topology. Meaning the GIS verifies the compliance with predefined geometric relations between features in certain feature classes. For example, there is a rule that there shall be no overlap of features nor any gaps between the representations of administrative areas.

Vector-geodata gets stored in many different ways. These are mainly dependent on the application they are used in. Nevertheless, there is at least one quite common but simple format, which is supported by almost every GIS system.

1.1.2.5 The Shape-Format

Initially, the Shape-format was introduced by the company ESRI as a simple data structure for the exchange of vector geodata (ESRI 1998). In the meantime, many other providers of GIS-applications have adopted it to provide a simple interface for the import and export of geodata or to provide a modest data structure for small projects. The Shape-format does not support topology. Each feature class can only hold features of one geometric type. Information on the spatial reference system for the coordinates used in a dataset is not obligatory but at least possible. Many providers of geodata utilise this format to provide data product-independently. A shape-feature class consists at least of three obligatory files with the same name but with different suffixes:

- ∗.shp: the main file, holding geometries
- ∗.dbf: attribute table in dBase-format
- ∗.shx: index file for the link between geometries and attributes

Additional information get stored in further optional files like:

- *.sbn/*.sbx: spatial index (generated automatically)
- *.prj: spatial reference system
- *.shp.xml: metadata

1.1.2.6 Geodatabases

For the efficient use of geodata in (larger) projects, almost every GIS supports some kind of geodatabase. Geodatabases are database management systems (DBMS), which support the handling of spatial data. Some of them even offer functions for geospatial analysis directly within the DBMS. Most geodatabases can store raster geodata as well as vector data. Furthermore, they provide features to organise data like in folder structures and take care of spatial reference systems and topologies.

1.1.2.7 Spatial Reference Systems (SRS)

The spatial reference of geodata consists of coordinates that are related to the earth's surface by some kind of coordinate system. For this, a mathematical model of the earth's shape is required, which the coordinate system can be linked to. Usually, according to the earth's form, this model is a "flattened" (oblate) ellipsoid (of revolution), mostly defined by the parameters of its semi-major axis and its inverse flattening. Sometimes an additional gravitational model is applied to account for divergences between the ellipsoid and the geoid – the earth's real appearance (Snyder 1987).

The geodetic datum describes the linkage between the geoid and the idealised shape of the ellipsoid. It consists of the ellipsoids parameters and those for its orientation related to a known precisely measured point or a network of precisely measured locations on the earth's surface.

The internationally most common datum for geodata is the World Geodetic System 1984 (WGS84). In Europe, the European Terrestrial Reference System 1989 (ETRS89) defines the reference for coordinates of current geodata. It is based on the Geodetic Reference System 1980 (GRS80) that consists of a reference ellipsoid and a gravity field model like the WGS84.

To locate positions on the earth's surface by coordinates geographical or projected coordinate systems are applied to the modelled surface (Fig. 1.3).

1.1.2.8 Geographical Coordinate Systems

Geographical coordinates relate to a grid composed of vertical and horizontal circles around the earth – the so-called parallels and meridians – as a base for coordinates measured in degrees referred to as latitude and longitude.

Latitude describes a location's distance to the equator measured parallel to the earth's axis. The longitude measures its distance parallel to the equator related to the base meridian, mostly defined by the meridian that crosses the location of the Royal Observatory in Greenwich.

1.1.2.9 Projected Coordinate Systems

For easier reading of maps and plans and less complex computing of distances and areas, projected coordinate systems provide a flat rectangular grid (Cartesian coordinate system) as a reference for measurements in metric units.

For Europe, the Universal Transversal Mercator (UTM) Projection (ETRS89-TMzn, EPSG-Code 3038-3051) is the official reference system for conformal pan-European mapping with scales

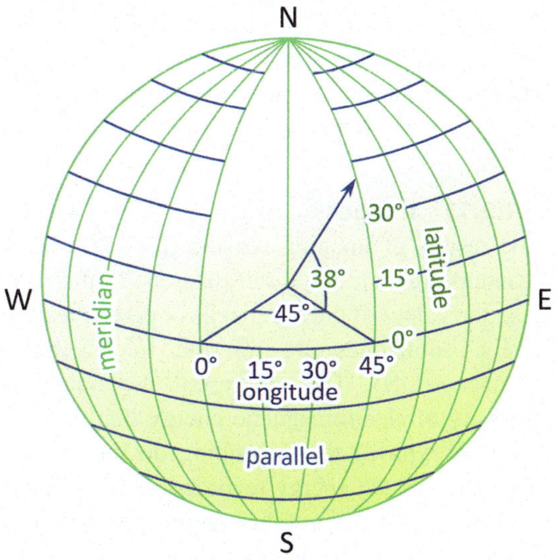

Fig. 1.3 Illustration of a geographical coordinate system. (Source: Author)

larger than 1:500,000. Less detailed maps are recommended to be drawn using Lambert conformal conic (ETRS89-LCC, EPSG-Code 3034) for conformal pan-European mapping at scales smaller or equal to 1:500,000 or using Lambert Azimuthal Equal Area Projection (ETRS89-LAEA, EPSG-Code 3035) for true area spatial representations in pan-European spatial analysis and reporting. All three of them are linked to the geoid – represented by the GRS80 – through the ETRS89 (European Commission 2014a).

The UTM-System covers Zones of 6° width by superimposing the so-called prime meridian of the zone with the vertical line at x = 500,000 m of the coordinate system. This practice of so-called false easting avoids calculations with negative values west of the prime meridian within a zone. The counting of zones starts at the International Date Line with the first prime meridian at 177° west of Greenwich. Hence zone 32 covers the zone three degrees west and east about the meridian 9° east of Greenwich.

Y-coordinates refer to the zero-latitude thus representing a location's absolute distance to the equator in meters.

1.1.2.10 Application of Geodata Models and Formats

Besides the technical properties of geodata, they can also be distinguished by their content. Typical fields of applications for raster data are for example, imagery, height models, land use classes, population data, atmospheric parameters like temperature, precipitation etc.

1.1.2.11 Imagery

The results of imaging sensors like cameras or scanners are stored in raster datasets. In this context, the values of the raster cells or pixels, respectively, sometimes are referred to as digital numbers (DN). They represent the quantised intensity of electromagnetic energy that the sensor was exposed to. Depending on the amount and range of the energy recorded, they are positive integer numbers of different bit depths defining the number of gradations between the lowest and the highest signal value. This defines the radiometric resolution expressed in bits of binary numbers. Standard bit depths are 8 bits representing 256 levels for consumer cameras and up to 16 bits representing 65,536 levels used with professional sensors.

Further aspects of digital imagery are explained in Sect. 1.5.

1.1.2.12 Digital Elevation Models

There are different kinds of models representing continuous surfaces. These digital elevation models (DEM) are differentiated as:

- DSM: Digital surface model, describing the height of the earth's surface, including all objects in the landscape.
- DTM: Digital terrain model, representing the terrain without vegetation or human-made objects.
- DHM: Digital height model also referred to as normalised DSM (nDSM) having the heights of all objects on the bare DTM (resulting from the calculation DSM-DTM).

The values of DEMs are usually of some floating-point data type to allow negative values as well as decimal numbers.

The raster-model is very common to represent this kind of geodata, but there is also a special vector data model for surfaces. Triangulated irregular networks (TIN) express surfaces by triangular areas resulting from a network formed by lines connecting mass-points of known heights.

1.1.2.13 Network Datasets

Another special vector based model for geodata is a network dataset. It is a collection of different vector feature classes and tables containing the all necessary information for performing network analyses: The network itself (holding attributes for the impedance of the edges), possible turns, barriers etc. Further information on this kind of geodata can be found in Part I, Sect. 3.3 in Chap. 3.

1.1.3 Geodata Interoperability (by Andreas Redecker)

For the exchange of geodata, it is vital to have data structures and methods that follow standardised rules. With these providers can advertise the properties of their data in a mutually intelligible form to potential users on the one hand. On the other hand, agreed formats and data structures allow the exchange of the data between different systems that internally might operate with individual, i.e. proprietary data models.

Many geodata is highly dynamic, and the exchange of that information can be very time-dependent. Therefore besides the exchange of files geodata more often are provided as services. That means that a user can directly use a provider's data by accessing it via a network. After receiving a standardised request, the provider's system will transfer the desired information to the user in a standardised format. This can be metadata about the data provided as well as the desired data itself.

Besides proprietary protocols standardised request and transfer methods are commonly used especially within public infrastructures. The central organisation that defines most of the standards to describe and transfer geodata is the Open Geospatial Consortium (OGC, http://www.opengeospatial.org/).

For geodata services, special formats support the delivery of spatially or thematically limited extracts of a provided dataset. Some of them even support the streaming of the data to be able to transfer large amounts, especially with raster data.

The most important standards that allow real-time access to (distributed) geodata over the internet are the OGC standards WCS, WFS and WMS.

1.1.3.1 WFS

A Web Feature Service allows interacting with geodata in a geodatabase on the level of single features (vector data). It supports request for:

- metadata about the service (in XML-format)
- description of datasets (in XML-format)
- delivery of feature data (geometry and attributes in GML-format)
- manipulation of the features (edit, create, delete, lock)

1.1.3.2 WCS

A Web Coverage Service provides access to raster data. Depending on its configuration level it offers services for:

- metadata about the service (in XML-format)
- description of certain datasets (in XML-format)
- delivery of raster data (raster formats)
- complex requests
- data processing
- data manipulation

1.1.3.3 WMS

The Web Mapping Service standard allows requesting geodata by stating the extent and choice of layers or requesting attribute information for single objects from a geodata service supporting this standard (for raster and vector data). Depending on the request it returns:

- metadata about the service (in XML-format)
- a raster image with a map (in a common raster format)
- attribute information (in XML-format)

Whereas WCS and WFS are designed to deliver data for further processing, WMS is intended to provide maps for display (Fig. 1.4).

1.1.3.4 GML

The XML-based Geography Markup Language was defined by the OGC as a universal format for the storage and transfer of geodata. Besides feature (vector) data, it can also be used to represent coverages (raster) and sensor data.

1.1.3.5 WKT/WKB

The markup language Well Known Text is used to describe vector-geodata in a human-readable, easy transferable way. It is supported by many applications that comply with OGC standards. Its

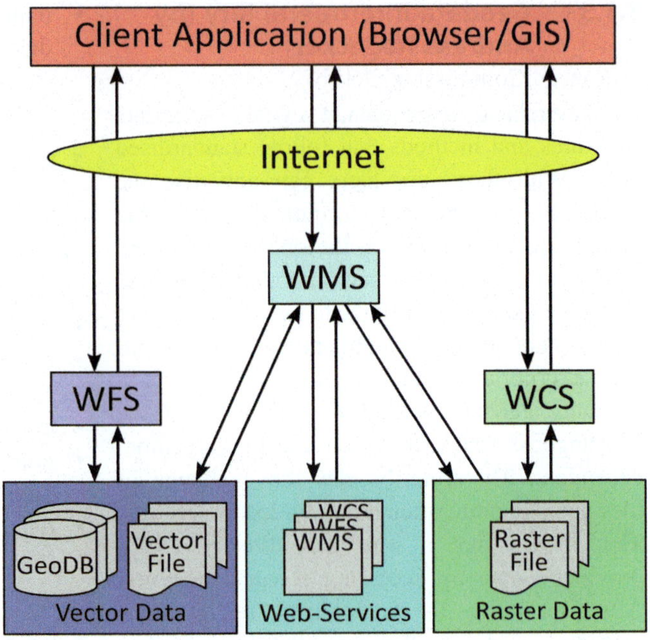

Fig. 1.4 Overview scheme of OGC-web-services. (Source: Author)

binary counterpart Well Known Binary is used to handle geospatial data within databases.

1.1.3.6 KML/KMZ

The Keyhole Markup Language is an XML-based format for the transfer of 2D and 3D geodata within internet-based applications like maps and earth browsers. KMZ files contain zip-compressed KML content. Initially developed for the use in Google Earth it became an OGC standard later on.

1.1.3.7 GPX

For the exchange of records from GPS-receivers, the GPS Exchange Format was developed by the company TopoGraphix. It represents waypoints, routes and tracks as coordinates with attributes in an open XML scheme. It can be handled by many applications.

1.1.4 Metadata (by Andreas Redecker)

Information about the characteristics of geodata and geodata services is important for the reliability of most analyses. General descriptions about the objects held in the geodata as well as information about the spatial reference, resolution, attributes, geometric accuracy, origin, copyright and many other aspects make up the so-called metadata. Usually, it is held in a special .xml-file delivered with the data itself. International standards for the description of geographical information are defined by the ISO (International Organization for Standardization):

- ISO 19115:2003 Geographic information – Metadata
  It defines the schema required for describing geographic information and services. It provides information about the identification, the extent, the quality, the spatial and temporal schema, spatial reference, and distribution of digital geographic data. (ISO 2018a)
- ISO/TS 19139:2007 Geographic information -- Metadata -- XML schema implementation
  It defines Geographic MetaData XML (gmd) encoding, an XML Schema implementation derived from ISO 19115. (ISO 2018b).

Standardised metadata are the key to the Infrastructure for Spatial Information in the European

Community (INSPIRE) that is aimed to easily share and use spatial data within the EU (European Commission 2014b).

1.2 International Data Sources (by Vít Pászto, Karel Macků, Andreas Redecker, and Nicolai Moos)

1.2.1 Eurostat

Eurostat represents the main official statistical body of the European Union with its headquarters in Luxembourg. The main task of Eurostat is to provide high-quality statistics about and for Europe (Eurostat 2018a). Thanks to these statistics, we can compare individual countries and/or various regions in a comprehensive way based on factual information. Most of the data that Eurostat collects comes from national statistical offices, which are obliged to report selected statistical indicators to Eurostat. In this sense, Eurostat serves as a common European statistical office for all member countries. For more information about Eurostat mission, goals and history, please, go to the official website – https://ec.europa.eu/eurostat.

1.2.1.1 Eurostat Spatial Data

The main body collecting spatial data and information within Eurostat is called Geographic Information System of the COmmission (GISCO). This unit is responsible for maintaining the geographical databases, creating and publishing maps and map applications. Besides the data management, GISCO also cooperates with other Eurostat units and publishes research texts on various topics (e.g. Rural-urban typology, Urban Europe etc.). GISCO also leads their own activities, such as GEOSTAT initiative and Merging statistics and geospatial information in the European statistical system. More details on GISCO activities and data is available at – https://ec.europa.eu/eurostat/web/gisco.

Talking about datasets, GISCO provides reference geodatasets (geographically covering EU) in five main themes:

- Administrative/Statistical Units – this section contains geodata about administrative hierarchical units NUTS (2003–2016), Urban audit data (2001–2014), countries-level units (2006–2016), census geodata (2011) and communes (LAU2 units, 2006–2013). Most of the datasets are provided in Esri Geodatabase format, Shapefile, and some other data models.
- Population Distribution/Demography – the section includes three main projects, namely GEOSTAT population grids (2006 and 2011), Urban Clusters (2006 and 2011), and DEGURBA (Degree of Urbanisation 2001 and 2014). Except for Urban clusters, which is provided as a raster (TIFF, geoTIFF), all datasets are provided in vector format (Esri Geodatabase, Shapefile).
- Transport Networks – this section contains two major datasets – airports (2006 and 2013) and ports (2009 and 2013). Both geodata sources are provided in Esri Geodatabase and Shapefile.
- Land cover – as indicated by the name of this section, it includes data on Land Use/Cover frame Statistical Survey (LUCAS) with the reference year 2009. Again, geodata is prepared in Esri Geodatabase and Shapefile. Besides, there are links to related datasets of Corine Land Cover (CLC) and Urban Morphological Zones (UMZ) provided by the European Environmental Agency (EEA, now being translated to Copernicus programme).
- Elevation – in this part, the main focus is on geodata referring about digital elevation model (DEM) and its derived products (e.g. slope, aspect, coloured relief). This section contains Digital Elevation Model data (in two different coordinates systems), data on Aspect, Slope, Coloured relief, Hillshade, and Hydrography. All the datasets are available in raster format (GeoTIFF).

1.2.1.2 Eurostat Statistical Data

As a counterpart to the spatial part of Eurostat data, there is a statistical part containing great number of tabular data with the possibility to link them together with spatial data. On the homepage of Eurostat, the first option to search for a data is a tab "Data", which redirects the user straight to available databases (https://ec.europa.eu/eurostat/data/database). There exist several options on how to search for a data using Data navigation tree:

- Database by theme
- Tables by theme
- Tables by EU policy
- Cross-cutting topics
- New items (sorted by code)
- Recently updated items (sorted by code)

Besides Data navigation tree, the user can perform search "database by the theme" also via context menu; this option brings additional links to respective EU policy indicators. In the context menu on the Eurostat website, it is possible to browse the database by alphabet order (Statistics A-Z).

Also, there are special data products and services available at the Eurostat webpage – Population Census 2011, Experimental Statistics, Bulk Download, Web Services, Microdata, Metadata and Data validation service. Although this dataset provides valuable information on specific topics or using specific (technical) approaches, only the main database will be further explored.

Searching and Downloading Data from Eurostat Main Database
Using any means of data search, it will bring the user to the list of main tables or databases, in which the specific topics are listed. It is worth to note that the Eurostat database contains hundreds of tables in various topics. Therefore, it is not possible to list them all in this book. In all cases, the tables are logically organised, and it is very intuitive to download data. In Fig. 1.5, there is an example of an expanded Data navigation tree with individual tables on the Economy and finance theme (main GDP aggregates in particular).

Basic information about the selected indicator is available by clicking blue "i" icon, yellow

Fig. 1.5 Expanded Data decision tree on the Eurostat data website. (Source: (c) European Union 1995–2018)

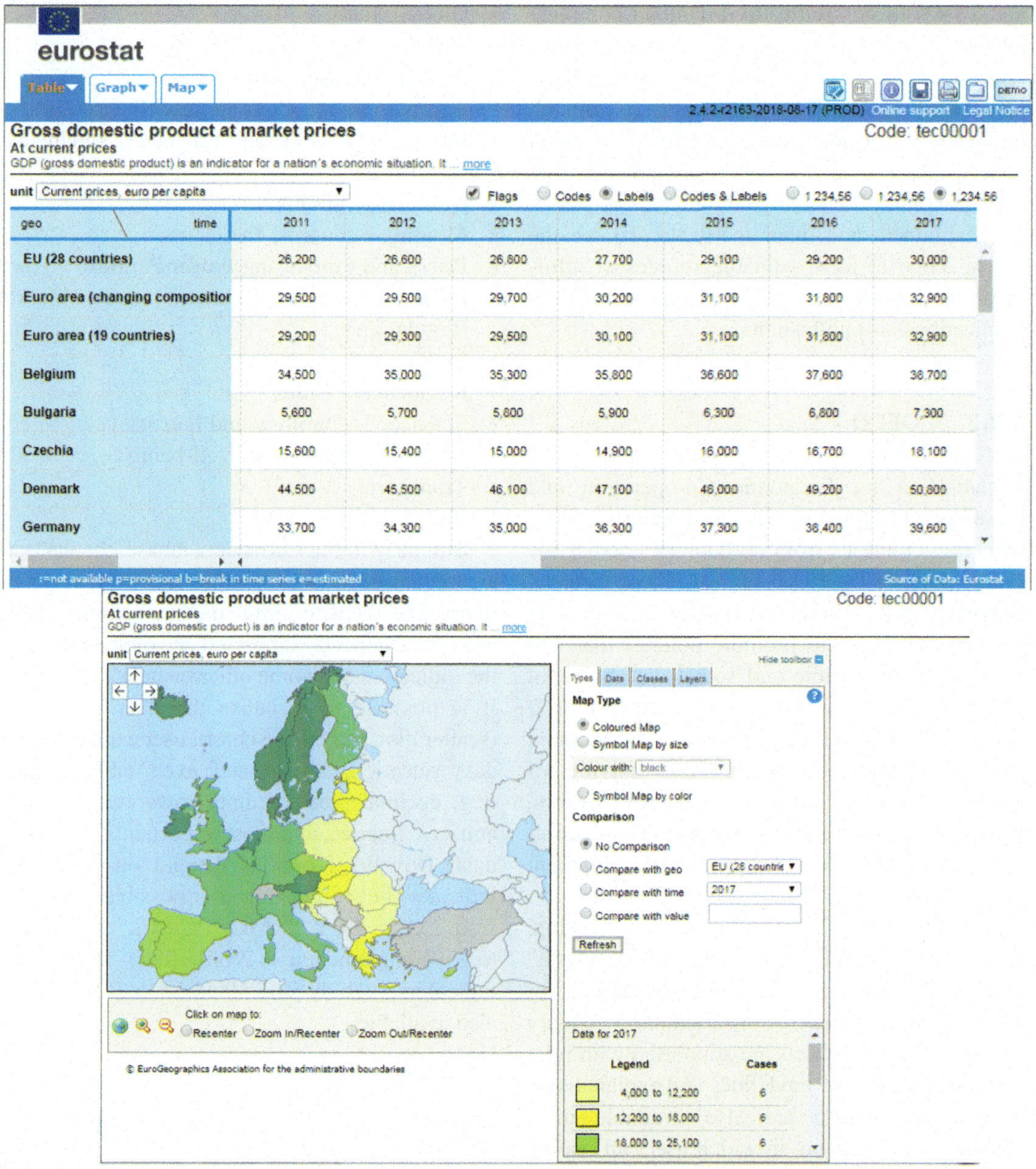

Fig. 1.6 Interactive interface for data selection, customisation and download on the Eurostat data website. (Source: (c) European Union, 1995–2018)

"zip" icons stand for downloading a data in a TSV format, and the first "marker" icon takes the user to the interactive interface with graphs, table, and map (if available). In this interactive environment (Fig. 1.6), it is possible to customise selection in tables, change the visual representation (graph or map) and download data selected data (by clicking "floppy disk" icon). Once submitted to download, a several data formats will be available to choose – XLS with or without footnotes (with and without short descriptions), HTML (with and without short descriptions), XML, PDF (with and without short descriptions), and TSV as a possibility to download complete table.

Besides the datasets listed above, it is important to note that Eurostat produces research publications, manuals and guidelines, working papers, yearbooks, brochures and leaflets, methodologies, books, digital publication, Statistics Explained, and other materials. To see the whole scope of these valuable sources of information, check https://ec.europa.eu/eurostat/publications/all-publications.

1.2.2 OECD

Organisation for Economic Co-operation and Development (OECD) is international body gathering countries from 36 countries across the globe (most of the EU countries, USA, Canada, Mexico, Chile, Turkey and others). The main goal of OECD is "to promote policies that will improve the economic and social well-being of people around the world" (OECD 2018). OECD creates a platform for international cooperation, sharing experiences and problem solving of social, economic, and environmental topics. Besides policy issues, OECD analyses and compare data with a focus on future development predictions.

To support policy and decision making, OECD runs special data portal "OECD Data" (https://data.oecd.org), where one can search for a data using hypertext, or browse data by county or topic. All the datasets are supported with relevant methodological guidelines and explanations. The statistical data are also available in a standalone application, in which user can search, filter, customise, visualise and download statistical data covering various themes:

- General Statistics
- Agriculture and Fisheries
- Demography and Population
- Development
- Economic Projections
- Education and Training
- Environment
- Finance
- Globalisation
- Health
- Industry and Services
- Information and Communication Technology
- International Trade and Balance of Payments
- Labour
- National Accounts
- Monthly Economic Indicators
- Prices and Purchasing Power Parities
- Public Sector, Taxation and Market Regulation
- Productivity
- Regions and Cities
- Science, Technology and Patents
- Social Protection and Well-being
- Transport

Similarly to the Eurostat database, the datasets are organised into expanding tree-system on an interactive website. After data selection, a table with data appears supported by explanations of the indicators and some other metadata (Fig. 1.7). It is possible to visualise the data as a chart (scatter plot, bar or line chart), user can customise data selection, layout and even table options (e.g. decimal places, empty rows etc.), manage and save queries, and most importantly download data. Download options depend on a selected indicator, but in general; it is possible to choose from XLS, CSV, XML, PC-axis, and others (e.g. complementary Word files). For some indicators, a bulk download as a RAR file is also available.

1.2.3 UN

As the United Nations (UN) is a well-known institution, just a brief note about its mission is to be mentioned. The UN is the international and intergovernmental organisation established in 1945, and its goal is constituted in the Charter of the United Nations, most importantly to protect human rights, freedoms, and a wide range of basic societal principles (e.g. healthcare, social equality and many others).

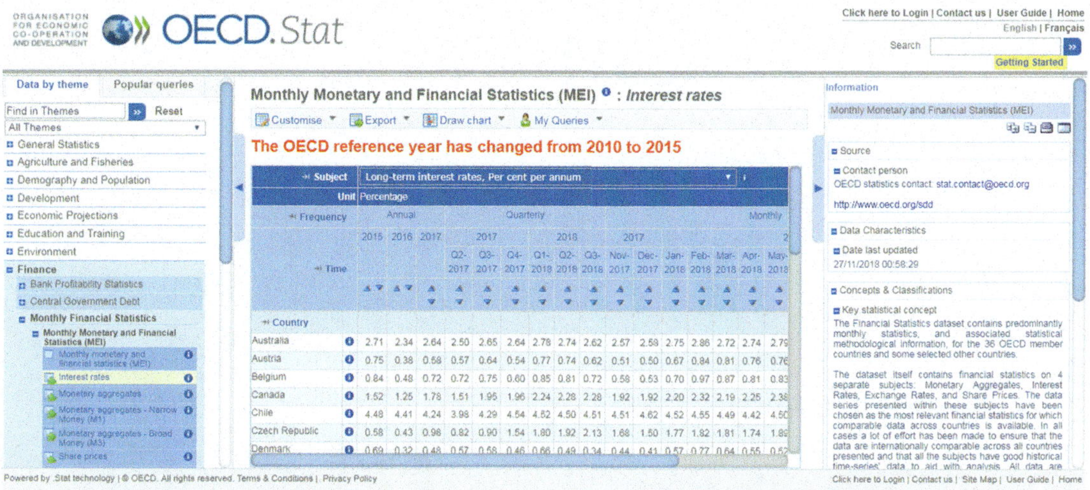

Fig. 1.7 Interface of OECD database after a selection. (Screenshot from OECD.Stat webpage)

The main body within the UN responsible for statistical data dissemination is its Statistical Division (UNSD). Statistical Division coordinates activities of international, national and other statistical organisations. Its primary focus is on data collection, processing and dissemination, methodology standardisation, and capacity development (UNSD 2018b). Thematically, UNSD covers topic such as development indicators (mainly Sustainable Development Goals – SDG), economy, environment, geospatial information, population and society. From a dataset perspective, UNSD lists the following main sources:

- UN Data
- Open SDG Data Hub
- Monthly Bulletin of Statistics Online
- SDG Indicators Global Database
- UN Comtrade
- UN ServiceTrade
- National Accounts Main Aggregates
- Joint Organisations Data Initiative
- Disability Statistics Database
- Population Census Datasets
- Population and Vital Statistics Report
- Demographic Yearbook System
- Minimum Set of Gender Indicators

The main data search tool provided by the UNSD is "UNdata" (http://data.un.org/), which serves as a primary search engine aggregates all the available UN data. According to UNSD (2018a), there are 32 databases with around 60 million records available. On that basis, only most general features of the UNdata portal will be mentioned. First, the user can search for data via full the text-search. The other option is to choose "Datamarts" feature, which brings the user to a data tree interface with the categorised dataset. In this environment, one can apply searching filters, select data (columns), order records, transpose rows and columns, share tables and also download data. When downloading, the user can choose between two main formats – XML and CSV. Moreover, detailed data description, including metadata, is available here.

It is also possible to look up for datasets that will be published by using the "Update Calendar" option. These options are complemented with a glossary and API (Application Programming Interface) helping users to understand indicators and/or use the data in their applications. Links to other UNSD specialised statistical databases as well as "popular statistics" are provided at the main webpage (Fig. 1.8).

Fig. 1.8 UNdata main page with full text search, links to other UNSD databases and a popular search. (Screenshot from UNdata webpage, Copyright © 2018 UNSD)

1.2.4 WTO

World Trade Organisation (WTO) refers to itself as the only global international organisation dealing with the rules of trade between nations (WTO 2018). The main goal of WTO is to participate in trade negotiations and to help conclude trade contracts. WTO offers rich information sources, including the legal text of WTO agreements, economic analysis, publications, glossaries and terminology database, and statistical data. Main statistical databases are organised into thematic groups:

- Tariffs
- Trade in Services
- Global value chains
- Merchandise trade
- Non-tariff measures
- Trade in services
- Trade and tariff maps

A fundamental instrument for statistical data access is WTO Data portal (http://data.wto.org/). In the interface of Data portal search, it is possible to choose from more than 200 indicators, around 300 reporting economies (country profiles), about 200 products/sectors, up to 300 partner economies, and time series with more than 70 years of history. The user can also filter the data based on topic, product classification, trade partner, and frequency. Moreover, user can exchange selector rows for columns (and vice versa) by dragging&dropping respective items and apply for changes in a resulting data table (Fig. 1.9). Once the selection is done, it is possible to download selected data and table composition as an XLS file and/or CSV. There is also an option to look into metadata with detailed information about the selected data. The user can also display the whole database inventory, where all the available indicators are listed and described.

1.2.5 World Bank

The World Bank was established in 1944 originally to offer low-interest loans for countries affected by World War II. Since then, The World Bank has grown into an organisation with 189 member countries. In general, The World Bank is a vital source of financial and technical assistance to developing countries around the world (World Bank Group 2018). As regards data sources, The World Bank offers the World Bank Open Data (https://data.worldbank.org/) portal as the main proxy for various information sources. The main tool for a data search is a full-text window with two browsing options – by country and indicator. Both options take the user to a list of countries and indicators. When the searched indicator is chosen, the interactive tool will appear (Fig. 1.10) and the user can select the display method (line or bar chart, map), linked indicators, time span, check metadata, visit another data&visualisation tool (e.g. DataBank) and download the data in CSV, XML or XLS.

Besides the main search interactive tool, The World Bank Open Data portal provides links to other data resources:

- Open Data Catalog
- DataBank
- Microdata Library
- World Development Indicators

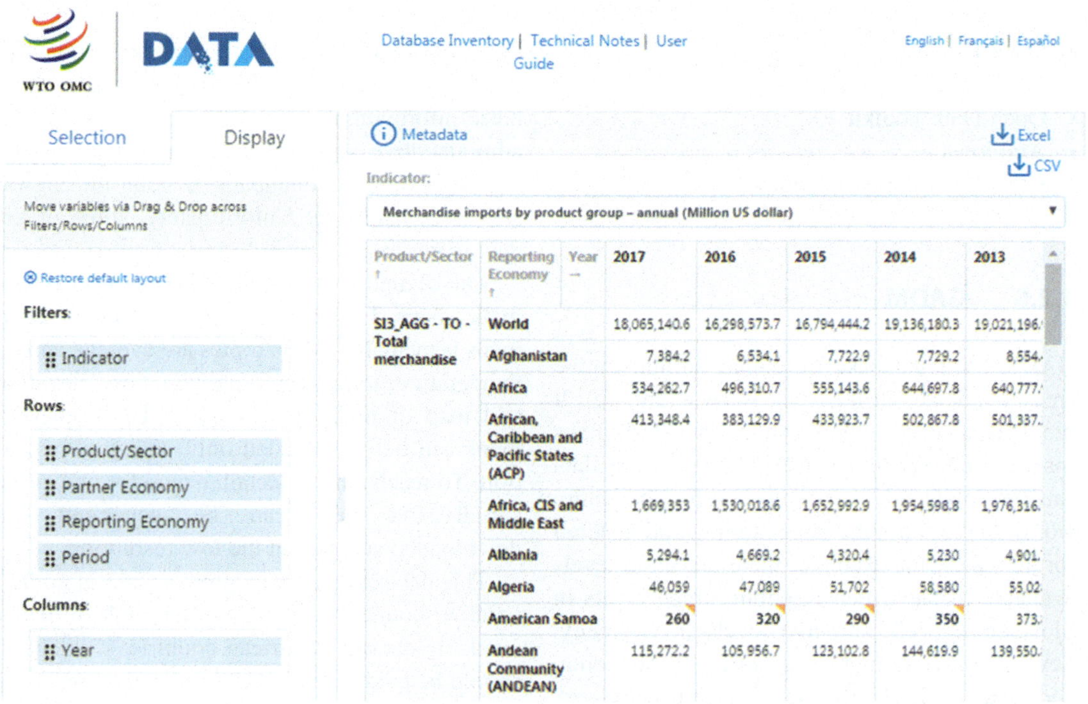

Fig. 1.9 WTO interactive selection tool. (Screenshot from WTO Data portal)

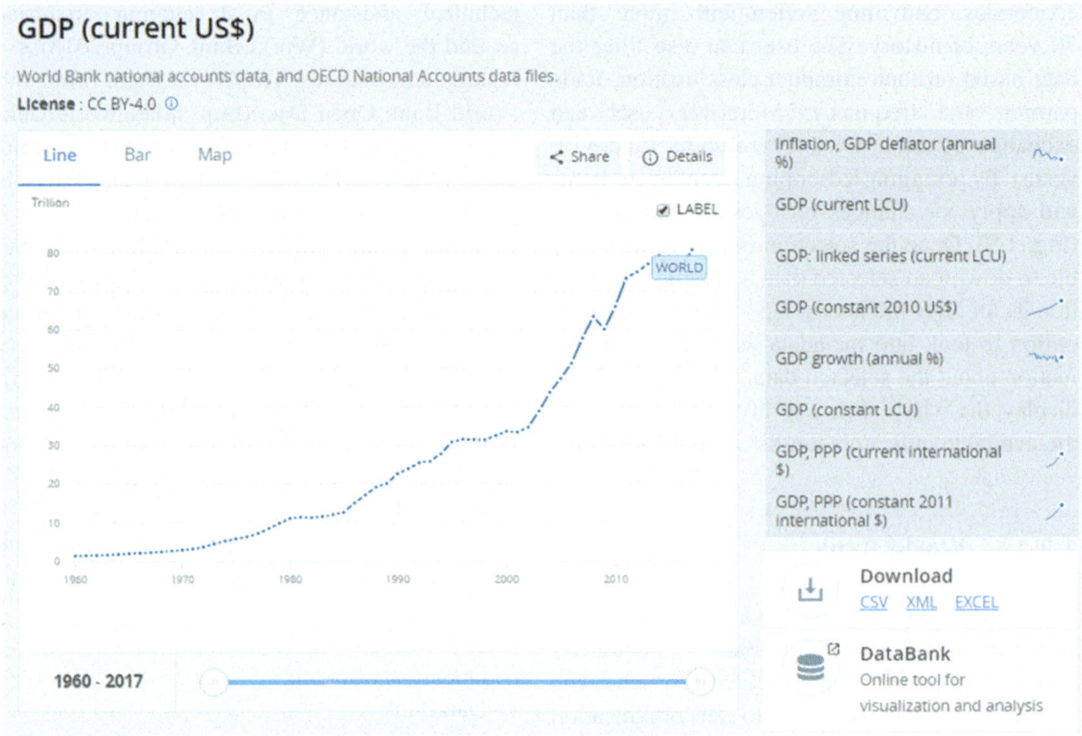

Fig. 1.10 World Bank interactive tool for data exploration. (Screenshot from The World Bank Open Data portal)

- Open Finances
- Projects & Operations
- Open Data Toolkit
- AidFlows
- Global Consumption Database

1.2.6 GADM

GADM is a database of global administrative areas available at the link www.gadm.org. It provides spatial administrative data and maps for all countries of the world. The spatial data can be download by country or for the entire world. There is not the same administrative detail for all states; for example, there are three levels available in the Czech Republic, three levels in Slovenia as well, and five levels in Germany. Several data formats are offered: shapefile, geopackages, KMZ and .rds (file for R software). The coordinate system of downloaded data is WGS 84. Regarding the attribute data, only basic information is provided – the name of the administrative, unit, a code and type (state, region, district, municipality), both in English and local language. Unfortunately, some of the data is missing.

The second part of the GADM project is thematic maps. For almost every country, a set of maps is available. Main topics are average annual temperature, total annual precipitation, elevation and map of night light activity. Unfortunately, maps can be downloaded only with low resolution. To reach a better detail, a map for one of the sub-division unit can be generated but downloaded still just in the low resolution.

This dataset could be a great source of administrative boundaries for countries with difficulty available data (e.g. African countries). Attention should be paid on classification level – since data is not provided by any government organisation, the user should always check if the classification follows the official administrative system of the country. The data are freely available for academic use and other non-commercial use. Redistribution or commercial use is not allowed without prior permission (GADM 2018).

1.2.7 Esri Open Data

Esri, as one of the most important GIS company worldwide, offers the collection of Data&Maps. This collection includes over 120 pre-symbolized vector data layers for North America, Europe, and the world. Datasets include several topographical data, demographic data, and transportation data. Access to the data is provided by the Esri Data & Maps Group on ArcGIS Online. Data can be downloaded in several GIS formats and can also be connected directly to Esri software products like ArcGIS or ArcGIS Online.

Second important od Esri data is their ArcGIS Open Data portal available at http://opendata.arcgis.com. This portal aggregates over 1000,000 datasets from over 5000 organisations worldwide. The idea of this portal is to offer the space and tools to share any spatial data as the open data. Data can be easily searched, visualised and downloaded in several GIS formats like KML, shh or GeoJSON, and can be accessed via several API (e.g. ArcGIS REST). Data covers many topics ranging from hydrology to criminality, depending on the users that published their datasets there.

1.2.8 OpenStreetMap

If the project needs free basic vector data for a certain area, the OpenStreetMap (OSM) project might be the first address to visit. OpenStreetMap was founded in 2004 as a free editable map of the world, inspired by the concept of Wikipedia where everybody who has something to contribute can participate and feed the OSM-databases from all over the world. To use these databases, one simply has to visit *openstreetmap.org* and browse through the maps in the interactive web map. If the files should rather be opened in a GIS, e.g. to do some calculations, there can also always be defined a freely selectable subset on the web map to download then and import it into any standard GIS. The downloaded dataset contains all features that OSM provides, as there are points of interest, rivers, streets, outlines of buildings etc.

If the project area is not yet clearly defined or the project area requires complete datasets of whole states, countries or continents, then it's worth to take a look on *geofabrik.de* where there are direct download links that contain the same features as listed above for the chosen administrative area.

The whole project until now counts more than two million registered users while numbers are growing, which is one of the reasons why OSM-data is not the most trustworthy kind of data one can get. As the number of participants is steadily increasing, so does the number of people who may incorporate wrong datasets into the OSM-database – no matter if by accident or on purpose – what leads to a not directly recognisable inaccuracy in few areas. These inaccuracies exist as long as somebody detects and fixes them. So, if the project requirements demand a completely credible dataset and not just something that helps to get an overview, feed some background map or do some basic analysis in teaching classes, one has to take into account that OSM-data and its crowd-based digital modelling of the world's surface cannot fully replace the national datasets provided by governments and official releases that are mostly more reliable and trustworthy.

1.2.9 Urban Atlas

Urban Atlas is a service in the frame of the EU Copernicus program, the world's largest single earth observation program, and provides pan-European reliable, inter-comparable and high-resolution land use and land cover data for functional urban areas (FUA) and their surroundings. In the first reference year 2006 Urban Atlas included 319 FUAs with more than 100.000

inhabitants (as defined by the Urban Audit) classified into 20 different classes (e.g. urban fabric, agricultural, industrial/commercial, green urban areas, etc.). Since the second reference year 2012 Urban Atlas comprises 800 FUAs in sum, as the surroundings of the FUAs with more than 50.000 inhabitants were added to the database as well as various new classes for selected FUAs, like a Street Tree Layer (STL), the building height of core urban areas in European capitals or wetlands.

The classification is conducted by using a combination of statistical image classification and visual interpretation of Very High Resolution (VHR) satellite imagery. Finally, the Urban Atlas product is enriched with functional information (road network, services, utilities, etc.), using additional data sources such as local city maps or online map services. The access to the Urban Atlas database can be reached via *land.copernicus.eu/local/urban-atlas*. After creating a free account, all datasets of the demanded city/area are available for download.

1.3 National Data Sources

This section focuses on the three countries, from which the Spationomy project partners were drafted. To keep the logic of the previous section, both main statistical and geospatial bodies and their data sources will be mentioned. As regards the statistical offices, the situation in Czechia and Slovenia is rather simple – both countries have one official statistical institution – while in Germany, every single federal state runs its own statistical office. It is worth to mention that the standardisation level of indicators is strictly followed, so the datasets should be mutually comparable. Nevertheless, there exists an office on a national level in Germany that collect selected statistical indicators. The latter office will be a subject in this chapter. It is worth to mention that each EU member state is obliged to provide statistical data within the European Statistical System (ESS) via their national statistical offices. Complete list of official statistical bodies of EU member states is given in Table 1.2.

When talking about geodata sources in the three countries, the status quo is much more diverse. In each country, there are several institutions dealing with some geodata; therefore, only the main geoportals collecting the most important geodatasets will be described in this part.

1.3.1 Czechia

1.3.1.1 Czech Statistical Office

Czech statistical office (https://www.czso.cz) is the central authority for providing statistics in Czechia. It is also the main body to report statistics to Eurostat. Every product from the office is based on statistical data. Therefore, only the main data source – Public database – will be here described. The public database is an interactive search engine for most of the statistical data that the Czech statistical office produces. Within the Public database, there are three options on how to obtain data:

- Statistics – this options is indicator-based, i.e. the user can choose the individual indicator via data tree interface. Then, it is possible to sort the data, filter it, display in table, chart or map, or download.
- All about territory – this choice let the user select a territory or region and display is summarising statistic (territory profile). As in the previous case, the data can be sorted, modified, displayed in table, chart or map, printed in a template or downloaded.
- Customised selection – this represents the most advanced method of data selection. From the user perspective, it is necessary to know which indicator, territory and period is desired since this selected is stepwise (Fig. 1.11). Once the selection is ready, the user can modify a layout, download selected data, or go back and change the selection settings.

Regardless of the method, it is possible to select indicators from these main thematic groups:

Table 1.2 European Union member states statistical offices

Country	Statistical office	Webpage
Austria	Statistik Austria	https://www.statistik.at
Belgium	National Statistical Institute (STATBEL)	https://statbel.fgov.be/en
Bulgaria	National Statistical Institute of Bulgaria	www.nsi.bg/en
Croatia	Central Bureau of Statistics	https://www.dzs.hr
Cyprus	Statistical Service of the Republic of Cyprus (CYSTAT)	http://www.mof.gov.cy/mof/cystat/statistics.nsf/index_en/index_en?OpenDocument
Czech Republic	Czech Statistical Office	https://www.czso.cz/csu/czso/home
Denmark	Statistics Denmark, Statistics Greenland	https://www.dst.dk/en, http://www.stat.gl
Estonia	Statistics Estonia	https://www.stat.ee/en
Finland	Statistics Finland	https://www.stat.fi/index_en.html
France	National Institute of Statistics and Economic Studies (INSEE)	https://www.insee.fr/en
Germany	Federal Statistical Office (DESTATIS)	https://www.destatis.de/EN
Greece	Hellenic Statistical Authority (ELSTAT)	http://www.statistics.gr/en/home
Hungary	Hungarian Central Statistical Office	www.ksh.hu/?lang=en
Ireland	Central Statistics Office	https://www.cso.ie/en
Italy	The Italian National Institute of Statistics (ISTAT)	https://www.istat.it/en
Latvia	Central Statistical Bureau of Latvia	https://www.csb.gov.lv/en
Lithuania	Statistics Lithuania	https://www.stat.gov.lt/en
Luxembourg	National Institute of Statistics and Economic Studies (STATEC)	https://statistiques.public.lu/en/index.html
Malta	National Statistical Office	https://nso.gov.mt
Netherlands	Statistics Netherlands	https://www.cbs.nl/en-gb
Poland	Statistics Poland	http://stat.gov.pl/en
Portugal	Statistics Portugal (INE)	https://www.ine.pt/xportal/xmain?xpgid=ine_main&xpid=INE&xlang=en
Romania	National Institute of Statistics	http://www.insse.ro/cms/en
Slovakia	Statistical Office of the Slovak Republic	https://slovak.statistics.sk
Slovenia	Statistical Office of the Republic of Slovenia (SURS)	https://www.stat.si/statweb/en/home
Spain	National Statistical Institute (INE)	https://www.ine.es/en/welcome.shtml
Sweden	Statistics Sweden	https://www.scb.se/en
United Kingdom	Office for National Statistics (ONS)	https://www.ons.gov.uk/

Source: http://www.unece.org/stats/links.html, authors survey

- Dwellings
- Prices, Inflation
- Tourism
- Foreigners
- Transportation, Inf. and Communication
- Energy
- Financial data, public budgets
- GDP, National Accounts
- Information Technologies
- Business Cycle Surveys
- Crime, Accidents, Fires
- Culture, sport
- Forestry
- Labour Costs and Earnings
- Retail Trade, Hotels and Restaurants
- Population
- Business Register Data
- Industry
- Living conditions, Household Income and Expenditure
- Population Census
- Services

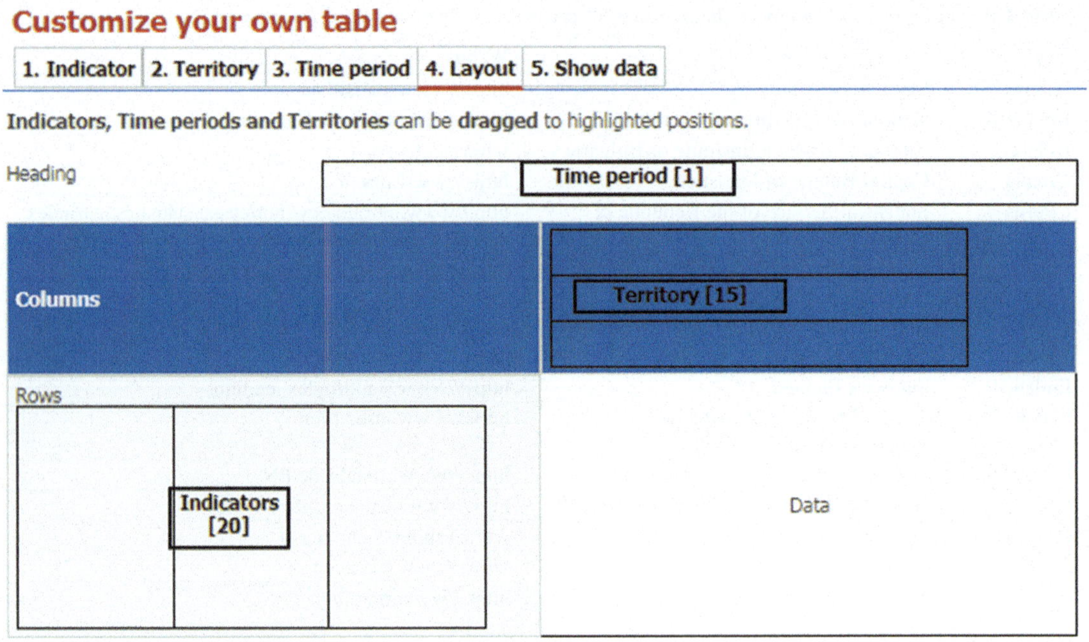

Fig. 1.11 Selection process via Customised selection option in the Public database. (Screenshot from Public database, Czech Statistical Office)

- Social Security
- Construction
- Territory, residential structure
- Science, Technology and Innovation
- Elections
- Education
- External Trade
- Employment, Unemployment
- Health Care, Incapacity for Work
- Agriculture
- Environment

Most of the indicators could be downloaded as XLS, PDF, XML and PNG (for maps) format. All datasets are complemented with metadata, and methodological guidelines are also available at the Czech Statistical Office.

1.3.1.2 Czech Office for Surveying, Mapping and Cadastre

Czech Office for Surveying, Mapping and Cadastre (ČÚZK) is the main state institution responsible for the production of spatial data. The main tasks of this office are, e.g. to complete administration of Czech cadastre, mapping of Czech Republic in all scales, the creation of Fundamental Base of Geographic Data (ZABAGED), implementation of geodetic surveys or standardisation of geographic names.

ČÚZK offers access to all map and data products by Geoportal. It is a web interface to access the spatial data produced and updated by activities of the Czech Office for Surveying, Mapping and Cadastre (ČÚZK 2018a). The Geoportal is available at the web page www.geoportal.cuzk.cz. The Geoportal offers services of data sharing according to rules of the EU INSPIRE Directive. It allows to search for spatial data and other products, to access services based on the spatial data and to obtain the products via e-shop. Most of them are charged according to the amount of data user request. An overview of all products is also available on the web page, few of them is described on the following lines.

Orthophoto of Czechia

It is a periodically updated dataset of aerial images covering the whole republic. An orthophoto is a geo-referenced ortho-photographic display of the Earth surface. Orthophotos show the photographic image of the Earth surface transformed in the way that image shifts generated during the acquisition of aerial images are removed. Since 2010 the photography has been carried out by a digital camera, which has caused an additional increase of product quality up to the spatial resolution of 0.2 m per pixel. This aerial images can serve as a suitable base map for use by for planning, project preparation, environmental protection, risk management and other applications done by organisations, state institutions and local governments (ČÚZK 2018b).

ZABAGED

The ČÚZK (ČÚZK 2018d) describes ZABAGED dataset as following: The geographic base data of the Czech Republic (ZABAGED®) is a digital vector model of the territory of the Czech Republic. ZABAGED® is a part of the surveying information system and belongs to information systems of the public service. It is maintained as a seamless database for the entire territory of the CR in a centralised information system managed by the Land Survey Office. Planimetric section of ZABAGED® contains two dimensional (2D) spatial information and descriptive information on settlements, roads, utility networks and pipelines, hydrology, administrative units and protected areas, vegetation and surface, terrain relief.

Both the orthophoto and ZABAGED database can be accessed as a Web Map Service (WMS). This can be easily added to any GIS software and then used for free. In total, ČÚZK offers almost 30 topics available as a free WMS, which is of course only for viewing, but sometimes this preview can sufficient as a base map for the project. A list of all services is available at the website in category Network services.

Registry of Territorial Identification, Addresses and Real Estates

Registry of Territorial Identification, Addresses and Real Estates (RÚIAN) is under operation since July 1st 2012 as an integral part of the whole system of public administration basic registries. The administrator and operator of RÚIAN are ČÚZK. The main benefit of the entire set of basic registries is to create such a set of reference data, which is obligatory for the performance of public administration agendas. In this case, it means the administration of descriptive and localisation data about territorial elements, territorial inventory units, teleological territorial elements and address data and their mutual relations (ČÚZK 2018c).

A part of the RÚIAN project IS the public remote access, through which RÚIAN and data are freely available via the internet for viewing or downloading in RUIAN exchangeable format (VFR – derived from GML format). Free remote access is available publicly at http://vdp.cuzk.cz/, unfortunately only in Czech. There several features categories can be downloaded – administrative units and boundaries (regions, districts, municipalities, etc.) and detailed spatial information at the municipality level – parcels, address points, streets and buildings. This detailed information is beneficial for different economic applications, local government management and planning or development.

1.3.2 Slovenia

1.3.2.1 Statistical Office of the Republic of Slovenia

Statistical Office of the Republic of Slovenia (SURS) represents the main institution in Slovenia responsible for collecting, managing, and distribution of statistical data about the country. According to SURS (2018), SURS "is professionally independent government service with autonomy as regards professional and methodological issues. The mission of the Slovene statistical office is to provide to users statistical data on the status and trends in the economic, demographic and social fields, as well as in the field of environment and natural resources."

As for a statistical data sources, SURS website (https://www.stat.si/StatWeb/en) offers four main option to access a statistical data – via dynamic search tool (although sensitive only for Slovenian indicator names), A to Z browsing, main database (SI-STAT), and preset themes:

- Agriculture, Forestry and Fishery
- Construction
- Culture
- Development and Technology
- Earnings and Labour Cost
- Education
- Energy
- Enterprises
- Environment
- Foreign Economic Relations
- GDP and National Accounts
- Industry
- Labour Market
- Population
- Prices and Inflation
- Quality of Life
- Regional Overview
- Social Protection
- Territory
- Tourism
- Trade and Services
- Transport
- European data

All the indicators in the listed themes are available in the main database (SI-STAT), which offer broader functionality for searching, selecting, filtering, displaying and downloading a data. In the database system, data is organised in four main categories – Fields of statistics (e.g. demography, economy and others), Census data, Cross-sectional reviews, and Archive for discontinued tables. By choosing a specific topic within the category, a list of individual indicators appear, and the user can then select a particular settings of selected indicator (e.g. when choosing Gross Domestic Product in an Economy section, several variations of Gross Domestic Product are available; including a selection of a year and respective data description). A data download is available in PC-axis format, XLS, TXT, CSV, and as an HTML.

1.3.2.2 The Surveying and Mapping Authority of the Republic of Slovenia

The Surveying and Mapping Authority of the Republic of Slovenia comprises the Main office, the Real estate office, the Mass real estate valuation office, the Geodesy office and twelve regional surveying and mapping administrations. These have been set up for the reasons of effective operation and the accessibility of services implemented by the Surveying and Mapping Authority of the Republic of Slovenia (Surveying and Mapping Authority of the Republic of Slovenia 2018a).

The offices cooperate with the regional surveying and mapping administrations to implement the following tasks to:

- prepare the national land survey service annual program
- organise the work of the regional surveying and mapping authorities
- implement the development assignments about surveying and mapping activities;
- provide the implementation of international obligations of national land survey service.

There is the e-Surveying data portal on the following link: http://egp.gu.gov.si/. After a quick registration, a user can access the web portal, where all available themes (17) are listed. Themes are, for example, remote sensing data, basic topographic maps, digital elevation model, register of geographical names or land cadastre. All the data can be very easily downloaded. Several interesting layers supporting the synergy of an economy and spatial data are on offer here — for example, data from Public Infrastructure Cadastre. This is a centralised database of public infrastructure objects and networks (roads, railways, water supply, sewage network, etc.). Each element in the database has the information about its type, location, identification number and ownership. The infrastructure network owners or managers are obliged to provide up-to-date information.

As well as in the Czech Republic, also Slovenian most detailed cadastral data can be freely downloaded on this portal. The following information is kept in the Land cadastre: the parcel identification code, border, surface, owner, land under the building, land evaluation. The relation to the Register of Spatial Units, Building Cadastre and Land Registry is also provided. Information on ownership of physical persons is not available to the public (personal data protection rules). Personal data about ownership can be provided by the Data Issuing Department of the Surveying and Mapping Authority of the Republic of Slovenia only when the end-user has a special right to use this personal data defined in law (Surveying and Mapping Authority of the Republic of Slovenia 2018b). Attribute and spatial data can be downloaded separately; the elementary unit for download is a municipality.

1.3.3 Germany

1.3.3.1 The Federal Statistical Office

The Federal Statistical Office (DESTATIS) is responsible for providing and disseminating statistical information and based on the federal structure and administration in Germany, DESTATIS implements federal statistical surveys in cooperation with the statistical offices of the 16 federal states (DESTATIS 2018). This implies the importance of DESTATIS since it acts as the main coordinator, ensuring that the data are collected by federal states according to standards, methodology and is delivered in time.

Besides actual news and information on DESTATIS homepage (https://www.destatis.de), it is possible to search for a specific data by looking at Facts & Figures tab on the main webpage, but more importantly, DESTATIS runs database application "GENESIS". Apart from the dynamic full-text search, it is possible to browse statistics by theme, which are grouped into nine main themes:

- Territory, population, labour market, elections
- Education, social security. Benefits, health, justice
- Housing, environment
- Economic sectors
- Foreign trade, enterprises, crafts
- Prices, earnings, income, consumption expenditure
- Public finances, taxes, public service personnel
- Economic accounts
- National and international indicator systems

Each of the themes contains several sub-themes in which individual indicators are available. Similar to other databases, the tree structure for data search is employed in the database interface. It is necessary to go through the tree structure down to the level with an individual table with an indicator (usually fifth level). In some cases, the tables are further split into the lower level of the hierarchy, for example:

4 Economic sectors – 47 Financial and other services – 473 Insurance – 47311 Statistics of insurance companies, pension funds – 47311-0001 Insurance companies' key figures: Germany, years, economic activities

Once the table with the desired table is selected, it is possible to generate results (as a table or chart) with respective indicators in a table (Fig. 1.12). Download options are XLS, XLSX, CSV, and HTML.

1.3.3.2 Federal Agency for Cartography and Geodesy (BKG)

"The Federal Agency for Cartography and Geodesy is the central service provider of topographic data, cartography, and geodetic reference systems for the German government." (BKG 2019a).

Its main tasks are to ensure a uniform coordinate system for the entire territory of the Federal Republic of Germany and to provide up-to-date spatial data of Germany via the internet. For this BKG integrates the official spatial data records of BKG and all sixteen federal states (Laender), as well as those of third-party suppliers. Their data is first edited and standardised by BKG before being made available in digital form.

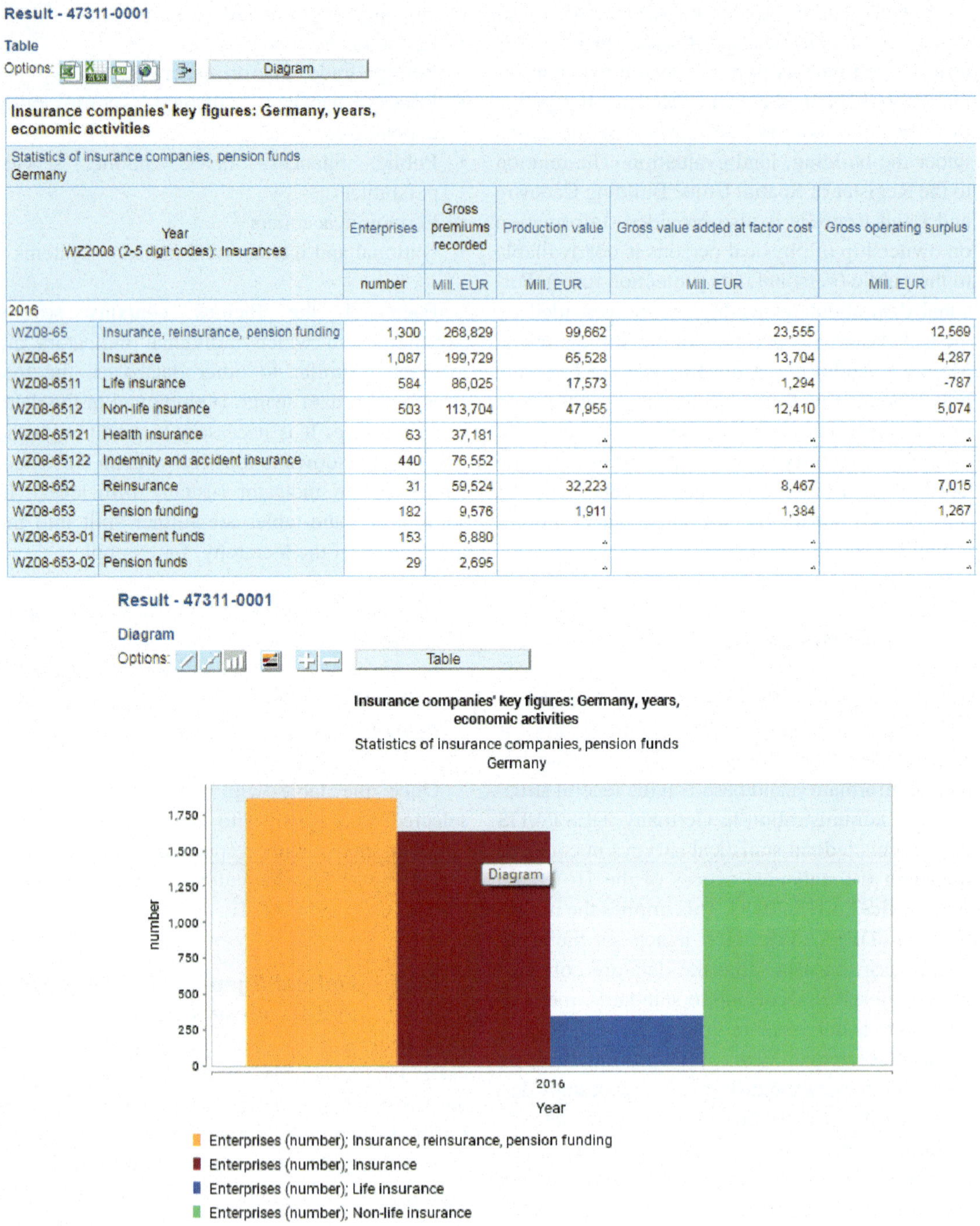

Result - 47311-0001

Table

Options: Diagram

Insurance companies' key figures: Germany, years, economic activities

Statistics of insurance companies, pension funds
Germany

Year WZ2008 (2-5 digit codes): Insurances		Enterprises	Gross premiums recorded	Production value	Gross value added at factor cost	Gross operating surplus
		number	Mill. EUR	Mill. EUR	Mill. EUR	Mill. EUR
2016						
WZ08-65	Insurance, reinsurance, pension funding	1,300	268,829	99,662	23,555	12,569
WZ08-651	Insurance	1,087	199,729	65,528	13,704	4,287
WZ08-6511	Life insurance	584	86,025	17,573	1,294	-787
WZ08-6512	Non-life insurance	503	113,704	47,955	12,410	5,074
WZ08-65121	Health insurance	63	37,181	.	.	.
WZ08-65122	Indemnity and accident insurance	440	76,552	.	.	.
WZ08-652	Reinsurance	31	59,524	32,223	8,467	7,015
WZ08-653	Pension funding	182	9,576	1,911	1,384	1,267
WZ08-653-01	Retirement funds	153	6,880	.	.	.
WZ08-653-02	Pension funds	29	2,695	.	.	.

Fig. 1.12 Resulting table and chart obtained in GENESIS database. (Screenshot from GENESIS database, © Federal Statistical Office)

Furthermore, the authority supports the establishment and expansion of spatial data infrastructure, which in turn enables all citizens to search for and take advantage of the spatial data offered by the federal government. BKG represents Germany's interests in international collaborative entities and projects addressing the fields of geodesy and geoinformation. It also advises its customers and offers customer-oriented solutions. (BKG 2019b).

The BKG operates the "Service Centre of the Federal Government for Geo-Information and Geodesy" on the web (http://www.geodatenzentrum.de). Besides News and some descriptive content, it provides access to Web Applications, Online Shops and Open Data.

Under the category "Web Applications" the service centre provides web-map-applications and JAVA-applets as clients for the access to and the use of geodata provided by the BKG. The service "Maps of BKG" ("Karten des BKG") allows an overview and browsing of geodatasets maintained by the BKG. The menu item "TopoPlusOpen Download" leads to a Java-Application that allows downloading tiles of the world-wide TopoPlusOpen-map.

The category "Online Shops" gives access to three specialised online shops. These allow ordering geodata, access geodata services, or to buy printed maps that are not free of charge and therefore not available for download.

In the section "Open Data" all datasets are available for download or are provided as WMS or WFS. They can be used free of charge according to different licenses specified in the metadata of the datasets. On the page "Free Data and Services of BKG" the following Products are on offer (Table 1.3).

The page "INSPIRE Themes" gives an overview of INSPIRE conformal services within the common spatial data infrastructure in Europe that are available free of charge. For each dataset, a description of its contents, downloads of PDFs holding the INSPIRE Data Specification and detailed documentation as well as the WMS- and WFS-URLs are provided. Datasets for the following INSPIRE themes are available:

- INSPIRE Hydro:
  - Physical Waters
  - Network
- INSPIRE Transport Networks:
  - Road Transport Network
  - Rail Transport Network
  - Water Transport Network
  - Air Transport Network
- INSPIRE Administrative Units
- INSPIRE Protected Sites
- INSPIRE Geographical Names
- INSPIRE Land Cover

In addition to the "Service Centre" described above, the BKG operates the web-portal https://www.geoportal.de for the Spatial Data Infrastructure Germany (GDI-DE). The GDI-DE is an initiative of the German federal government, the states and its municipalities. It constitutes the German part of the European spatial data infrastructure implemented via the EU Directive INSPIRE (GDI-DE 2019).

Besides comprehensive information on the GDI-DE the portal guides the way to geodata related resources of many different entities within the German federal and decentralised administration (Geoportal → Service → Viewer und Portale). Direct links to the GDI-pages of the states are provided via a member list on the sub-homepage "GDI-DE".

1.4 Other Statistical Data Sources

This section highlights microdata sources covering most of the European countries. According to Eurostat (2018b), microdata is records containing information on individual persons, households or business entities. In many cases, due to the individual nature of such datasets, microdata is not publicly accessed to protect personal or other sensitive information about the entity. Moreover, microdata is usually collected as a sample of a given population, therefore are in a very specific topic, demographic or business sample, and not representing the whole population (e.g. entire business sector, or country/region). Similarly to microdata, commercial datasets possess the same characteristics as regards free access. Based on the business nature of commercial datasets, these are usually provided upon a purchase, which somehow limits their wide-range usage

Table 1.3 Open data available at the BKG

Product	Metadata	Download format	Service
TopPlusOpen	PDF	Web application for downloading of extracts and tiles of the web map tile archive	WMS/ WMTS
Digital landscape model 1:250000	PDF/ TXT	Shape-file	WFS
Digital landscape model 1:1000000	PDF/ TXT	Shape-file	WMS/ WFS
CORINE land cover	PDF/ TXT	Shape-file	WMS/ WFS
Digital terrain model grid width 200 m	PDF/ TXT	GRID-ASCII XYZ-ASCII	WMS
Digital terrain model grid width 1000 m	PDF/ TXT	GRID-ASCII XYZ-ASCII	WMS
Digital topographic map 1:250000	PDF/ TXT	TIFF	WMS
Digital topographic map 1:500000	PDF/ TXT	TIFF	WMS
Digital topographic map 1:1000000	PDF/ TXT	TIFF	WMS
Administrative map of Germany (Federal States, administrative regions, districts)	PDF/ TXT	PDF	
Administrative map of Germany (Federal States)	PDF/ TXT	PDF	
Administrative map of Germany (national border, Federal States, districts)	PDF/ TXT	AI, EPS, SVG	
Administrative areas and NUTS regions at different scales, with and without population numbers	PDF/ TXT	Shape-file	WMS/ WFS
Geographical names 1:250000	PDF/ TXT	Shape-file, CSV	WMS/ WFS
Geographical grids for Germany in UTM- and Lambert-projection	PDF/ TXT	Shape-file, CSV	
Landscapes of Germany	PDF/ TXT		WMS
Archives	PDF/ TXT	Raster & vector files	

(especially by the scientific community often with the constrained budget). However, it is necessary to mention some of the important data sources that are classified as microdata or commercial data.

1.4.1 Eurostat Microdata

1.4.1.1 Community Innovation Survey

According to Eurostat (2018c), Community Innovation Survey (CIS) is a survey of innovation activity in enterprises as part of the EU science and technology statistics voluntarily (i.e. different countries contribute to the individual survey years). CIS uses harmonised questionnaire for all EU member states and as such presents unique and reliable source of data regarding innovation activities of enterprises of different size, age, and industry (Vaculík et al. 2017). As noted by Vaculík et al. (2017), the advantage of the CIS is the long-term experience with methodological issues related to the innovation activities involving data on technical types of innovation (product and process) as well as on the long underestimated findings on non-technical innovations (marketing and organizational). The datasets are available for research purposes only upon request. First, the research organisation has to be recognised as a research entity; then it is possible to apply for data itself based on a research proposal. More information about the dataset could be found at https://ec.europa.eu/eurostat/web/microdata/community-innovation-survey.

1.4.1.2 Eurostat Microdata – Other Sources

On the main Eurostat webpage about Microdata access (Eurostat 2018b), following microdata surveys and dataset are listed:

- European Community Household Panel
- European Union Labour Force Survey
- European Union Statistics on Income and Living Conditions
- Structure of Earnings Survey
- Adult Education Survey
- European Road Freight Transport Survey
- European Health Interview Survey
- Continuing Vocational Training Survey
- Community Statistics on information Society
- Micro-Moments Dataset
- Household Budget Survey
- Labour Force Survey
- Statistics on Income and Living Conditions

The latter two surveys are samples free for public in order to allow general public and researchers to become familiar with such microdata and to prepare software programmes (e.g. statistical computing tools) for the possible use of "full" microdata.

1.4.2 Global Entrepreneurship Monitor

Global Entrepreneurship Monitor (GEM) represents an international project on entrepreneurship data collection and research. The project monitors two aspects of entrepreneurship – individual behaviour and attitudes to entrepreneurship, and general conditions and the entrepreneurial context in each participating country. The former aspect is monitored by Adult Population Survey (APS), which looks at the characteristics, motivations and ambitions of individuals starting businesses, as well as social attitudes towards entrepreneurship (Global Entrepreneurship Monitor 2018). The latter aims at the national context in which individuals start businesses (Global Entrepreneurship Monitor 2018), and is based on experts reports within National Expert Survey (NES). For both datasets, a national team responsible for surveys needs to be formed, mostly voluntarily. Therefore, freely available datasets vary from country to country and from time to time. Detailed information about the project can be found at www.gemconsortium.org.

1.4.3 Amadeus – Bureau Van Dijk

Amadeus database provides information about companies across Europe and is maintained by a Moody's analytics company Bureau van Dijk (BvD). Amadeus database is one of the BvD's international databases with comprehensive information on private companies. The database includes basic information about the company (e.g. postal address, IDs, number of employees, industry category), company financials and their indicators, detailed corporate structures, and many more. All data are collected yearly, and according to BvD's webpage about the product (Bureau van Dijk 2018), the Amadeus database contains information on about 21 million companies across Europe. It has to be noted that besides the Amadeus database (and global database), BvD also maintains special database about Asia-Pacific region, insurance companies, intellectual properties, as part of international datasets. National datasets covering specific countries and specialist products are also provided by the company. Although very information-rich, databases offered by BvD are commercial and need to be purchased.

1.5 Earth Observation Data (by Carsten Jürgens)

Earth observation data, also called remote sensing data, are characterised to be pictorial representations (images) in raster format of the earth's surface acquired by sensors on board of associated platforms. Earth observation systems are characterised by up-to-date image data capture that is useful for various applications.

The principle of remote sensing relies on the electromagnetic energy, which is the transmitter of any information between the earth's surface objects and the image generating sensor on board of a platform (Jürgens 2003). Passive sensors are characterised by their dependency on emitted or reflected electromagnetic energy. Since the sun illuminates the earth during the daytime, passive sensors can capture the reflected portion of the incoming radiation. Therefore these sensors rely on the sun's illumination during daytime and are affected by cloud cover, which obscures the earth's surface (Campbell and Wynne 2011). In contrary, active systems have their own source of illumination and allow to capture images also during night time (Albertz 2013). Active sensors are LiDAR/Laser Scanning and RADAR, the latter case is also relatively unaffected by clouds. Due to the long microwave wavelengths used, clouds can be penetrated, and one can get image data of the earth's surface without any missing area due to cloud coverage.

1.5.1 Platforms

There is a variety of different imaging sensors that need to be mounted on-board of a flying platform. Platforms are subdivided into earth-orbiting satellites, airplanes, helicopters and unmanned aerial vehicles (UAV). Airplanes, helicopters and UAV's are able to acquire image data on demand and therefore are very flexible. For airborne systems, a flight plan has to be prepared to assure that the system covers the complete area of interest during the planned flight. Often aerial systems capture stereo images which can be used for 3D-interpretation. Therefore an overlap of at least ca. 50% is needed between adjacent images and around ca. 15–35% between flight lines. Aerial systems can be started almost at any time and adjusted to special needs by equipping them with specific sensors according to those needs. The flying height can be adapted to the mapping scale requested for an image campaign. Earth observation satellites instead move on fixed near-polar sun-synchronous orbits with fixed revisit rates. Since the sensor on a satellite platform cannot be changed like on an airborne platform, the user has to decide which earth observation satellite system is most applicable for a specific task. This means in contrary to flexible platforms like airplane, helicopter and UAV's, which can carry different sensors, in the case of satellites, one has to decide which system serves the user needs best according to its specific sensor system characteristics.

1.5.2 Sensor Types

As indicated earlier, there are active and passive sensors. The active sensors are characterised by their ability to acquire images during day and night and in the case of RADAR their robustness against clouds and unfavourable weather conditions due to their long microwave wavelengths (1 mm–1 m) that can penetrate clouds. The passive sensors operate in the so-called optical domain of the electromagnetic spectrum with much shorter wavelengths, ranging from the visible part of the electromagnetic spectrum (blue, green, red) to the short, middle, thermal infrared and passive microwave wavelengths (see Fig. 1.13).

Some specialised sensors are able to capture thermal emissions and passive microwave emissions during the night as well.

1.5.3 Types of Resolution

Earth observation sensor systems are characterised by different types of resolution.

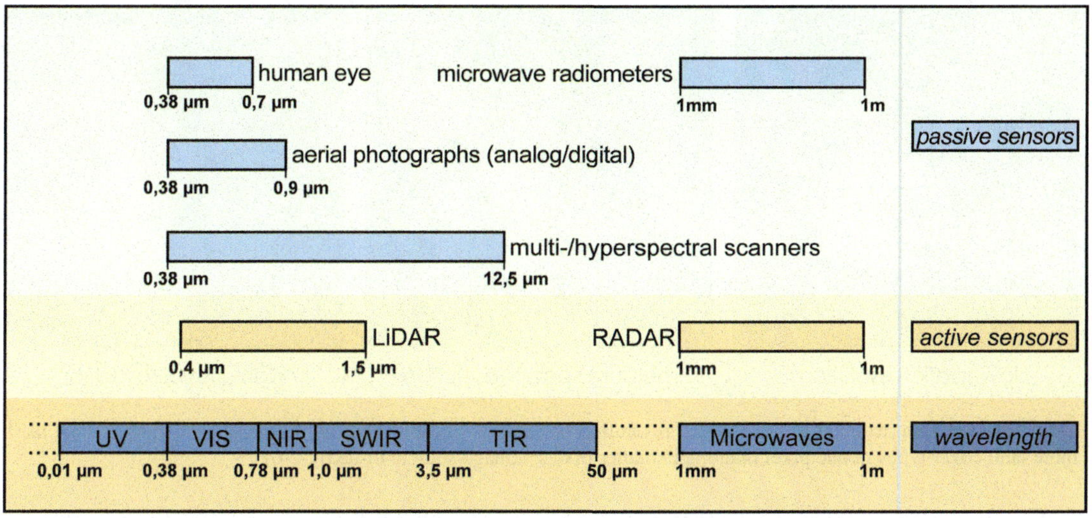

Fig. 1.13 Electromagnetic spectrum and selected wavelengths used for earth observation systems

One has to distinguish between four types of resolution, namely (Lillesand et al. 2015):

– *Temporal resolution*, which is the revisit rate or repetition rate, meaning in which time interval acquires the system another image of exactly the same area on the ground. In the case of earth observation satellites, this is dependent on the orbit parameters.
– *Spatial resolution*, which refers to the ground sampling rate or pixel size on the ground. One can imagine a pixel as the footprint on the ground, which gives a single „echo"of reflected energy, neglecting on how it is composed of different materials. The smaller a pixel is, the more details one can identify in the image. This is especially meaningful in urban areas with a lot of material variety within short distances, which requires larger scales for better discrimination of objects and materials. In contrary to that, in the countryside with fields and forests, for example, one does not necessarily need very small pixels due to the relatively homogeneous land cover. So coarser or medium spatial resolution pixels are acceptable for studies in such landscapes associated with smaller scales. Spatial resolutions vary in the range of cm and dm pixel size for airborne systems and ca. 0,30 m and ca. 1000 m for satellite systems (see Fig. 1.14).
– *Spectral resolution* refers to the width and number of different spectral bands. Digital images are captured and stored in separate spectral bands, meaning that individual portions of the electromagnetic spectrum are separated from other parts of the spectrum. The more individual bands are available the smaller is the individual bandwidth of each band, which refers to a high spectral resolution.
– *Radiometric resolution* refers to the sensitivity of a sensor to capture the photons of the reflected energy. The radiometric resolution is better/higher in systems that can differentiate more and therefore finer levels of reflectance. Early satellite systems typically had a radiometric resolution of 8 bit, meaning that 256 different intensities of reflected energy (0–100%) could be separately stored for each pixel. Modern systems have improved sensitivity (e.g. 12 bit) and can separate finer intervals. This allows to separate 4096 individual radiation intensities and therefore much more detailed discrimination of the reflected energy.

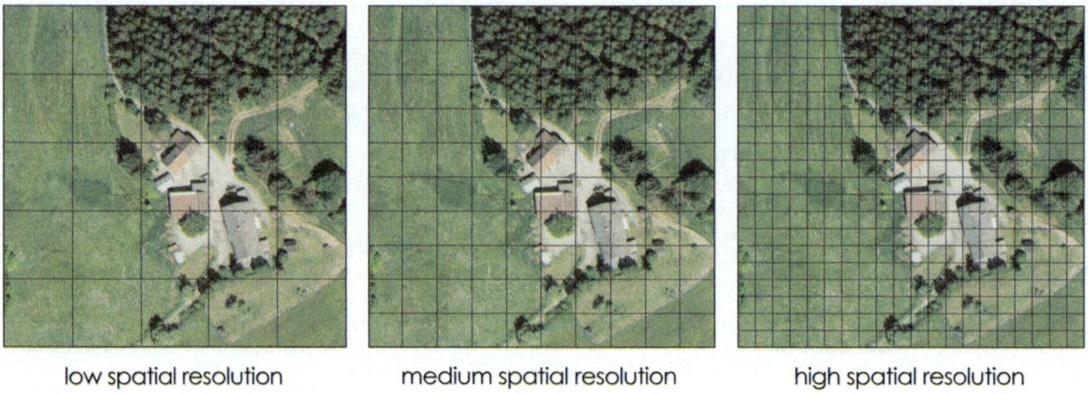

Fig. 1.14 Different examples of the spatial resolution of earth observation sensors. Larger pixels capture most likely more land cover types in one pixel than finer/smaller pixels. (Image source: dl-de/by-2-0)

For practical work with image data, one has to decide with which type of data one will work. This implies to define the scale of the investigation, and then one can decide on the necessary pixel size, which determines the level of detail. The pixel size mostly defines also the areal extent of images. As a rule of thumb, aerial images by UAV's and planes or helicopters cover a lot less ground per image than satellite images. And satellite images with a very high spatial resolution which means small pixels, cover less area per image than satellite images with coarser resolution. For large areas of interest, one has to consider to eventually stitch together images of different orbits or flight paths to cover the ground completely in an image mosaic.

One more aspect is the spectral characteristic of a sensor. Can that provide the information one is looking for? For instance, if one is interested in plant characteristics, the sensor should at least be able to capture near-infrared information, since in this portion of the electromagnetic spectrum essential information about plant conditions is located. In addition to that one has to consider the revisit rate, meaning how often one can get another „fresh "image of the same area of interest, which is necessary in the case of heavy cloud cover on an earlier image. Satellites have fixed orbits and depending on the systems constellation, satellites can reach daily coverage or acquire images of the same area of interest after a couple of days. All airborne platforms can acquire images on demand, as long as one has cloud-free conditions. Satellite images in the optical domain are much more affected by clouds. Due to fixed orbits, one cannot change the time of the satellites overpass. Since clouds disturb the direct observation of the ground, the user of these image data has to search in the satellite archives of the image provider for cloud-free images for the specific area of interest and specific period of the year.

An alternative chance for cloud-free images offers satellite constellation systems, which consist of a group of identical satellites that use the same orbit and fly one after the other in a fixed equal distance. With this technique (e.g. used by 5 German RapidEye satellites) the temporal resolution is reduced and the chance of a cloud-free image is increased (see Fig. 1.15).

Airborne and some satellite systems have the ability to be tilted along a track or across a track. This allows in the case of across-track, that the satellite points to an area that is not in nadir direction underneath the satellite, but aside of that. Doing the across-track tilt from two different orbits (with a time lag of a few days between the orbits and respective images) to the same target area, one gains two images with different perspectives that could be used as a so-called stereo pair. This allows the images to be used for a 3D representation of the earth. In the along-track version, one uses sensors with

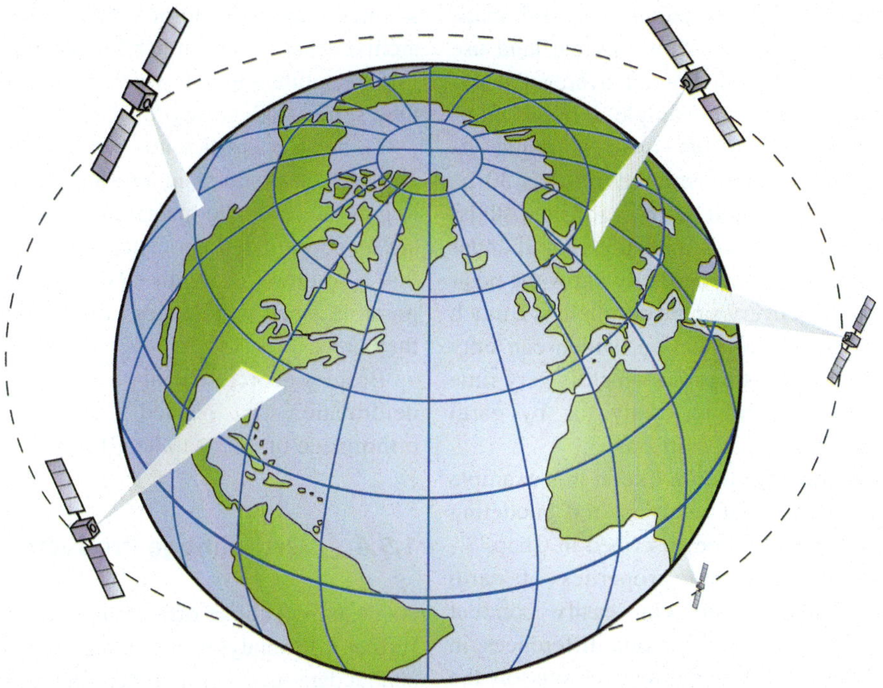

Fig. 1.15 Satellite constellation of five satellites

different viewing direction along the orbit pass: one sensor looks forward, another looks backwards. By using images of these two directions, also allows to use two different perspectives on the same area on the ground, but with almost simultaneous image capture and also allows a 3D representation of the ground. The 3D representation of the ground enables to generate digital height and digital surface models, which are requested in a number of applications (Crespi and Jürgens 2016). Due to the high repetition rates of modern earth observation systems, one can also update the 3D information in short intervals, e.g. in the case of changes mining of built-up structures on the ground.

Of course, one can also extract 2D information from the image data. Due to the fact that earth observation image data capture the specific ground situation at the time of overpass, the land cover in the resulting image reflects this specific situation. The changing characteristics of plants during the growing season or the vegetation period affects their reflective responses that can be imaged by sensor systems. Due to such seasonal effects and agricultural activities, the land cover/land-use varies between images of the same year and between years as well. Sometimes this information of change is the needed information, e.g. if one does a change detection study (Jürgens 2000 or Henits et al. 2016) for a multi-year period to document the real land cover changes in a certain region. Sometimes these changes disturb an analysis, especially if one seeks to document the average situation. For harvested fields, for instance, one is unable to determine the crop of that year. Or if one tries to average statistical information retrieved from the image data, harvested fields could be hindering. Also, artificial surfaces can vary if new objects are constructed in an area.

The described disadvantages of images representing the snap-shot situation on the earth's surface during image capture, is also an advantage, due to the fact that images are not generalised representations of the real situation at a certain point in time. All real objects are recorded, for instance, cars, trucks, ships etc. Therefore all images are historical records or documents. Another advantage compared to areal statistics is the fact that the images themselves and maps derived from the images show the real land use in the specific location and not an averaged situation like in statistics. For example, if one is interested in cornfields of an area, it might be interesting that an administrative unit has 30 ha of cornfields. This would be an aggregated statistical information for the administrative unit under investigation. However, to exactly know the position and extent of every single cornfield is a much more sophisticated information, which can only be gained by field inspection, which is very time consuming and therefore costly, or by earth observation images in a short time.

Area-specific information like in this example is required for spatial analysis and modeling approaches, which will be described in Chap. 3.

Due to the snap-shot properties of earth observation images, one can easily conduct time series analysis to find out differences in land use/land cover for defined regions on the globe. The oldest satellite-based earth observation data are available since 1972 from the first LANDSAT satellite Jürgens (1998). Since this year more and more different earth observation satellites came into orbit so that there is a great variety of different systems available nowadays. The latest development is the European fleet of Sentinel satellites. Those images are free of charge and have a high temporal revisit rate, a high spatial, spectral and radiometric resolution in addition to large area coverage. In addition to those freely available earth observation data from space, there are a lot of commercial providers who sell very high spatial resolution data. Typically prices of these images are calculated per square kilometer.

In the airborne domain, many countries have archives with aerial images starting with greyscale analogue World War II images. Approximately in the mid 1990ies, the analogue aerial images were transferred to colour images. In most countries, the analogue image capture techniques for aerial photographs ended around the turn to the third millennium and were transferred into digital image acquisition systems. In summary, in the airborne domain, there are even archives that cover a much longer time span than in the satellite domain, which is of importance for time series analysis approaches.

To be able to efficiently work with image data and derived products in a GIS system, one needs to geo-reference the image data. By default, there are no coordinates associated with image data, so that a common use with other spatial data is not possible since the images „do not know where they are".

Besides georeferencing, corrections for terrain deformation is needed to have so-called orthoimage products to be GIS-ready.

1.5.4 Orthoimage Products

As described above orthoimages are essential for further GIS-analysis in combination with further spatial data sets. Aerial images in raw format have various disadvantages, due to relief effects, radial distortions and the central perspective of the camera. This results in displaced positions of objects in the images. After proper image correction and georeferencing, the images have properties like (image) maps. Similar deformations affect satellite images, resulting in similar corrections to produce ortho satellite images. One should never consider a raw stereo image or single image for any mapping purpose or use in GIS. This rule applies to all raw earth observation images.

The following illustration (see Fig. 1.16) shows a distorted regular grid in a raw image and the corrected representation in the corresponding orthoimage. One can easily imagine how many distortions and scale variations could be in the raw image material.

1.5.5 Use of Earth Observation Image Data

These orthoimage data sets can then be used to generate up-to-date information based on the situation captured in those images. Images are the

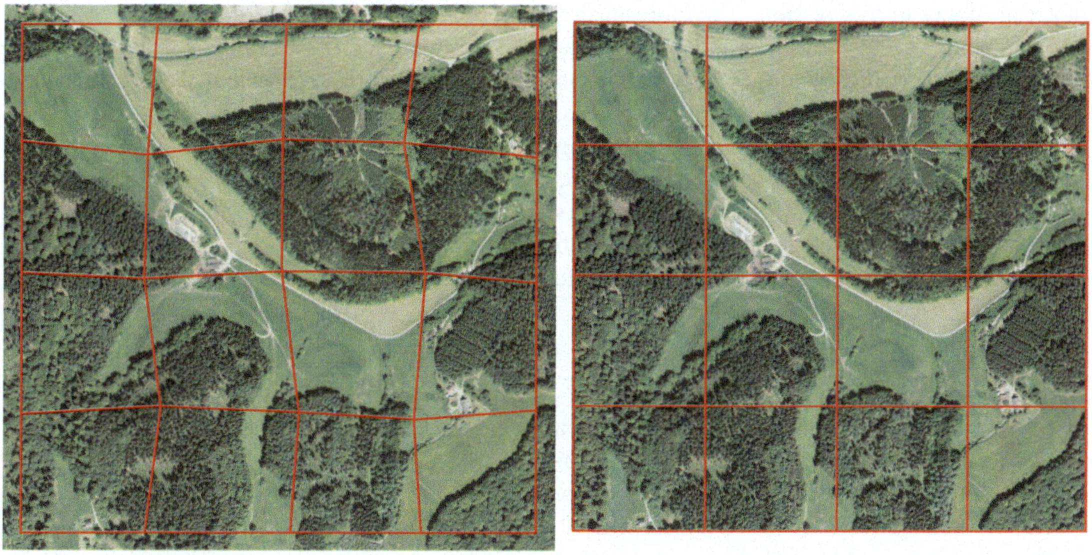

Fig. 1.16 Comparison between a raw aerial image and an orthoimage: the distorted regular grid in a raw aerial image and the corrected representation in the orthoimage shows the deformations and scale implications in the uncorrected raw image. (Image source: dl-de/by-2-0)

basic data sets for timely information in information systems. With specialized image processing techniques the extraction of specific information as possible. Each image processing step can generate another thematic GIS layer, depending on the type of image processing analysis.

By application of classification algorithms, one can generate an up-to-date land cover/land-use map of an area of interest. For the extraction of that kind of thematic information out of the potentially ambiguous image data, one has to define a proper nomenclature to describe the land cover/land-use classes for a clear separation among each other (Thunig et al. 2011). Normally a hierarchically structured nomenclature serves most needs and allows to merge classes of lower hierarchy to a class of the next level hierarchy if needed in generalisation processes. Automated and semi-automated classification procedures then use the prior knowledge regarding the land cover/land-use in the area of interest and extract the information regarding the defined classes based on the implemented algorithms.

Very often one needs a second or even third image of the same season to account for the seasonal variances in the crop and plant surfaces. The additional images can help to get reliable information in the sense that the classified areas are most likely true. For the accuracy assessment, one uses field data, which was not used for the classification and is tested against the classification result to determine its reliability. As a result, one can calculate the overall accuracy, users and producers accuracy. These give clear identification of how reliable each result for each individual class of the nomenclature is. It is also possible to get a map representation of the reliability of individual parcels. For further spatial analysis, the spatial reliability of data sets is a crucial point, since further analysis using the classification results or a resulting thematic map could benefit from that information if it is reliable and also could suffer from it in the case of poor quality.

1.5.6 Available Earth Observation Satellites

Due to the fact that aerial images normally are to be obtained at national agencies, these data are not considered in this section. Therefore in this section, you will find a selection of earth observation resources, including further information on the parameters of individual imaging systems. Due to the many and dynamically changing earth observation systems, one cannot describe all systems in detail. The following links will support your search for the optimal earth observation data source for your specific application.

1.5.6.1 Sentinel Satellite Fleet

The European Space Agency (ESA) is operating a fleet of different next-generation earth observation satellites that offer images at no cost to the user. The Sentinel satellites (https://sentinel.esa.int/) are characterised by a high temporal revisit rate due to a concept of twin missions for each type of Sentinel satellite in one orbit. One can find all technical parameters and technical guides describing the systems in depth for proper use of the images.

1.5.6.2 Landsat Satellites

The first earth observation satellite started in 1972 with Landsat-1. The Landsat fleet of satellites (https://landsat.usgs.gov/) was increased with time, and today Landsat-8 is in operation. The technical parameters (e.g. resolution) have changed several times since 1972 and can be investigated on the web portal. The US Geological Survey (USGS) operates this web page and offers the longest continuously-acquired medium resolution data archive, which can be used to time series analysis and change detection.

1.5.6.3 European Space Imaging

The company European Space Imaging (http://www.euspaceimaging.com/) is selling spatially very high-resolution satellite images from a group of commercial satellites. Each system is described in detail and potential applications are outlined.

1.5.6.4 Additional Resources

For further reading one is encouraged to visit the following web pages to gain a deeper understanding of remote sensing.

1.5.6.5 ESA

The European Space Agency (ESA) offers a web portal (http://www.esa.int) that gives one a full view on the space activities of ESA. One section is dedicated to remote sensing of the earth and gives detailed information on specific missions.

1.5.6.6 ESA EOPORTAL

This portal (https://eoportal.org) lists available earth observation resources and describes the individual satellite systems.

1.5.6.7 ESA EDUSPACE

This portal (http://www.esa.int/SPECIALS/Eduspace_EN/SEM7YN6SXIG_0.html) offers in-depth background information about earth observation principles, history, and specific satellites as well.

1.5.6.8 SATIMAGINGCORP

This portal (https://www.satimagingcorp.com/) is dedicated to showing real-world applications in various fields. In addition to that one can find image examples of satellites in high and medium resolution. Descriptions of many earth observation satellite systems are also available.

1.5.6.9 Satellite Image Archives

The following links are some examples for satellite image archives for image search and free download:

https://scihub.copernicus.eu/dhus/#/home
https://lta.cr.usgs.gov/SPOT_Historical
https://www.eodc.eu/services/data-services/
http://landcover.org/
https://earthexplorer.usgs.gov/

References

Albertz, J. (2013). *Einführung in die Fernerkundung. Grundlagen der Interpretation von Luft- und Satellitenbildern* (4th ed.). Darmstadt: Wissenschaftliche Buchgesellschaft.

BKG. (2019a). *Tasks and organization*. Retrieved January 10, 2019, from https://www.bkg.bund.de/EN/About-BKG/Tasks-and-Organization/tasks-and-organization.html

BKG. (2019b). *Maps and coordinates – BKG's range of tasks*. Retrieved January 10, 2019, from https://www.bkg.bund.de/EN/About-BKG/Tasks-and-Organization/Tasks-of-BKG/tasks-of-bkg.html

Bureau van Dijk. (2018) *Amadeus* [online – cit. 2018-12-10]. Available from: https://www.bvdinfo.com/en-gb/our-products/data/international/amadeus

Campbell, J. B., & Wynne, R. H. (2011). *Introduction to remote sensing* (5th ed.). New York: Guilford Press.

Crespi, M., & Jürgens, C. (2016). 3D remote sensing and urban remote sensing. *International Journal of Remote Sensing, 37*(15), 3437–3438. https://doi.org/10.1080/01431161.2016.1205314.

ČÚZK. (2018a). *ČÚZK: Geoportál*. https://geoportal.cuzk.cz/(S(2vb00ejqnx1uuiyzl4xvtokw))/Default.aspx?mode=TextMeta&text=about&side=about&menu=2. Accessed 28 Nov 2018.

ČÚZK. (2018b). *Orthophoto of the Czech Republic*. https://geoportal.cuzk.cz/(S(2vb00ejqnx1uuiyzl4xvtokw))/Default.aspx?mode=TextMeta&text=ortofoto_info&side=ortofoto&menu=23. Accessed 28 Nov 2018.

ČÚZK. (2018c). *ZABAGED*. https://geoportal.cuzk.cz/(S(2vb00ejqnx1uuiyzl4xvtokw))/Default.aspx?mode=TextMeta&text=dSady_zabaged&side=zabaged&menu=24. Accessed 28 Nov 2018.

ČÚZK. (2018d). *RÚIAN*. https://geoportal.cuzk.cz/(S(2vb00ejqnx1uuiyzl4xvtokw))/Default.aspx?mode=TextMeta&text=dSady_RUIAN&menu=31. Accessed 28 Nov 2018.

DESTATIS. (2018). *Our mission – Statistics* [online]. Wiesbaden, 2018 [cit. 2018-11-27]. Available from: https://www.destatis.de/EN/AboutUs/OurMission/StatisticsInfo.html

ESRI. (1998). *ESRI shapefile technical description*. White paper. Retrieved November 28, 2018, from https://support.esri.com/en/white-paper/279

European Commission. (2014a). *INSPIRE data specification on coordinate reference systems – Technical guidelines*. Retrieved January 2, 2019, from https://inspire.ec.europa.eu/id/document/tg/rs

European Commission. (2014b). *INSPIRE – An efficient way to share European spatial data!* Retrieved January 2, 2019, from https://inspire.ec.europa.eu/documents/inspire-efficient-way-share-european-spatial-data

Eurostat. (2018a). *About – Overview* [online]. Luxembourg [cit. 2018-11-27]. Available from: https://ec.europa.eu/eurostat/about/overview

Eurostat. (2018b). *Access to microdata – Overview* [online]. Luxembourg [cit. 2018–12–04]. Available from: https://ec.europa.eu/eurostat/web/microdata/overview

Eurostat. (2018c). *Community innovation survey* [online]. Luxembourg [cit. 2018–12–04]. Available from: https://ec.europa.eu/eurostat/web/microdata/community-innovation-survey

GADM. (2018). *GADM data*. https://gadm.org/data.html. Accessed 28 Nov 2018

GDI-DE. (2019). *GDI-DE: Networking data*. Retrieved January 10, 2018 from https://www.geoportal.de/EN/GDI-DE/gdi-de_artikel.html?lang=en

Global Entrepreneurship Monitor. (2018). *Data* [online]. London [cit. 2018-12-10]. Available from: https://www.gemconsortium.org/data

Henits, L., Juergens, C., & Mucsi, L. (2016). Seasonal multitemporal land-cover classification and change detection analysis of Bochum, Germany, using multitemporal Landsat TM data. *International Journal of Remote Sensing, 37*(15), 3439–3454. https://doi.org/10.1080/01431161.2015.1125558.

ISO. (2018a). *ISO 19115:2003. Geographic information – Metadata*. Retrieved January 2, 2019 from https://www.iso.org/standard/26020.html

ISO. (2018b). *ISO/TS 19139:2007 Geographic information – Metadata – XML schema implementation*. Retrieved January 2, 2019 from https://www.iso.org/standard/32557.html

Jürgens, C. (1998). Satellitenfernerkundung – ein Überblick über derzeitige und zukünftige Aufnahmeinstrumente. *Geoökodynamik, Jg.19* (Heft: 1–2): 119–137.

Jürgens, C. (2000). Change Detection – Erfahrungen bei der vergleichenden multitemporalen Satellitenbildauswertung in Mitteleuropa. *Photogrammetrie Fernerkundung Geoinformation*, 1/2000: 5–18.

Jürgens, C. (2003). Geo-Fernerkundung – Was ist das? *Praxis Geographie, 3/2003*: 4–7.

Lillesand, T. M., Kiefer, R. W., & Chipman, J. (2015). *Remote sensing and image interpretation* (7th ed.). New York: Wiley.

OECD. (2018). *About the OECD* [online]. Paris, 2018 [cit. 2018-11-27]. Available from: http://www.oecd.org/about/

Snyder, J. P. (1987). *Map projections – A working manual. U.S. geological survey professional paper 1395*

SURS. (2018). *About statistical office* [online]. Ljubljana, 2018 [cit. 2018-11-27]. Available from: https://www.stat.si/statweb/en/AboutUs/AboutStatOffice

Surveying and Mapping Authority of the Republic of Slovenia. (2018a). *Assignments and objectives | Surveying and mapping authority of the RS*. http://www.gu.gov.si/en/gu_vstopnastran/assignments_and_objectives/. Accessed 29 Nov 2018

Surveying and Mapping Authority of the Republic of Slovenia. (2018b). *e-Surveying data*. http://egp.gu.gov.si/egp/?lang=en. Accessed 29 Nov 2018

Thunig, H., Wolf, N., Naumann, S., Siegmund, A., Jürgens, C., Uysal, C., Maktav, D. (2011). Land use/land cover classification for applied urban planning – The challenge of automation. https://doi.org/10.1109/JURSE.2011.5764762

UNSD. (2018a). *UNdata – A world of information* [online]. New York, 2018 [cit. 2018-11-27]. Available from: http://data.un.org/

UNSD. (2018b). *Welcome to UNSD – About us* [online]. New York, 2018 [cit. 2018-11-27]. Available from: https://unstats.un.org/home/about/

Vaculík, M., Pászto, V., & Švarcová, B. (2017). Spatial distribution of innovation activities in Czech Republic, 2010–2012. *Journal of International Studies Vol, 10*(1).

World Bank Group. (2018). *What we do* [online]. Washington, DC, 2018 [cit. 2018-11-27]. Available from: https://www.wto.org/english/thewto_e/thewto_e.htm

World Trade Organisation. (2018). *UNdata – The WTO* [online]. Geneva, 2018 [cit. 2018-11-27]. Available from: https://www.wto.org/english/thewto_e/thewto_e.htm

Quantitative Methods

2

Polona Tominc and Vesna Čančer

Abstract

In today's world of digitization and the emergence of large amounts of data, it is extremely important that we know how to extract the information that this data captures and that represents foundations we need in organizations for efficient decision-making. Therefore, for graduates, regardless of the field of study, the knowledge of quantitative methods and the ability of "statistical thinking" is extremely important. Quantitative methods and quantitative approach to research and analysis of data imply important advantages over qualitative approach, but their combination can certainly in many areas mean an effective approach to obtaining information that is hidden within data.

In this chapter, we first consider some selected methodological approaches to data analysis, in particular the basic principles of inferential statistics, which enables us to generalize the results of a random sample, with a certain degree of probability, to a statistical population. We also show how to use the logistic regression methodology on a case that shows the importance of combining the economic and geographic aspects of research, i.e. the "spationomy" approach.

In the next part of the chapter the multi-criteria decision-making is presented, that has proven useful in solving spatial decision problems. We develop and apply the multi-criteria model for the protection of agricultural land for food self-sufficiency. Based on the literature review, the theoretical backgrounds of multi-criteria decision-making, together with the use of multi-criteria decision making in land-use evaluation and management are introduced. In addition, the multi-criteria model for the protection of agricultural land for food self-sufficiency is developed, taking into account the characteristics of the protection of agricultural land and public data base information in Slovenia. For this purpose, we followed the frame procedure for multi-criteria decision making by using the group of methods based on assigning weights to criteria – in this research, we used the Simplified Multi-attribute Rating Technique and the Analytic Hierarchy Process. Selected geographical and economic factors were structured in the criteria hierarchy. In synthesis, the additive model was used in order to select the most favorable solution. The aggregate values obtained with an additive model were completed by considering synergies and redundancies among criteria by a fuzzy measure – discrete Choquet integral. The results enable suggesting measures for the protection of agricultural land for food self-sufficiency.

P. Tominc (✉) · V. Čančer
Faculty of Economics and Business, University of Maribor, Maribor, Slovenia
e-mail: polona.tominc@um.si; vesna.cancer@um.si

V. Pászto et al. (eds.), *Spationomy*, https://doi.org/10.1007/978-3-030-26626-4_2

Keywords
Quantitative methods · Inferential statistics · Logistic regression · Decision theory · Multi-criteria decision-making · Food self-sufficiency · Protection of agricultural land · Prescriptive approach · Weight · Synergy

2.1 Introduction

Due to rapid changes in the digitalization of every-day life that results in the high velocity of data generation, it becomes more and more important that graduates, regardless the study field, have competence in data and business analysis. Data analysis is the process of examining data sets in order to draw conclusions about the information they contain; increasingly with the aid of specialized systems and software. Data analysis technologies and techniques are widely used in commercial industries to enable organizations to make more-informed business decisions and by scientists and researchers to verify or disprove scientific models, theories and hypotheses.

We are in the era of "big data", which is characterized by the high-volume, high-velocity and high-variety information assets that demand cost-effective, innovative forms of information processing for enhanced insight and decision making (Gartner IT Glossary 2018). Hidden information in big data can help companies unlock the strategic value of this information by allowing a company to understand where, when and why the company's customers buy; to optimize workforce planning and operations; to improve inefficiencies in company's supply chain; to predict market trends and future needs and to become more innovative and competitive; to name just a few opportunities. It means that data visualization increasingly becomes a top-need for organizations that are data-driven. There is a need today to develop data visualization abilities within companies, as unique digital assets in the business, for maximum impact and consistent execution against strategic business practices and goals, are implemented (Big Data Quarterly 2016; Šebjan and Tominc 2015). These abilities of organizations can be undoubtedly built based on the competences of their employees.

Quantitative research competence and the so called "statistical thinking" have become indispensable for able citizenship, and for the competences needed for successful placement in the labour market.

Recent decades saw a great increase in the use of quantitative research methods, not only in the natural science, but particularly in the social sciences. Although the research process itself is in general the same, the quantitative and qualitative researches differ in terms of methods for data collection, the procedures and methodologies used for data analysis and the style of communication of the findings (Kumar 2005).

There are some important characteristics and advantages of using quantitative methods. An important feature is that the variables that quantitative methods include in the analysis are both numerical and nominal; data analysis is, in principle, quick, especially with the help of software; the results can be generalized to a statistical population if the data base represents a random sample; very often, quantitative analysis-based findings are also accepted as more credible by policy and decision makers.

In this chapter we first introduce the selected set of quantitative statistical methodologies, that address the basic principle of inferential statistics with the principles of generalizing sample statistics to the parameters of the statistical population – thus, to show how to use sample data to estimate population parameters. In the next section of the chapter we present the use of the binary logistic regression. The goal of logistic regression is to explain categorical variable, divided into two groups, on the interval of explanatory variables (interval scaled, ratio scaled, categorical) (Janssens et al. 2008).

In the third part of this chapter the multi-criteria decision-making is presented; it has proven useful in solving spatial decision problems. The main goal of the third section of this chapter is to develop and apply the multi-criteria model for the protection of agricultural land for food self-sufficiency. The issue of food self-sufficiency (i.e. covering domestic food needs with domestic production) and related food security is namely becoming increasingly important both in the world and in the European Union and, of course, in individual European Union Member States, including Slovenia (Court of Audit 2013).

Based on the literature review, the theoretical backgrounds of multi-criteria decision-making, together with its' use in land-use evaluation and management are introduced. In continuation of this chapter, the multi-criteria model for the protection of agricultural land for food self-sufficiency is developed, taking into account the characteristics of the protection of agricultural land and public data base information in Slovenia. For this purpose, we followed the frame procedure for multi-criteria decision making by using the group of methods based on assigning weights to criteria (Čančer and Mulej 2013) – in this research, we used the Simplified Multi-attribute Rating Technique and the Analytic Hierarchy Process. Selected geographical and economic factors were structured in the criteria hierarchy. In synthesis, the additive model was used in order to select the most favorable solution. The aggregate values obtained with an additive model were completed by considering synergies and redundancies among criteria by a fuzzy measure – discrete Choquet integral. The results enable suggesting measures for the protection of agricultural land for food self-sufficiency.

2.2 Multivariate and Univariate Statistics

Multivariate statistical methods are extremely popular in contemporary research, both in business, economics and management studies, as well as in the geoinformatics and spatial studies in general. Multivariate statistics provides analytical tools when complex research models are addressed, with many independent and dependent variables, all or majority of them being associated with each other to varying degree. The domains of multivariate and univariate statistics can be characterized by the number of dependent and independent variables involved in the research, which are defined as follows (Tabachnick and Fidell 2013):

- Independent variables are the conditions that differ, and to which the subjects or objects of analysis are exposed; or are the characteristics that the subjects or objects of the analysis bring into the research.
- Dependent variables are predicted by independent variables. That is why independent variables are sometimes called predictor variables, while the dependent variables are sometimes called response or outcome variables.

The dependent and independent variables are defined within the context of an individual research, therefore the dependent variable in one research may be in the role of independent variable in the other.

Univariate, bivariate and multivariate statistical methods are usually used simultaneously, especially in the interdisciplinary research (Hu and Zhang 2017) - in several situations, the cross-disciplinary scientific approach, which combines management sciences and natural sciences, is important. Examples can be found in natural resources management (Robinson et al. 2012), in the context of real-estate markets (Benson et al. 2000), or when performing decision-making on the basis of geo-analysis models (Yue et al. 2015).

Univariate statistics refer to the research situation, where single dependent variable is involved, but there may be several independent variables involved (for example testing the difference between the sample mean and population mean, testing the difference between population means in k different groups, that may represent independent or paired groups, one-way analysis of variance and several others). Bivariate statistics usually refers to the analysis of relationship between two variables, usually by Pearson correlation coefficient, Spearman rank correlation coefficient or Chi-square analysis. Multivariate statistical methods are the extension of univariate and bivariate statistics. Within multivariate methodological approach several dependent and independent variables are simultaneously analysed and complex interrelationship among variables are revealed and assessed in statistical inference (Tabachnick and Fidell 2013).

2.3 Descriptive and Inferential Statistics

Inferential statistical methods use sample statistics to make predictions about the population parameters (Agresti and Finlay 2009), with the quality of the inference being dependent on how well the sample represents the population characteristics. In general, sampling is a process of selecting a few units (a sample) from a bigger group (a sampling population), to become a basis for estimating or predicting the prevalence of an unknown piece of information, situation or outcome regarding the sampling population (Kumar 2005). Ideally samples are selected by some random process, so that they represent the population of interest; two main factors that affect the inferences drawn from a sample are the size of a sample and the extent of variation in the sampling population.

While the descriptive statistics describes samples of subjects or objects of the research in terms of variables or combination of variables, inferential statistical techniques test hypotheses about differences in populations on the basis of measurements made on samples. If differences are reliable, descriptive statistics (sample statistics) can be used to provide estimations of statistical parameters in the population. Descriptive and inferential statistics are rarely an either-or proposition, since we are usually interested in both describing and making inferences about the data, but the restrictions and constraints that should be taken into account are different. For example, if simple description of the sample is the major goal, many assumptions which are necessary for inference within the multivariate statistical methods, may be relaxed (Tabachnick and Fidell 2013).

2.4 Statistical Inference: Estimation

A single number calculated from a sample, is called a point estimate (a sample statistic):

$\overline{Y}$ – a sample mean value
p – a sample proportion

A sample mean value (statistic) is used to estimate the population mean value (parameter) of the variable analysed, μ, while the sample proportion (statistic) is used to estimate the population proportion (parameter), π (the proportion of statistical units with a certain characteristics).

The sampling distribution of a sample statistic is the probability distribution that specifies probabilities for the possible values the statistic can take (Agresti and Finlay 2009). Each sample statistic has a sampling distribution (sample mean, sample proportion, sample median and so forth). A sampling distribution speciefies probabilities not for individual observations but for possible values of a statistic computed from the observations; it is important in inferential statistics for predicting how close a statistic falls to the parameter it estimates – sampling distribution describes how statistic varies (ibid).

According to the central limit theorem (McClave et al. 2011), the sampling distribution of the random sample mean for large samples is approximately a normal distribution[1]. The approximate normality of the sampling distribution applies no matter what the shape of the population distribution of the observed variable – for large random samples the sampling distribution of $\overline{Y}$ is approximately normal even if the population distribution is highely skewed, U shaped or highely discrete (such as binary distribution) (ibid).

How large the sample size must be to make the above conclusion of the central limit theorem, largly depends on the skewness of the population distribution of the observed variable – if the

[1] The normal distribution, characterized by the bell-shaped curve, is the most important probability distribution for statistical analysis. It is described in details for example in Agresti and Finlay (2009, p. 78).

population distribution is bell shaped, the sampling distribution is bell shaped as well for all sample sizes. More skewed distributions require larger sample sizes, but for more cases the sample size of 30 is sufficient, although it may not be large enough to make precise inference (ibid). The result applies also to proportions, since a sample proportion is a special case of the sample mean, for observations coded as 0 and 1 (ibid).

The mean of sampling distribution equals mean of the sampled population ($\overline{Y}$ is the unbiased estimate of μ) and the standard deviation of sampling distribution equals standard error of the mean:

$$se_{\overline{Y}} = \frac{\text{stand.dev.of sampled population}}{\text{square root of sample size}}$$

If the sample size, n, is large, the sample standard deviation, s, is a good estimator of a standard deviation of sampled population, σ:

$$s = \frac{1}{n-1}\sum_{i=1}^{n}\left(y_i - \overline{Y}\right)^2$$

A large samples confidence interval for μ, with the confidence level $(1 - \alpha)$ %, equals

$$\overline{Y} \pm z_{\alpha/2} se_{\overline{Y}}, \qquad (2.1)$$

where $z_{\alpha/2}$ is the adequate z-value of a standardized normal distribution.

Commonly used confidence levels are (Tol 2009; Bathelt et al. 2004):

$$90\% \Rightarrow z_{\alpha/2} = 1.645$$
$$95\% \Rightarrow z_{\alpha/2} = 1.96$$
$$99\% \Rightarrow z_{\alpha/2} = 2.56$$

Similarly holds for the population proportion. If the sample size is large (the condition is satisfied if $np \geq 15$ and $n(1 - p) \geq 15$) and a random sample is selected from the target population, the sampling distribution of sample proportion is approximately normal (McClave et al. 2011). The mean of sampling distribution equals population proportion (p is the unbiased estimate of π) and the standard deviation of sampling distribution equals standard error of the proportion:

$$se_{\pi} = \sqrt{\frac{p(1-p)}{n}}$$

A large samples confidence interval for π, with the confidence level $(1 - \alpha)$ %, equals

$$p \pm z_{\alpha/2} se_{\pi}$$

where $z_{\alpha/2}$ is the adequate z-value of a standardized normal distribution.

Example 2.1: Some Characteristics of the Real Estate Market

In the city of Maribor, the real estate agency wanted to estimate the market value of the apartments, available on the market, based on the past transactions and characteristics of apartments that were sold in the past. Among the influential factors having an impact on the market value (price in EUR/s.m) also the distance from the city centre was considered. The sample of 1117 past transactions was analysed; the sample was obtained in years 2013 and 2014 (Živkovič and Tominc 2016). The descriptive statistics of the sample data for variables "price per sq. m" and "the distance from the city centre" is as follows (Table 2.1).

The frequency histogram, presented at Fig. 2.1 shows the frequency distribution for variable describing the distance of the real estate from the city centre. This Figure also presents the normal distribution, adjusted to the empirical frequency distribution, showing the differences among the empirical distribution (sample data) and theoretical distribution (adjusted normal distribution). Fig. 2.2 presents the frequency histogram for variable describing the price of the real estate (in EUR per sq.m.; frequency distribution and adjusted normal distribution exhibit a higher degree of similarity. The statistical tests for testing the hypothesis about the normal distribution of variables in the population (Kolmogorov-Smirnov and Shapiro-Wilk test) exceed the

Table 2.1 Sample descriptive statistic for real estate data

	The distance from the city centre (m)	Price in EUR/sq. m
Sample size	1116	1116
Sample mean	1646.16	857.81
Median	1510.00	863.96
Mode	1160.00	1000
Sample standard deviation	917.88	195.10
Sample variance	842,509.78	38,064.74
Range	6740	1033.50
Min	40.00	348.08
Max	6780.00	1381.58

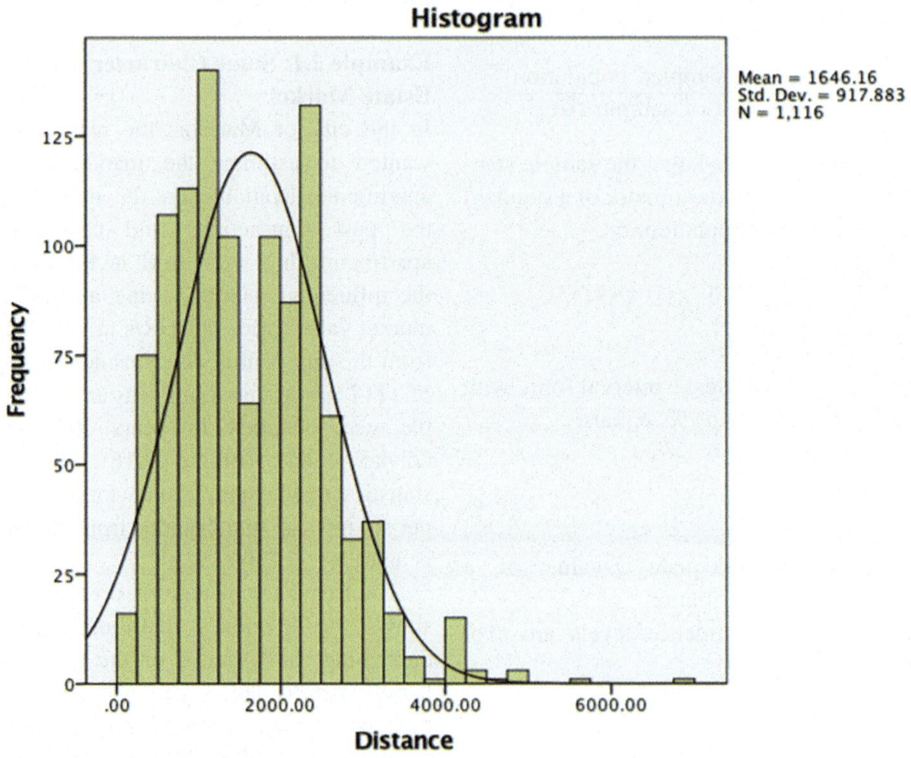

Fig. 2.1 Frequency distribution for the "Distance from the city centre" (in m)

scope of this chapter, and the reader can get acquainted with them in the literature (Tabachnick and Fidell 2013).

Using the above mentioned procedure and Eq. (2.1), the 95% confidence interval for the population mean value for variable "distance from the city centre" is calculated:

$$\overline{Y} \pm z_{\alpha/2} se_{\overline{Y}} = 1.80 \pm 1.96 * \frac{5.36}{\sqrt{1117}}$$
$$= 1.80 \pm 0.314$$

Therefore, we can be 95% confident, that the population mean value for mean distance from the city centre for the apartments sold in the

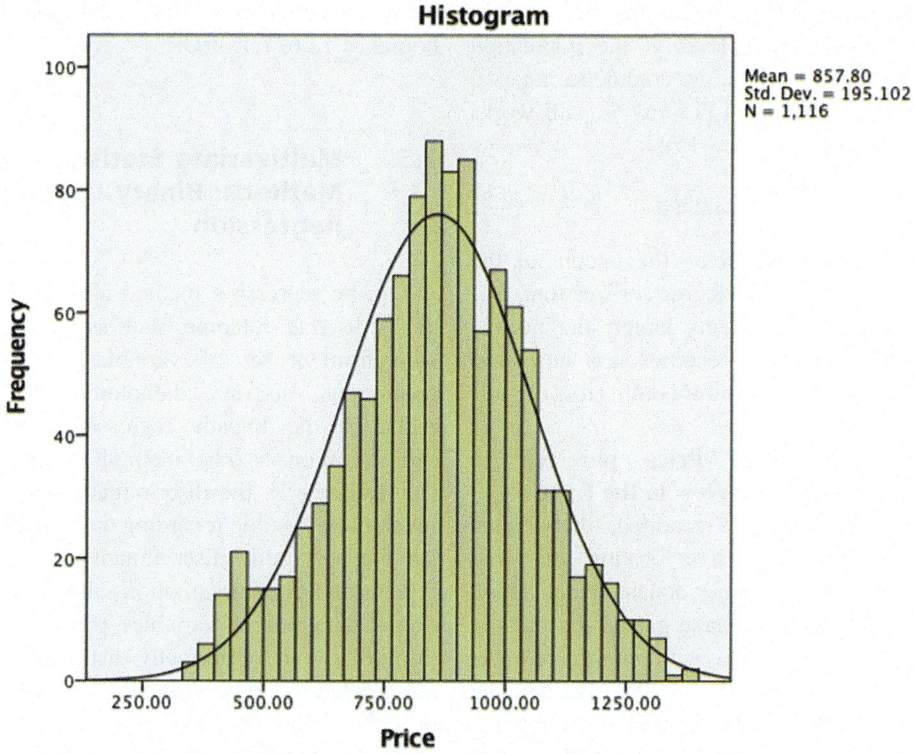

Fig. 2.2 Frequency distribution for the "Price" (in EUR/sq.m)

market in 2013 and 2014 in Maribor, lies between the lower bound of 1.486 km and the upper bound of 2.114 km.

Similarly, we can calculate the 95% confidence interval for the population mean value for variable "price in EUR per sq. m". We can be 95% confident, that the population mean value for mean price in EUR per sq. m, for the apartments sold in the market in 2013 and 2014 in Maribor, lies between the lower bound of 846.385 EUR per sq. m and the upper bound of 868.935 EUR per sq. m.

2.4.1 The t-distribution

For obtaining a confidence interval for any sample size (also less than 30), the assumption of normal population distribution must be fulfilled – in this case the sampling distribution of $\overline{Y}$ is normal even for small sample sizes (Agresti and Finlay 2009). If the population standard deviation, σ, is known, the exact standard error of the sample mean is known as well:

$$se_{\overline{Y}} = \frac{\sigma}{\text{square root of sample size}}$$

and the following equation can be used also when n < 30:

$$\overline{Y} \pm z_{\alpha/2} se_{\overline{Y}}$$

But in practice, the population standard deviation and therefore the exact standard error of mean, are not known. Substituting the sample standard error, s, for σ, to get the estimated standard error, then introduces extra error. To account for this increased error, the t-distribution[2] for obtaining confidence intervals, is used. T-distribution is robust against violation of the normal population distribution assumption (a statistical method is called robust with respect to particular assumption

[2] T-distribution is described in details for example in Agresti and Finlay (2009, p.118).

if it performs adequately even when the assumption is violated (ibid)). Even if the population distribution is not normal, the confidence interval with the confidence level (1 − α) %, still works quite well (especially if n > 15):

$$\overline{Y} \pm t_{n-1;\alpha/2} se_{\overline{Y}}, \qquad (2.2)$$

where $t_{n-1;\alpha/2}$ represents the t-score of the t-distribution with (n–1) degrees of freedom, df.

As the sample size gets larger, the normal population distribution becomes less important because of the central limit theorem (ibid).

Example 2.2: Mean Price per sq. m for Apartments with a View to the Lake
In August 2017, it was reported, that Zurich remains the most expensive location for Swiss property at CHF12,250 per square metre. However, houses in Lucerne have gained the most in value over the past decade, with one square metre costing CHF 8500, up 82% on 2007. In particular demand are homes in lake regions, according to a report published on August 17, 2017 by the federal institute of technology ETH Zurich. The area with the second-highest increase in property prices over the past 10 years was Horgen, overlooking Lake Zurich, where a square metre has climbed by 80% to CHF 11,000 (Swissinfo 2017).

Suppose, that a random sample of 20 apartments that have a view to the lake were obtained. The sample mean value for price per sq. m equals 11,110 EUR, with the sample standard deviation that equals 3250.25 EUR.

Using the above described procedure and Eq. (2.2), the 95% confidence interval for the population mean value for variable "price per sq. m" for apartments with the view to the lake, is calculated:

$$\overline{Y} \pm t_{n-1;\frac{\alpha}{2}} se_{\overline{Y}} =$$
$$= 11,110 \pm t_{19;0.025} * \frac{3,250.25}{\sqrt{20}}$$
$$= 11,110 \pm 2.093 * 726.78$$

Therefore, we can be 95% confident, that the population mean value for mean price per sq.m for the apartments sold in the market, lies between the lower bound of 9588.85 EUR and the upper bound of 12,631.15 EUR.

2.5 Multivariate Statistical Methods: Binary Logistic Regression

A logistic regression method allows one to predict a discrete outcome, such as group membership from a set of variables, that may be continuous, discrete, dichotomous or a mix. Although the logistic regression answers the same questions as other methods for classification (for example as the discriminant analysis), it is much more flexible regarding any other multivariate method: unlike discriminant analysis, logistic regression has no assumptions about the distribution of the predictor variables, predictor variables do not have to be normally distributed and linearly related to the dependent (grouping) variable, or of equal variance within each group (Tabachnick and Fidell 2013). Unlike a multiple regression method, logistic regression does not require a linear relationship between the dependent and independent variables. Also the error terms (residuals) do not need to be normally distributed and homoscedasticity is not required. As already mentioned, the dependent variable in logistic regression is not measured on an interval or ratio scale.

But some other assumptions still apply. The dependent variable may have two or more categories – binary or multinomial logistic regression analysis may be employed. Binary logistic regression, that is presented in this chapter, requires the dependent variable to be binary. Logistic regression also requires the observations to be independent of each other (the observations should not come from repeated measurements or matched data). Logistic regression also requires there to be little or no multicollinearity among the independent variables. This means that the independent variables should not be too highly correlated with each other. Although this analysis does not require the dependent and independent variables to be related linearly, it requires that the independent variables are linearly related to the logit or log odds (defined later by Eq. 2.3). Logistic regression typically requires a large sample size as well. A general guideline is that one needs at minimum of 10 cases with the least frequent outcome for each independent variable in the model. For example, if you have 5 independent variables and the expected probability of your least frequent outcome is 0.10, then you would need a minimum sample size of 500 (10∗5/0.10) (ibod).

Logistic regression is often used in analyses that use data of geographic location in different fields of research (for example Rodrigues et al. 2014; Bhat et al. 2009).

In the binary logistic regression model, the outcome variable, *Y*, is the probability of having one outcome or another based on a nonlinear function of the best linear combination of predictors; with two outcomes (ibid):

$$Y_i = \frac{e^u}{1 + e^u}$$

where Y_i is the estimated probability that the i-th case ($i = 1,2, \ldots, n$) is in one of the categories and u is the linear regression equation:

$$u = A + B_1X_1 + B_2X_2 + \cdots + B_kX_k$$

with constant A, coefficients B_j and predictors X_j for k predictors ($j = 1,2, \ldots, k$).

Taking the logarithm leads to the logit or log odds in the form

$$\ln\left(\frac{Y}{1-Y}\right) = A + \sum B_jX_j, \qquad (2.3)$$

For estimating the coefficients of the logit function the maximum likelihood procedure is used. This is an iterative procedure that starts with the arbitrary values of coefficients and determines the direction and size of change in coefficients that will maximize the likelihood of obtaining the observed frequencies. Then residuals of the predictive model nased on those coefficients are tested and another determination of direction and size of change in coefficients is made. The procedure continues until the coefficients change very little. Maximum likelihood estimates are then those parameter estimates that maximize the probability of obtaining the observed outcome frequencies (Hox 2002).

The procedure od applying the binary logistic procedure is presented in the following example, where the suitability of the model is assessed, using the model Chi-square, Cox and Snell R square, Nagelkerke R square and Classification table (Janssens et al. 2008); the meaning and the significance of the logistic regression coefficients is assessed as well.

Example 2.3: Characteristics of NUTS-2 Regions in Slovenia

We would like to find out whether several indicators associated with spatial, economic and social view-points are characteristical for the NUTS-2 regions in Slovenia. In the NUTS (Nomenclature of Territorial Units for Statistics) codes of Slovenia (SI), the three levels are:

- NUTS 1, no subdivisions
- NUTS 2, Macro regions (two)
- NUTS 3, Statistical regions (12)

The data for 212 municipalities of Slovenia was obtained from the Statistical Office of Slovenia (Statistical office RS 2017) for the following indicators:

- Population density (population per sq. km, 2nd half of 2016)
- Natural change (natural increase per 1000 population, 2016)
- Net migration (total net migration per 1000 population, 2016)
- Registered unemployment rate, 2016
- Tertiary education (number of students per 1000 population, 2015/2016)
- Convicted adults and juveniles per 1000 population (2015)
- Waste (in kg per capita, 2016)
- Persons employed per enterprise in the municipality, 2015
- Number of dwellings per 1000 population
- Average monthly net earnings (in EUR, 2016)

Table 2.2 Mean values and standards deviations for selected indicators of spatial, social and economic characteristics of municipalities in NUTS-2 regions of Slovenia

Indicators	Region	N	Mean	Std. Dev.
Population density	Western region	73	142.60	166.57
	Eastern region	139	100.39	97.06
Natural increase per 1000 population	Western region	73	0.85	3.94
	Eastern region	139	−1.04	4.64
Total net migration per 1000 population	Western region	73	1.36	7.08
	Eastern region	139	0.88	9.94
Registered unemployment rate	Western region	73	8.26	1.83
	Eastern region	139	12.12	3.73
Number of students per 1000 population	Western region	73	40.05	5.42
	Eastern region	139	36.65	6.35
Convicted adults and juveniles per 1000 population	Western region	71	2.00	0.85
	Eastern region	129	3.28	1.86
Generated municipal waste (kg per capita)	Western region	73	473.86	121.65
	Eastern region	139	399.19	172.96
Persons employed per enterprise in the municipality	Western region	73	3.38	1.31
	Eastern region	139	3.30	1.58
Number of dwellings per 1000 population	Western region	73	411.86	92.97
	Eastern region	139	410.55	71.95
Average monthly net earnings	Western region	73	960.60	86.99
	Eastern region	139	909.43	73.05

Municipalities were grouped according to the NUTS 2 macro regions in Eastern region (139 municipalities, value 1) and Western region (73 municipalities, value 0).

The descriptive statistics for variables included into the analysis is presented by Table 2.2.

The descriptive statistics in Table 2.2 shows, that in the municipalities of the Eastern region there is on average lower population density, lower (negative) natural increase of population, higher registered unemployment rate, lower net migrations, lower number of students, higher number of convicted adults and juveniles per 1000 inhabitants, lower generated waste, almost equal number of persons employed by enterprises in the municipality and number of dwellings per 1000 inhabitants and on average lower monthly net earnings, as compared with the Western region of Slovenia.

The logistic regression results are presented in the Tables 2.3, 2.4, 2.5, 2.6, 2.7 and 2.8 below (discriminant classification technique, that could be useful in such research, cannot be used, since the data significantly differ from the multivariate normal distribution).

Table 2.3 Case processing summary

Unweighted Cases		N	Percent
Selected Cases	Included in Analysis	200	94.3
	Missing Cases	12	5.7
	Total	212	100.0
Unselected Cases		0	.0
Total		212	100.0

Table 2.4 Dependent variable encoding

Original Value	Internal Value
Western region	0
Eastern region	1

Table 2.5 Omnibus tests of model coefficients

		Chi-square	df	Sig.
Step 1	Step	117.631	10	.000
	Block	117.631	10	.000
	Model	117.631	10	.000

The model has been estimated on the basis of 200 cases, due to 12 missing cases, as presented by Table 2.3.

Table 2.6 Model summary

Step	−2 Log likelihood	Cox & Snell R Square	Nagelkerke R Square
1	142.564a	.445	.611

a. Estimation terminated at iteration number 6 because parameter estimates changed by less than .001.

Table 2.7 Classification table

			Predicted		
			region		
	Observed		Western region	Eastern region	% Correct
Step 1	region	Western region	51	20	71.8
		Eastern region	14	115	89.1
	Overall Percentage				83.0

a. The cut value is .500

Table 2.8 Variables in the logistic regression equation

		B	S.E.	Wald	df	Sig.	Exp (B)
Step 1	X1 - Population density	−0.007	0.002	8.910	1	0.003	0.993
	X2 - Natural increase per 1000 population	−0.012	0.058	0.041	1	0.839	0.988
	X3 - Total net migration per 1000 population	0.019	0.026	0.522	1	0.470	1.019
	X4 - Registered unemployment rate	0.677	0.140	23.434	1	0.000	1.968
	X5 - Number of students per 1000 population	−0.028	0.043	0.417	1	0.519	0.973
	X6 - Convicted adults and juveniles per 1000 population	0.831	0.223	13.816	1	0.000	2.295
	X7 - Generated municipal waste (kg per capita)	−0.001	0.001	0.609	1	0.435	0.999
	X8 - Persons employed per enterprise in the municipality	0.063	0.132	0.227	1	0.634	1.065
	X9 - Number of dwellings per 1000 population	−0.005	0.003	2.466	1	0.116	0.995
	X10 - Average monthly net earnings	−0.007	0.003	4.259	1	0.039	0.993
	Constant	2.096	4.293	0.238	1	0.625	8.131

The Dependent variable encoding table (Table 2.4) is linked to the coding of the "region" variable, with 0 for Western region and 1 for Eastern region. It is important, that the dichotomous dependent variable is encoded by 0 and 1, in order to be able to perform the logistic regression and to answer the relevant research question, namely, what is the contribution of a selected independent variable to the differences among both regions.

The output is comprised of two blocks, "Block 0" corresponds to the zero-model (the model is initial model, created only on the basis of constant) and "Block 1" is related to the full-model (that includes all independent variables). Here the tables of the "Block 1″ results are presented.

With the purpose to determine the suitability of the model, the results for the model Chi-square, Cox and Snell R square, Nagelkerke R square and Classification table, are presented in Tables 2.5, 2.6 and 2.7.

The model Chi-square which may be found in Table 2.5, suggests that the full-model is more suitable than the zero-model, with at least one of the regression coefficients of the independent variables is significantly different from zero. The number 10 for degrees of freedom corresponds to the number of estimates in the full-model as compared to the zero-model (regression coefficients for 10 independent variables). Also the Step, Block and the Model Chi-squares are all equal in the current application (blocks of variables are included within the stepwise method).

Cox & Snell R square in Table 2.6 is comparable to R square in the linear regression, in a sense that a higher value corresponds to a better fit of the model, but the Cox & Snell R square cannot accept the maximum value of 1. Nagelkerke R square in Table 2.6 does meet this requirement, since it falls within a range from 0 to 1. This criterion is 0.611, which is on the high side and points to the quite high suitability of the full-model.

In the classification table, presented in Table 2.7, it appears that the full-model classifies 115 + 51 units as being correct (83%), whereas 14 municipalities were incorrectly included into the Western region based on the values of independent variables, and similarly, 20 were incorrectly included into the Eastern region. The score of 83% of correct classification points to the quite high suitability of the full-model, as well.

The next step consists of the assessment of the logistic regression coefficients. On the basis of the Table 2.8, the probability of response variable (Eastern region), using the Eq. (2.3), is:

$$\begin{aligned}\mathrm{u} = 2.096 - 0.007\mathrm{X}_1 - 0.012\mathrm{X}_2 + 0.019\mathrm{X}_3 +\\ +0.677\mathrm{X}_4 - 0.028\mathrm{X}_5 + 0.831\mathrm{X}_6 - 0.001\mathrm{X}_7 +\\ +0.063\mathrm{X}_8 - 0.005\mathrm{X}_9 - 0.007\mathrm{X}_{10}\end{aligned}$$

The Wald statistics, which is Chi-square distributed, and the p-values linked to it, are used to determine whether the regression coefficient differs significantly from zero or not. Results in Table 2.8 reveal, that regression coefficients linked to Population density, Registered unemployment rate, Convicted adults and juveniles per 1000 population and Average monthly net earnings, differ significantly from zero. The interpretation of the regression coefficients is examined in terms of "odds" and "log odds".

The "log odds" is defined as follows:

$$u = A + B_1X_1 + B_2X_2 + \cdots + B_kX_k = Logit = Log\ odds = Log\ \frac{Probability\ (Event)}{Probability\ (No\ event)}$$

(In our case the probability that the municipality belongs to the Eastern region.)

The regression coefficient B shows the change in the "log odds" for a change in the independent variable X by one unit. The above equation may be transformed into the function of "odds":

$$\begin{aligned}Odds &= \frac{Probability\ (Event)}{Probability\ (No\ event)}\\ &= e^{A+B_1X_1+B_2X_2+\cdots+B_kX_k}\end{aligned}$$

The Exp(B), sometimes called "odds ratio", reveals that the change of an independent variable for one unit, will result in decrease (in the case of negative B) or in the increase (in the case of positive B) of the "odds", by a factor expressed by Exp(B).

Therefore, the explanation of the significant regression coefficients is as follows.

- The negative and significant regression coefficient B linked to the Population density, that equals −0.007 and the "odds ratio" that equals 0.993 indicate, that for the unit increase in the Population density we see the 0.7% decrease in "odds" of being in the Eastern region. Municipalities with the higher Population density are less likely to be a part of the Eastern region.
- The positive and significant regression coefficient B linked to the Registered unemployment rate, that equals 0.677 and the "odds ratio" that equals 1.968 indicate, that for a unit increase in the Registered unemployment rate we see the 96.8% increase in "odds" of being in the Eastern region. Municipalities with higher Registered unemployment rate are more likely to be a part of the Eastern region.
- The positive and significant regression coefficient B linked to the Convicted adults and juveniles per 1000 population, that equals 0.831 and the "odds ratio" that equals 2.295 indicate, that for a unit increase in the rate of Convicted adults and juveniles per 1000 population, we see the 129.5% increase in "odds" of being in the Eastern region. Municipalities with higher rate of Convicted adults and juveniles per 1000 population, are more likely to be a part of the Eastern region.

– The negative and significant regression coefficient B linked to the Average monthly net earnings, that equals −0.007 and the "odds ratio" that equals 0.993 indicate, that for the unit increase in the Average monthly net earnings, we see the 0.7% decrease in "odds" of being in the Eastern region. Municipalities with the higher Average monthly net earnings, are less likely to be a part of the Eastern region.

The analysis leads to the conclusion that the Eastern region is scored lower regarding several spatial, economic and social indicators, with significantly lower results especially regarding the population density and average monthly net income of inhabitants, while significantly higher rates of registered unemployment and convicted adults and juveniles per 1000 population.

2.6 Multi-criteria Decision Making

Multi-criteria decision making (MCDM) based on assigning weights, describes the set of approaches for making decisions with respect to several, more or less conflicting criteria. They should be used when intuitive decision making is not enough for several reasons: e.g. conflicting criteria or disagreement between decision makers about what criteria are relevant or more important and what alternatives and preferences are acceptable.

The approach used in this chapter follows a prescriptive approach to decision-making (Raiffa 1994): instead of regarding people as perfect rational individuals, we develop systematic decision-making procedures supportive of decision-making. Based on Belton and Stewart's (2012) process of MCDM:

- From problem identification and structuring,
- Through model building and its application for informing and thinking,
- To the action plan for problem solving,

It follows the frame procedure for MCDM based on assigning weights that has already been introduced and well-verified in practice (Čančer and Mulej 2013):

Problem definition and structuring. We identify the problem, describe it thoroughly and determine criteria and alternatives. An MCDM problem can be structured in the form of the hierarchy of criteria and alternatives.

Criteria weighting and measuring local alternatives' values. Weights express the relative importance of criteria. Their values are from 0 to 1. As decision makers may have difficulties when determining the weights of criteria directly, expressing the importance of criteria can be supported with several methods. By using the computer support, the criteria's importance can be expressed by using the methods based on:

- ordinal (e.g. SMARTER),
- interval (e.g. SWING and SMART) and
- ratio scale (e.g. AHP), or
- by direct weighting.

When using the SMARTER method (Edwards and Barron 1994), we rank the attributes in the order of importance for the attribute changes from their worst level to the best level. We start from the most important attribute. The weights can be obtained by the centroid method. In SMART (Edwards 1977), we assign 10 points to the least important attribute change from the worst criterion level to its best level, and then give points (more than or equal to 10, but less than or equal to 100) to reflect the importance of the attribute change from the worst criterion level to the best level relative to the least important attribute change. When using the SWING method (Von Winterfeldt and Edwards 1986) for expressing the criteria's importance, we assign 100 points to the most important attribute change from the worst criterion level to the best level, and then assign points (less than or equal to 100, but more than or equal to 10) to reflect the importance of the attribute change from the worst criterion level to the best level relative to the most important attribute change. In SMART and SWING, the weight of the jth criterion, w_j, is obtained by:

$$w_j = \frac{t_j}{\sum_{j=1}^{m} t_j}, \tag{2.4}$$

where t_j – the points given to the jth criterion, and m – the number of criteria. When the criteria are structured in two levels, the weight of the sth attribute of the jth criterion, w_{js}, is obtained in SMART and SWING by:

$$w_{js} = \frac{t_{js}}{\sum_{s=1}^{p_j} t_{js}}, \tag{2.5}$$

where t_{js} – the points given to the sth attribute of the jth criterion, and p_j – the number of the jth criterion sub-criteria.

The AHP method is based on pairwise comparisons of criteria's importance. Expressing judgments on criteria's importance is based on the 9-point scale: 1 – a criterion is equally important as the compared criterion, 3 – a criterion is moderately more important than the compared criterion, 5 – a criterion is strongly more important than the compared criterion, 7 – a criterion is very strongly more important than the compared criterion, and 9 – a criterion is extremely more important than the compared criterion. Even numbers are interpreted by means of odds: 2 means from equally to moderately more, 4 from moderately more to strongly more etc. The local weights of criteria are calculated by using eigen vectors and eigen values (for details see Saaty 1999).

The values of alternatives with respect to criteria on the lowest hierarchy level (attributes) can be measured by:

- Value functions,
- Pairwise comparisons, and
- Directly.

When using value functions, the lower and upper bounds should be defined, and then appropriate formula (for example, for linear, piecewise linear or exponential value function) should be applied to obtain the local values of alternatives.

Pairwise comparisons of alternatives' values with respect to the criterion on the lowest hierarchy level is one of major advantages of the AHP method, especially when nominal data are available. Expressing preferences to alternatives is based on the 9-point scale; the strengths of preferences can be described as follows: 1 – an alternative is equally preferred as the compared alternative, 3 – an alternative is moderately more preferred than the compared alternative, 5 – an alternative is strongly more preferred than the compared alternative, 7 – an alternative is very strongly more preferred than the compared alternative, and 9 – an alternative is extremely more preferred than the compared alternative. The procedure for calculating the local values of alternatives is equal to calculating the criteria weights.

Synthesis, ranking and sensitivity analysis. In the Multi -Attribute Value (or Utility) Theory (MAVT or MAUT) and the methodologies that were developed on its bases (e.g., SMART, AHP), the additive model is used in synthesis: the aggregate (final) value of each alternative is the sum of the weighted values of the alternative with respect to criteria:

$$v(X_i) = \sum_{j=1}^{m} w_j v_j(X_i), \tag{2.6}$$

for each $i = 1, 2, \ldots, n$,

where $v(X_i)$ is the value of the ith alternative, w_j is the weight of the jth criterion and $v_j(X_i)$ is the local value of the ith alternative with respect to the jth criterion.

When the criteria are structured in two levels, the aggregate alternatives' values are obtained by

$$v(X_i) = \sum_{j=1}^{m} w_j \left(\sum_{s=1}^{p_j} w_{js} v_{js}(X_i) \right), \tag{2.7}$$

for each $i = 1, 2, \ldots, n$,

where p_j is the number of the jth criterion sub-criteria, w_{js} is the weight of the sth attribute of the jth criterion and $v_{js}(X_i)$ is the local value of the ith alternative with respect to the sth attribute of the jth criterion.

The use of the additive model (2.6) or (2.7) is not appropriate when there is an interaction among the criteria. If the criteria can interact with each other, not only the weights of each criterion but also weighting on subsets of criteria should be considered (Moaven et al. 2008; Sridhar et al. 2008). Marichal (2000) defines and describes three kinds of interaction among criteria that could exist in the decision-making problem: correlation, complementary, and preferential dependency.

The aggregate values obtained with an additive model can be completed by considering synergies and redundancies among criteria by a fuzzy measure – discrete Choquet integral. Let us consider a finite set of alternatives X and a finite set of criteria K in an MCDM problem. In order to have a flexible representation of complex interaction phenomena between criteria, it is useful to substitute to the weight vector w a non-additive set function on K allowing to define a weight not only on each criterion, but also on each subset of criteria. For this purpose, the concept of fuzzy measure has been introduced. A suitable aggregation operator, which generalizes the weighted arithmetic mean, is the discrete Choquet integral.

Proposed in capacity theory (Choquet 1953), the concept of the Choquet integral was used in various contexts, among them in non-additive utility (value) theory (Grabisch 1995; Marichal 2000). Let us adapt its definition to the MAVT. Following (Grabisch 1995) and (Marichal 2000), this integral is viewed here as an m-variable aggregation function; let us adopt a function-like notation instead of the usual integral form, where the integrand is a set of m real values, denoted $v = (v_1, \ldots, v_m) \in \mathfrak{R}_n$. The (discrete) Choquet integral of $v \in \mathfrak{R}_n$ with respect to w is defined by

$$C_w(v) = \sum_{j=1}^{m} v_{(j)} \left[w\left(K_{(j)}\right) - w\left(K_{(j+1)}\right) \right], \quad (2.8)$$

where (.) is a permutation on K – the set of criteria, such that $v_{(1)} \leq \ldots \leq v_{(m)}$, where $K_{(j)} = \{(j), \ldots, (m)\}$. Similarly, to consider the interactions among the second-level criteria, the alternative's value with respect to the jth first level criterion can be substituted with the Choquet integral

$$C_w\left(v_j\right) = \sum_{s=1}^{p_j} v_{(js)} \left[w\left(K_{(js)}\right) - w\left(K_{(js+1)}\right) \right]. \quad (2.9)$$

With ranking, we can select the most appropriate alternative(s), eliminate the alternative(s) with the lowest aggregate value, or compare the alternatives with respect to their aggregate values.

Several types of sensitivity analysis enable decision makers to investigate the sensitivity of the goal fulfilment to changes in the criteria's weights (e.g. gradient and dynamic sensitivity), and to detect the key success or failure factors for the goal fulfilment (e.g. performance sensitivity). Jankowski (1995) addressed another advantage of performing sensitivity analysis, namely the problems of imprecision, uncertainty and inaccurate determination of decision makers' preferences are addressed in MCDM techniques by sensitivity analysis of decision recommendations.

2.7 Use of Multi-criteria Decision Making in Land-Use Evaluation and Management

Many spatial decision making problems, such as location or site selection, the selection of optimal utility routes in urban or rural areas, and land use decision making, are complex problems and thus require considering multiple, more or less conflicting criteria when choosing the best alternative(s). From 1980s onwards, multi-criteria evaluation methods were introduced to spatial decision making and Geographical Information Systems (GIS) (Malczewski 1999; Massam 1980; Voogd 1983), together with file exchange modules and computer programs (see, e.g., Jankowski 1995).

As Burian et al. (2012) exposed, new methods and approaches were developed and applied in spatial planning in the last 15 years (for example, GIS, Global Positioning Systems (GPS), and remote sensing techniques). In addition, Burian et al. (2015) presented a new approach to automatic optimal land use scenario modelling using the developed ArcGIS Urban Planner extension, functional in the assessment of land suitability and the detection of optimal areas suitable for urban development. The land suitability was assessed with respect to physico-geographical factors and socioeconomic factors with the weights, set at modelling final scenarios.

GIS based multi-criteria analysis has been used for supporting decisions about several problems. Rikalovic et al. (2014) applied it in industrial site selection. Ogryzek and Rzasa (2017) used it in the revitalization of space, applied on a local revitalization program. To assess revitalization projects of various types of urban public space, Palicki (2015) created valuation models of different types of public space based on multi-criteria analysis methodology taken into account the multidimensionality of assessment criteria (space and economic, social, economic, urban and cultural indicators); the PROMETHEE method with the software D-sight was used. In addition, Torrieri and Batà (2017) used the Integrated Spatial Multi-criteria Decision Support System (ISMDSS), i.e., the integration of GIS and multi-criteria analysis, in strategic environmental assessment, to support the preparation of environmental assessment reports and the construction of scenarios for the adoption of urban plans. The table of effects included the agriculture, forestry, tourism, industry, noise, waste, and atmosphere and water indicators.

Moreover, Rinner (2007) proposed to use principles of geographic visualization in conjunction with multi-criteria evaluation methods to support expert-level spatial decision making and presented a case study where the AHP was used to calculate composite measures of urban quality of life for neighborhoods. The study of land analysis for urban extension presented by Chen (2014) introduces the multi-criteria decision analysis based on the additive model. Due to data limitations, this study was based on simplified factors, such as population, employment and average income. The lowest weight was assigned to average income, and the highest to population (Chen 2014). Yang et al. (2008) combined suitability modeling with remote sensing, landscape ecological analysis and GIS to develop a spatial analyzing system for urban expansion land management; to address the uncertainties during the evaluation process, grey relational analysis (GRA) was combined with the AHP. Land evaluation was done with respect to the indicators of biological and water resources, soil resources, and society and economy. Among the first level criteria, the lowest weight was assigned to society and economy, and the highest to soil resources.

Javadian et al. (2011) focused on environmental suitability analysis of educational land use by using AHP and GIS. The criteria were access range, compatibility and slope. A GIS-based MCDM using pairwise comparisons of the AHP was also applied in investigating suitable locations for a new recreational park (Lawal et al. 2011). Along with the criteria academic area buildings, student residential zone and pedestrian paths, the model included also the criteria road networks, slope and land use (Lawal et al. 2011).

The research of Nyeko (2012) explores application of MCDM and GIS to find the best spatial allocation of land to future agriculture and forest development. Factors concerning allocation to agriculture were rainfall, road, settlement, population, water, normalized difference vegetation index, and land use. The AHP was used to determine the criteria's weights. Nyeko (2012) stated that the objectives of land allocation to agriculture are to increase productivity of land and the scale of farming and to protect the environmental including controlling soil erosion and soil degradation. The biophysical parameters used in suitability analyses of land for agriculture land use were the Normalized Difference Vegetation Index (NDVI) used as measurement of biomass density and soil fertility, settlement and road network maps used in determining accessibility, rainfall map used in assessing adequacy of rainfall received, and reference land use map used in setting the constraints. The most important criterion was rainfall, and the least important criteria were NDVI and land use.

Pažek et al. (2018) presented a multi-criteria DEXi model for the assessment of individual less favored areas and farming systems with respect to criteria of sustainability and farming potential. The criteria were as follows: farm description, social structure, amount of natural handicap payment, amount of agri-environment payments, amount of direct payments (Pažek et al. 2018). This multi-attribute DEXi model can be applied to a smaller number of farms, emphasizing the qualitative aspects of decision making based on judgments when measuring values of alternatives, which is proven suitable when there is a lack of numeric data.

Besides the multi-criteria applications in spatial decision making based on the approaches of multi-criteria decision making, the literature review offers also the applications based on multi-criteria optimization. Baja et al. (2007) developed spatial modelling procedures for agricultural land suitability analysis using compromise programming and fuzzy set approach within a GIS environment. However, GIS is not needed in each spatial decision making decision support. For example, the Multi-criteria Analysis Shell for Spatial Decision Support (MCAS-S) (Australian Government, Department of Agriculture and Water Resources ABARES 2017) is a software tool that enables support of multi-criteria analysis in spatial decision making without GIS programming, removing the technical obstacles to non-GIS users.

2.8 The Development of the Multi-criteria Model for the Protection of Agricultural Land for Food Self-Sufficiency

2.8.1 Problem Definition and Structuring

Protection of agricultural land and food self-sufficiency are indispensable in reducing the risk associated with climate change, increasing demand for ecological goods and services, the threats of natural and social catastrophes, and economic crisis.

Since 2000, the European Landscape Convention (Council of Europe 2000) has been developing a new perspective for the perception of a landscape socio-ecological system where people become a crucial part of this system, providing services and benefits to human well-being due to sustainable transformations and ecological resource preservation (Mele and Poli 2017).

As stated in the Resolution on the strategic orientations for the development of Slovenian agriculture and food industry by 2020 (UL RS 2011), the basic task of agriculture is to ensure adequate supply of safe food. According to the United Nations, the countries in the climate zone which includes Slovenia need about 3000 square meters of farmland (fields, meadows and orchards) per habitant for the required amount of food, which amounts to approximately 600,000 hectares (UL RS 2011). Court of Audit of the Republic of Slovenia found that in 2010 the competent authorities in Slovenia were not successful in protecting agricultural land (Court of Audit 2013), which reflects in food self-sufficiency.

The multi-criteria model for the protection of agricultural land for food self-sufficiency should thus include selected geographical and economic factors, structured in the criteria hierarchy. Because of the geographical and economic diversity of statistical regions in Slovenia, the global goal is to measure the capacity of alternatives – statistical regions for food self-sufficiency with respect to geographical and economic factors.

As the heterogeneity of the public, private, and voluntary dataset entails different standards, formats and scattered sources (Mele and Poli 2017), we limited on the latest available official data sources of the Statistical Office of the Republic of Slovenia (SURS). To include the data about the agricultural land in overgrowing (Glavan et al. 2017), we also used the available data of the ministry, responsible for agriculture and food (MKGP). The multi-criteria model for the protection of agricultural land for food self-sufficiency includes the first level criteria that are geographical and economic factors, second level criteria or attributes, and alternatives. The alternatives in the model are statistical regions, according to the Nomenclature of Territorial Units for Statistics, the so-called the NUTS 3 level, as follows: Pomurska (X_1), Podravska (X_2), Koroška (X_3), Savinjska (X_4), Zasavska (X_5), Posavska (X_6), Jugovzhodna Slovenija (X_7), Osrednjeslovenska (X_8), Gorenjska (X_9), Primorsko-notranjska (X_{10}),

Table 2.9 Relevant criteria

First-level criteria		Second level criteria		
Name	Symbol	Name and description	Symbol	Source
Geographical factor	z_1	Utilized agricultural area[a] (ha, 2016)	z_{11}	SURS (2018a)
		Agricultural land with limited opportunities for farming (% of utilized agricultural area, both in ha, 2016)	z_{12}	SURS (2018a)
		Agricultural land in overgrowing (ha, 2011)	z_{13}	MKGP (2011)
Economic factor	z_2	Number of agricultural holdings (2016)	z_{21}	SURS (2018a)
		Annual working units[b] (2016)	z_{22}	SURS (2018a)
		Economic size of an agricultural holding[c] (ESU, 2016)	z_{23}	SURS (2018a)
		Number of agricultural holdings with economic size >100.000 EUR	z_{24}	SURS (2018a)
		Employment in agriculture, forestry and fisheries (number of employment, 2016)	z_{25}	SURS (2018b)
		Average monthly net earnings per employee in agriculture, forestry and fisheries (EUR, 2017)	z_{26}	SURS (2018b)

[a]Utilised agricultural area (UAA) consists of arable land, kitchen gardens, grassland, orchards (intensive, extensive), olive plantations, vineyards and nurseries, used for agricultural production (irrespective of their ownership). Common grassland is not included. (SURS 2017)

[b]Annual working units (AWU): Expressing the extent of (AWU) is based on the ratio between the number of hours worked on the farm in one year and the extent of work done by one fully employed person in one year (1800 hours), which is being used by the national labor force statistics. (SURS 2018a)

[c]The economic size of an agricultural holding is assessed by summing up the products of standard gross margin (SGM) values of individual cost 10/11 items and the extent of their production. The economic size is expressed in ESU (European Size Unit), which equals EUR 1200. (SURS 2018a)

Goriška (X_{11}), Obalno-kraška (X_{12}). The hierarchy of the selected relevant criteria for which data can be obtained in the data sources of SURS and MKGP is presented in Table 2.9.

2.8.2 Criteria Weighting and Measuring Local Alternatives' Values

The weights of the first-level and then of the second-level criteria were determined hierarchically, so that the sum of the weights on the lower level with respect to the criterion on the higher level was one. Following the importance attributed to the geographical and socioeconomic factors in several studies (e.g., Burian et al. 2015; Pažek et al. 2018; Yang et al. 2008) we made the following pairwise comparison of the importance of the first level criteria with respect to the global goal: the geographical factor is twice[3] as important as the economic one. The weights of the second-level criteria of the geographical factor were determined directly, i.e. considering the available data of utilized agricultural area, agricultural land with limited opportunities for farming taking into account the evaluation that only 30% of agricultural land with limited opportunities for farming is arable land (Pažek et al. 2018) and agricultural land in overgrowing in Slovenia. The importance of the second-level criteria of the economic factor was expressed by using the SWING method (see Fig. 2.3), by following the results of several studies about the importance of socioeconomic factors to food self-sufficiency, and characteristics in Slovenia. According to Chen (2014) and Rikalovic et al. (2014), employment is the most important

[3] According to the AHP scale: 2 – from equally important to moderately more important.

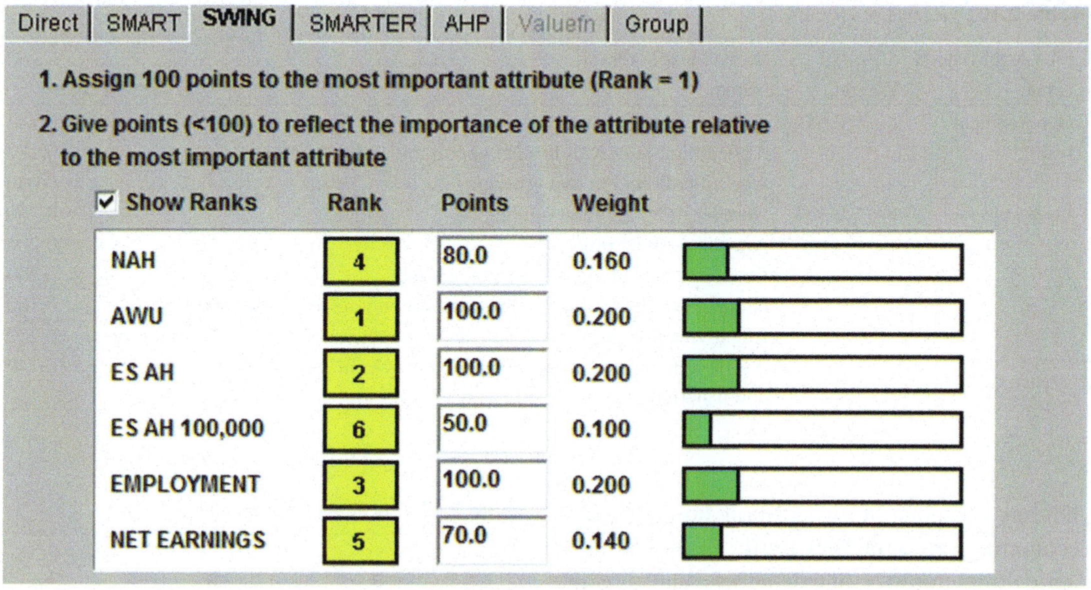

Fig. 2.3 Expressing the importance of the second-level criteria of the economic factor by using the SWING method. (Note: NAH – number of agricultural holdings, AWU – annual working units, ES AH – economic size of an agricultural holding, ES AH 100,000 – number of agricultural holdings with economic size >100,000 EUR, EMPLOYMENT – employment in agriculture, forestry and fisheries, NET EARNINGS – average monthly net earnings per employee in agriculture, forestry and fisheries)

attribute, therefore 100 points were given to the change from the lowest to the highest level. Similarly, 100 points were given to the change of the annual working units. Furthermore, 30 points less, i.e. 70 points, was given to the average monthly net earnings per employee in agriculture, forestry and fisheries.[4] Because insufficient size of agricultural holdings presents a serious problem in Slovenia (UL RS 2011), 100 point were given to the change from the lowest to the highest level of the economic size of an agricultural holding.[5] From this point of view, the change from the lowest to the highest level of the number of agricultural holdings is 20 points less important, and the change from the lowest to the highest level of the number of agricultural holdings with economic size >100,000 EUR is 50 points less important.[6] The weights in Fig. 2.3 were calculated by following (2.5). The obtained weights of criteria are presented in Table 2.10.

The latest public available data with respect to the second-level criteria (attributes) in the official data sources of SURS in 2018, together with the latest available data for the agricultural land in overgrowing of MGZT were considered when measuring the local alternatives' values – for this purpose, value functions were used. Decreasing linear value functions were used to measure alternatives' values with respect to agricultural land with limited opportunities for farming, in percentage of utilized agricultural area, and agricultural land in overgrowing, and increasing linear value functions were used to measure alternatives' values with respect to other second-level criteria: utilized agricultural area, number of agricultural holdings, annual working units, economic size of an agricultural

[4] The given points correspond to the ones of employment and payment presented in Chen (2014).

[5] This is also supported by the results of studies conducted in other countries (see, e.g., Mannaf and Uddin 2012).

[6] The sensitivity analysis was made to verify how changes in criteria' weights influence the aggregate alternatives' values.

Table 2.10 Weights of criteria

First-level criteria		Second level criteria	
Name	Weight	Name	Weight
Geographical factor	$w_1 = 0.667$	Utilized agricultural area	$w_{11} = 0.787$
		Agricultural land with limited opportunities for farming	$w_{12} = 0.173$
		Agricultural land in overgrowing	$w_{13} = 0.040$
Economic factor	$w_2 = 0.333$	Number of agricultural holdings	$w_{21} = 0.160$
		Annual working units	$w_{22} = 0.200$
		Economic size of an agricultural holding	$w_{23} = 0.200$
		Number of agricultural holdings with economic size >100.000 EUR	$w_{24} = 0.100$
		Employment in agriculture, forestry and fisheries	$w_{25} = 0.200$
		Average monthly net earnings per employee in agriculture, forestry and fisheries	$w_{26} = 0.140$

holding, number of agricultural holdings with economic size >100,000 EUR, employment in agriculture, forestry and fisheries, and average monthly net earnings per employee in agriculture, forestry and fisheries. In order to ensure the greatest distinction between the alternatives, the lower bound is equal to the lowest datum and the upper bound is equal to the highest datum for each second-level criterion. The obtained local alternatives' values are presented in Table 2.11.

Table 2.11 Local alternatives' values

Part 1

	X_1	X_2	X_3	X_4	X_5	X_6
v_{11}	0.846	1.000	0.164	0.773	0.000	0.285
v_{12}	1.000	0.640	0.000	0.222	0.003	0.537
v_{13}	0.144	0.019	0.848	0.535	1.000	0.738
v_{21}	0.670	1.000	0.125	0.923	0.000	0.377
v_{22}	0.583	0.992	0.163	1.000	0.000	0.396
v_{23}	0.641	1.000	0.146	0.646	0.000	0.256
v_{24}	0.536	1.000	0.062	0.430	0.000	0.185
v_{25}	0.560	0.974	0.178	1.000	0.000	0.384
v_{26}	0.637	0.592	0.395	0.616	0.000	0.577

Part 2

	X_7	X_8	X_9	X_{10}	X_{11}	X_{12}
v_{11}	0.562	0.709	0.309	0.158	0.303	0.074
v_{12}	0.218	0.369	0.527	0.000	0.054	0.140
v_{13}	0.227	0.308	0.664	0.689	0.000	0.160
v_{21}	0.694	0.638	0.292	0.081	0.390	0.156
v_{22}	0.598	0.651	0.269	0.019	0.338	0.080
v_{23}	0.357	0.513	0.272	0.038	0.280	0.089
v_{24}	0.244	0.422	0.319	0.010	0.202	0.064
v_{25}	0.655	0.707	0.317	0.098	0.393	0.093
v_{26}	1.000	0.781	0.801	0.774	0.892	0.418

Note: alternative's value with respect to: utilized agricultural area (v_{11}), agricultural land with limited opportunities for farming (v_{12}), agricultural land in overgrowing (v_{13}), number of agricultural holdings (v_{21}), annual working units (v_{22}), economic size of an agricultural holding (v_{23}), number of agricultural holdings with economic size >100.000 EUR (v_{24}), employment in agriculture, forestry and fisheries (v_{25}), average monthly net earnings per employee in agriculture, forestry and fisheries (v_{26}); alternatives: Pomurska (X_1), Podravska (X_2), Koroška (X_3), Savinjska (X_4), Zasavska (X_5), Posavska (X_6), Jugovzhodna Slovenija (X_7), Osrednjeslovenska (X_8), Gorenjska (X_9), Primorsko-notranjska (X_{10}), Goriška (X_{11}), Obalno-kraška (X_{12})

2.8.3 Synthesis, Ranking and Sensitivity Analysis

To obtain the alternatives' values with respect to geographical factor, to economic factor and the aggregate alternatives' values with respect to all criteria included in the criteria hierarchy (Table 2.9), the additive model was used. For instance, the aggregate value of Pomurska (X_1) is calculated by (2.7), considering the weights in Table 2.10 and the local alternative's values in Table 2.11:

$$\begin{aligned} v(X_1) = {} & w_1(w_{11}v_{11}(X_1) + w_{12}v_{12}(X_1)+ \\ & +w_{13}v_{13}(X_1)) + w_2(w_{21}v_{21}(X_1) + w_{22}v_{22}(X_1)+ \\ & +w_{23}v_{23}(X_1) + w_{24}v_{24}(X_1) + w_{25}v_{25}(X_1)+ \\ & +w_{26}v_{26}(X_1)) \end{aligned}$$

The results in Table 2.12 show that the highest aggregate value with respect to all criteria and

Table 2.12 The alternatives' values, obtained with the additive model

	Value with respect to geographical factor v_1	Value with respect to economic factor v_2	Aggregate value v	Rank
Pomurska (X_1)	0.845	0.607	0.766	2
Podravska (X_2)	0.899	0.936	0.911	1
Koroška (X_3)	0.163	0.179	0.169	9
Savinjska (X_4)	0.668	0.806	0.714	3
Zasavska (X_5)	0.041	0.000	0.027	12
Posavska (X_6)	0.346	0.367	0.353	7
Jugovzhodna Slovenija (X_7)	0.489	0.597	0.525	5
Osrednjeslovenska (X_8)	0.634	0.628	0.632	4
Gorenjska (X_9)	0.361	0.363	0.361	6
Primorsko-notranjska (X_{10})	0.152	0.153	0.153	10
Goriška (X_{11})	0.248	0.410	0.302	8
Obalno-kraška (X_{12})	0.089	0.142	0.107	11

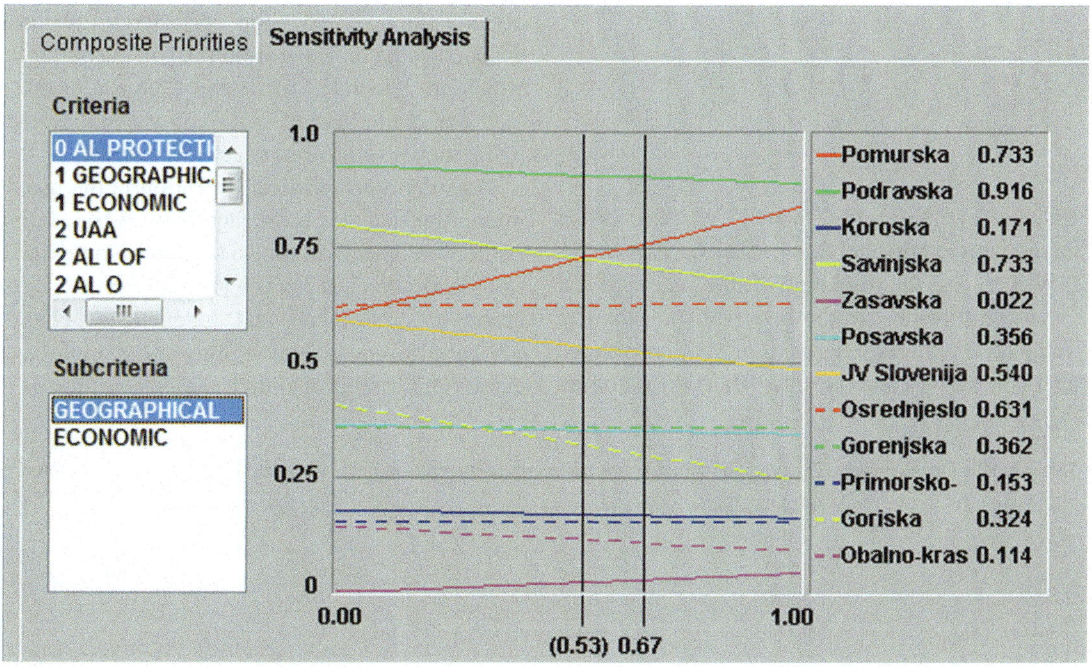

Fig. 2.4 The results of the sensitivity analysis with respect to the geographical factor – additive model

with respect to geographical factor was obtained by Podravska (X_2), followed by Pomurska (X_1), and the lowest aggregate value with respect to all criteria and with respect to geographical factor was obtained by Zasavska (X_5). With respect to economic factor, the highest value was obtained by Podravska (X_2), followed by Savinjska (X_4), and the lowest level was obtained by Zasavska (X_5). Sensitivity analysis (see Fig. 2.4) showed that changes of the first-level criteria does not influence the ranking of the alternative with the highest aggregate value Podravska (X_2); if the weight of geographical factors decreased for more than 0.14 (if it would be less than 0.53),

Savinjska (X_4) region would have ranking 2 instead of Pomurska (X_1). The results of the gradient sensitivity analysis showed that the ranking results are not sensitive to changes in the second-level criteria weights.

The aggregate values obtained with an additive model were completed by considering an interaction among criteria by a fuzzy measure – discrete Choquet integral. The results of correlation analysis let us report that there is positive correlation between utilized agricultural area and agricultural land with limited opportunities for farming, where correlation coefficient is 0.807, $p < 0.01$. Positive correlation can be overcome by using a weight on a subset of criteria $w_{11,12} < w_{11} + w_{12}$, note that $w_{11} + w_{12} = 0.960$ (Table 2.10), $w_{11,12} = 0.807$. For instance, for Pomurska (X_1), where $v_{13} < v_{11} < v_{12}$ (Table 2.11), we have

$$C_w(v_{11}, v_{12}, v_{13}) = v_{13}[w_{13,11,12} - w_{11,12}] + \\ + v_{11}\left[w_{11,12} - w_{12}\right] + v_{12}w_{12}, \quad (2.10)$$

where $w_{13,11,12} = 1$. Following (2.9), the Choquet integral for other alternatives can be expressed. However, as the ranking of the alternatives' values with respect to the geographical factor in Table 2.11 differs for each of the considered alternatives, we cannot use (2.10) as the common formula for calculating the Choquet integral in the example case; it should be expressed for each alternative by considering (2.9).

The values of the Choquet integral with respect to geographical factor are presented in Table 2.13.

The redundancy factor between utilized agricultural area and agricultural land with limited opportunities for farming decreased the aggregate values of Pomurska (X_1), Podravska (X_2), Osrednjeslovenska (X_8) and Goriška (X_{11}) region and changed the ranking of Pomurska (X_1) and Savinjska (X_4) region.

2.9 Conclusions

Quantitative methods in analyzing and processing data in order to obtain the information that organizations need for effective business decision-making, have a number of advantages, which are illustrated by cases of the use of these methods in the field of spationomy. On the other hand, they also have a range of restrictions.

An important practical limitation in the use of inferential statistics is the fact that it is based on a random sample; obtaining a random sample (i.e. a sample where each statistical units from a population, has an equal and independent chance (probability) of selection into the sample and even more, this probability can be calculated) is

Table 2.13 The alternatives' values, obtained by considering an interaction between criteria with the Choquet integral

Alternative	Choquet integral C with respect to geographical factor	Aggregate alternative value v	Rank
Pomurska (X_1)	0.737	0.694	3
Podravska (X_2)	0.803	0.847	1
Koroška (X_3)	0.163	0.169	9
Savinjska (X_4)	0.668	0.714	2
Zasavska (X_5)	0.041	0.027	12
Posavska (X_6)	0.346	0.353	7
Jugovzhodna Slovenija (X_7)	0.489	0.525	5
Osrednjeslovenska (X_8)	0.625	0.626	4
Gorenjska (X_9)	0.361	0.361	6
Primorsko-notranjska (X_{10})	0.152	0.153	10
Goriška (X_{11})	0.240	0.297	8
Obalno-kraška (X_{12})	0.089	0.107	11

in most cases expensive, measured in both time as in money. On the other hand, only a random sample of statistical units provides a reliable bases for generalization of sample statistics to the parameters of the statistical population.

The logistic regression, which we present in the second part of the chapter, is a very useful method for analyzing relations between variables since the assumptions upon which logistic regression is based can in principle be easy to ensure. However, the limitation of logistic regression is that it does not provide information about the influence of the selected variables on the response variable (as the results of logistic regression are often misinterpreted) but the odds ratio is a measure of the nature and the strength of an association (Agresti and Finlay 2009).

The results of the application of MCDM methods presented in the third part of this chapter enable suggesting measures for the protection of agricultural land for food self-sufficiency.

Based on the literature overview it can be concluded that the majority of multi-criteria applications in spatial decision making are based on the additive model assuming no interactions between criteria. Thus, aggregate alternatives' values are obtained as weighted arithmetic means (the sum of the weighted values of the alternative with respect to criteria). Pairwise comparisons within the AHP method are commonly used for criteria weighting and measuring local alternatives' values.

The crucial limitation is the availability of the up-to-date data in public data sources for the attributes of the geographical and economic factors delineated by statistical regions. This limitation influenced the selection of the second-level criteria, structured in the criteria hierarchy.

Further research possibilities are to extend the criteria hierarchy to sustainable development, which will result in interdisciplinary cooperation of several stakeholders when making decisions for the protection of agricultural land for food self-efficiency. Another improvement of the presented model is to include the criteria about which the data can be obtained with on-site analysis and questionnaires. A promising possibility is to connect the GIS and MCDM. Further research possibilities include also the use of MCDM for solving other complex decision making problems of spatial economy, such as location or site selection, the selection of optimal utility routes in urban or rural areas, the assessment of LFAs, the evaluation of urban or rural quality of life, landscape services, public space from different perspectives, and in the environmental assessment.

References

Agresti, A., & Finlay, B. (2009). *Statistical methods for the Social sciences*. Pearson: Prentice Hall.

Australian Government, Department of Agriculture and Water Resources ABARES. (2017). *Multi-criteria analysis (MCAS-S)*. Retrieved January 18, 2018, from http://www.agriculture.gov.au/abares/aclump/multi-criteria-analysis

Baja, S., Chapman, D. M., & Dragovich, D. (2007). Spatial based compromise programming for multiple criteria decision making in land use planning. *Environmental Modelling and Assessment, 12*(3), 171–184.

Bathelt, H., Malmberg, A., & Maskell, P. (2004). Clusters and knowledge: Local buzz, global pipelines and the process of knowledge creation. *Progress in Human Geography, 28*(1), 31–56.

Belton, V., & Stewart, T. J. (2012). *Multiple criteria decision analysis–An integrated approach*. Boston/Dordrecht/London: Kluwer Academic Publishers.

Benson, E. D., Hansen, J. L., & Schwartz, A. L., Jr. (2000). Water views and residential property values. *The Appraisal Journal, 3*(48), 260–271.

Bhat, C. R., Ipek, N., & Sener, I. N. (2009). A copula-based closed-form binary logit choice model for accommodating spatial correlation across observational units. *Journal of Geographical Systems, 11) 3*, 243–272.

Big Data Quarterly. (2016). Four steps to increasing Data Visualization Competence. Available on April 3rd, 2018, at http://www.dbta.com/BigDataQuarterly/Articles/4-Steps-to-Increasing-Data-Visualization-Competency-109990.aspx

Burian, J., Sorbiova, K., Tuček, P., & Tučkova, M. (2012). Possibilities of delimitation of city centers using GIS. Issue 71, conference proceedings World Academy of Science. *Engineering and Technology*, 81–90.

Burian, J., Šťastny, S., Brus, J., Pechanec, V., & Voženilek, V. (2015). Urban planner: Model for optimal land use scenario modelling. *Geografie, 120*(3), 330–353.

Čančer, V., & Mulej, M. (2013). Multi-criteria decision making in creative problem solving. *Kybernetes, 42*(1), 67–81.

Chen, J. (2014). GIS-based multi-criteria analysis for land use suitability assessment in City of Regina. *Environmental Systems Research, 3*(13), 1–23.

Choquet, G. (1953). Theory of capacities. *Annals of Institute of Fourier, 5*, 131–295.

Council of Europe. (2000). European landscape convention; Report and convention; Council of Europe: Strasbourg, France.

Court of Audit. (2013). *Audit report: The effectiveness of the protection of agricultural land as a condition for self-sufficiency*. Ljubljana: Court of Audit of the Republic of Slovenia.

Edwards, W. (1977). How to use multiattribute utility measurement for social decision making. *IEEE Transactions on Systems, Man, and Cybernetics, 7*, 326–340.

Edwards, W., & Barron, F. H. (1994). SMARTS and SMARTER: Improved simple methods for multiattribute utility measurement. *Organizational Behavior Human Decision Processes, 60*, 306–325.

Gartner IT Glossary. (2018). Available on February 2nd 2018 at https://www.gartner.com/it-glossary/?s=Big+data

Glavan, M., Malek, A., Pintar, M., & Grčman, H. (2017). Prostorska analiza kmetijskih zemljišč v zaraščanju v Sloveniji [Spatial analysis of the abandonment of agricultural land in Slovenia]. *Acta agriculturae Slovenica, 109*(2), 261–279. https://doi.org/10.14720/aas.2017.109.2.10.

Grabisch, M. (1995). Fuzzy integral in multicriteria decision making. *Fuzzy Sets and Systems, 69*, 279–298.

Hox, J. J. (2002). *Multilevel analysis*. Mhawah, NJ: Erlbaum.

Hu, J., & Zhang, Y. (2017). Discovering the interdisciplinary nature of Big Data research through social network analysis and visualization. *Journal of Scientometrics, 112*(1), 91–109.

Jankowski, P. (1995). Integrating geographical information systems and multiple criteria decision-making methods. *International Journal of Geographical Information Systems, 9*(3), 251–273.

Janssens, W., Wijnen, K., DePelsmacker, P., & VanKenhove, P. (2008). *Marketing research with SPSS*. Harlow: Pearsons Education.

Javadian, M., Shamskooshki, H., & Momeni, M. (2011). Application of sustainable urban development in environmental suitability analysis of educational land use by using AHP and GIS in Tehran. *Procedia Engineering, 21*, 72–70.

Kumar, R. (2005). *Research methodology: a step-by-step guide for beginners*. London/Thousand Oaks/New Delhi: Sage.

Lawal, D. U., Matori, A.-N., & Balogun, A.-L. (2011). A geographic information system and multi-criteria decision analysis in proposing new recreational park sites in Universiti Teknologi Malaysia. *Modern Applied Science, 5*(3), 39–55.

Malczewski, J. (1999). *GIS and multi-criteria decision analysis*. New York: Wiley.

Mannaf, M., & Uddin, T. (2012). Socioeconomic factors influencing food security: Status of maize growing households in selected areas of Bogra district. *Bangladesh Journal of Agricultural Research, XXXV* (1&2), 177–187.

Marichal, J. L. (2000). An Axiomatic Approach of the Discrete Choquet Integral as a Tool to Aggregate Interacting Criteria. *IEEE Transactions on Fuzzy Systems, 8*(6), 800–807.

Massam, B. H. (1980). *Spatial search*. Oxford: Pergamon Press.

McClave, J. T., Benson, P. G., & Sincich, T. (2011). *Statistics for business and economics*. Boston/Columbus: Pearson.

Mele, R., & Poli, G. (2017). The effectiveness of geographical data in multi-criteria evaluation of landscape services. *Data, 2*(9), 1–11.

MKGP–Ministry of Agriculture, Forestry and Food RS. (2011). *MKGP portal. Ministrstvo za kmetijstvo, gozdarstvo in prehrano*. Retrieved November 23, 2011, from, http://rkg.gov.si/GERK/

Moaven, S., Ahmadi, H., Habibi, J. Kamandi, A. (2008). A fuzzy model for solving architecture styles selection multi-criteria problem. In: *Proceedings of the second UKSIM European symposium on computer modeling and simulation IEEE computer society*.

Nyeko, M. (2012). GIS and multi-criteria decision analysis for land use resource planning. *Journal of Geographic Information System, 4*, 341–348.

Ogryzek, M., Rzasa, K. (2017). Geodetic Congress (BGC Geomatics), Baltic.

Palicki, S. (2015). Multi-criteira assessment of public space from the social perspective. *Real Estate Management and Valuation, 23*(4), 24–34.

Pažek, K., Irgolič, A., Turk, J., Borec, A., Prišenk, J., Kolenko, M., & Rozman, Č. (2018). Mutli-criteria assessment of less favoured areas: A state level. *Acta Geographica Slovenica, 58*(1), 98–104.

Raiffa, H. (1994). The prescriptive orientation of decision making: a synthesis of decision analysis, behavioral decision making, and game theory. In S. Rios (Ed.), *Decision theory and decision analysis: Trends and challenges* (pp. 3–13). Boston/Dordrecht/London: Kluwer Academic Publishers.

Rikalovic, A., Cosic, I., & Lazarevic, D. (2014). GIS based multi-criteria analysis for industrial site selection. *Procedia Engineering, 69*, 1054–1063.

Rinner, C. (2007). A geographic visualization approach to multi-criteria evaluation of urban quality of life. *International Journal of Geographical Information Science, 21*(8), 907–919.

Robinson, P., Genskow, K., Shaw, B., & Shepard, R. (2012). Barriers and opportunities for integrating social science into natural resource management: Lessons from national estuarine research reserves. *Environmental Management, 50*(6), 998–1011. https://doi.org/10.1007/s00267-012-9930-6.

Rodrigues, M., de la Riva, J., & Fotheringham, S. F. (2014). Modeling the spatial variation of the explanatory factors of human-caused wildfires in Spain using

geographically weighted logistic regression. *Applied Geography, 48*, 52–63.
Saaty, T. L. (1999). *Decision making for leaders*. Pittsburgh: RWS Publications.
Šebjan, U., & Tominc, P. (2015). Impact of support of teacher and compatibility with needs of study on usefulness of SPSS by students. *Computers in Human Behaviour, 53*, 354–365.
Sridhar, P., Madni, A. M., & Jamshidi, M. (2008). Multicriteria decision making in sensor networks. *IEEE Instrumentation & Measurement Magazine, February, 2008*, 24–29.
Statistical office RS, SI-STAT data portal. (2017). Available on March 27, 2018 from http://pxweb.stat.si/pxweb/Database/Municipalities/Municipalities.asp
SURS–Statistical office RS. (2017). *Methodological explanation: Farm structure survey, Slovenia*. Ljubljana: Republic of Slovenia, Statistical office.
SURS–Statistical office RS. (2018a). *SI-STAT data portal*. Retrieved August 8, 2018, from, http://pxweb.stat.si/pxweb/Database/Okolje/Okolje.asp.
SURS–Statistical office RS. (2018b). *SI-STAT data portal*. Retrieved August 8, 2018, from, http://pxweb.stat.si/pxweb/Database/Regije/Regije.asp.
Swissinfo. (2017). *Highest Swiss property prices recorded in Zurich*. Available on April 25, 2018 from: https://www.swissinfo.ch/eng/housing-market_highest-swiss-property-prices-recorded-in-zurich/43482440
Tabachnick, B. G., & Fidell, L. S. (2013). In etc (Ed.), *Using multivariate statistics*. Boston, Columbus: Pearson.
Tol, R. S. J. (2009). The economic effects of climate change. *Journal of Economic Perspectives, 23*(2), 29–51.
Torrieri, F., & Batà, A. (2017). Spatial multi-criteria decision support system and strategic environmental assessment: A case study. *Buildings, 7*(96), 1–17.
UL RS – Uradni list RS. (2011). Resolucija o strateških usmeritvah razvoja slovenskega kmetijstva in živilstva do leta 2020 – "Zagotovimo.si hrano za jutri" (ReSURSKŽ) (Resolution on the strategic orientations for the development of Slovenian agriculture and food industry by 2020 - Provide food for tomorrow, in Slovenian). Uradni list RS, 25/2011, 04. 04. 2011.
Von Winterfeldt, D., & Edwards, W. (1986). *Decision analysis and behavioral research*. Cambridge: Cambridge University Press.
Voogd, H. (1983). *Multicriteria evaluation for urban and regional planning*. London: Pion.
Yang, F., Zeng, G., Du, C., Tang, L., Zhou, J., & Li, Z. (2008). Spatial analyzing system for urban land-use management based on GIS and multi-criteria assessment modeling. *Progress in Naural Science, 18*, 1279–1284.
Yue, S., Wen, Y., Chen, M., Lu, G., Hu, D., & Zhang, F. (2015). A data description model for reusing, sharing and integrating geo-analysis models. *Environmental Earth Sciences, 74*(19), 7081–7099. https://doi.org/10.1007/s12665-015-4270-5.
Živkovič, R., Tominc, P. (2016). Market approach to evaluation of real estate values - model of adjustments of comparable sales´ prices (In Slovene: Ocenjevanje vrednosti stanovanj po načinu tržnih primerjav - model določanja prilagoditev cen primerljivih nepremičnin). Sir∗ius, No. 3, pp 66–84.

Spatial Analysis in Geomatics

3

Andreas Redecker, Jaroslav Burian, Nicolai Moos, and Karel Macků

Abstract
For processing geodata there are many different approaches of which all of them require their own specific input data and parameters to generate an outcome that suits the respective case of application. This chapter introduces the most common analyses that are conducted using a GIS. From basic tools like buffering certain vector geometries or merging operations of two different datasets to interpolating area wide raster datasets out of point data there is a huge variety of different toolsets that can be applied when using geodata. To understand why and how these toolsets are utilised, how they are parametrized and which other things are important to make proper use of all the different possibilities these toolsets are providing, this chapter sums up the analyses in reasoned groups and illustrates the many different approaches of spatial analyses through proper examples and depictions.

Keywords
Spatial analysis · Network analysis · GIS · Spatial statistics · Raster resolution · Geoprocessing · Data conversion

A. Redecker (✉) · N. Moos
Geography, Geomatics Group, Ruhr-University Bochum, Bochum, Germany
e-mail: andreas.redecker@rub.de; nicolai.moos@rub.de

J. Burian · K. Macků
Department of Geoinformatics, Palacký University Olomouc, Olomouc, Czech Republic
e-mail: jaroslav.burian@upol.cz; karel.macku@upol.cz

3.1 Simple Spatial Analysis (by Andreas Redecker)

This sub-chapter gives an overview of fundamental GIS methods for performing basic spatial analysis with feature data. Nevertheless, depending on the data involved and the workflow incorporating these methods, they can deliver highly valuable output. The process of manipulating geodata is called geoprocessing. To automate workflows all operators that are involved in an analysis can be combined with a geoprocessing model.

3.1.1 Selections

In many cases, not all features of a feature class are supposed to take part in an analysis. The selection of the desired objects can be performed based on the attributes of the features or incorporating their spatial characteristics. Depending on the GIS used, these two different methods can be applied successively or in one process.

3.1.1.1 Select by Attribute

This method is like selecting datasets in a database using a so-called WHERE-clause of the very common Structured Query Language (SQL). "The WHERE clause is used to extract only those records that fulfill a specified condition" (w3schools.com 2018) according to the feature's

V. Pászto et al. (eds.), *Spationomy*, https://doi.org/10.1007/978-3-030-26626-4_3

properties stored in the attribute table of a feature class (Fig. 3.1).

3.1.1.2 Linking Tabular Data

If the necessary properties for an attribute-based selection are not held in the attribute table of the feature class itself, they can be linked to it from external tabular data. For this, both tables need to contain a field (column) with matching entries. These must uniquely identify a feature in the attribute table as well as its corresponding data in the table to be linked.

3.1.1.3 Select by Location

The spatial approach for selecting features needs a second feature class whose features locations or extents determine which features of the original feature class will be selected. For this, the desired spatial relationship (e.g. intersect, contain, within a distance, etc.) and distance (optional) need to be specified (Fig. 3.2).

3.1.2 Single Feature Class Operations

To prepare features for further analysis or to better visualise results, two major operations are available to change the structure of single feature classes.

3.1.2.1 Buffer

A Buffer is a proximity function, describing an equidistant line around a feature. Therefore, the resulting geometry type of a buffer operation always is a polygon – no matter if the input was a point, line or polygon-type feature class. The distance value for the construction of buffers around the features in a feature class is either defined by a single value or derived from an individual property in the attribute table for every single feature (Fig. 3.3).

3.1.2.2 Dissolve

The Dissolve operation consolidates the features in a feature class. Based on similar values in a specified attribute field it merges the geometries

Expression:
Name LIKE 'fagus sylvatica'
AND
Age >= 50

ID	Name	Age
1	quercus robur	30
2	fagus sylvatica	79
3	acer pseudoplatanus	76
4	sorbus aucuparia	25
5	fagus sylvatica	37
6	quercus robur	88
7	fagus sylvatica	51

Fig. 3.1 Example of an attribute-based selection. (Source: Authors)

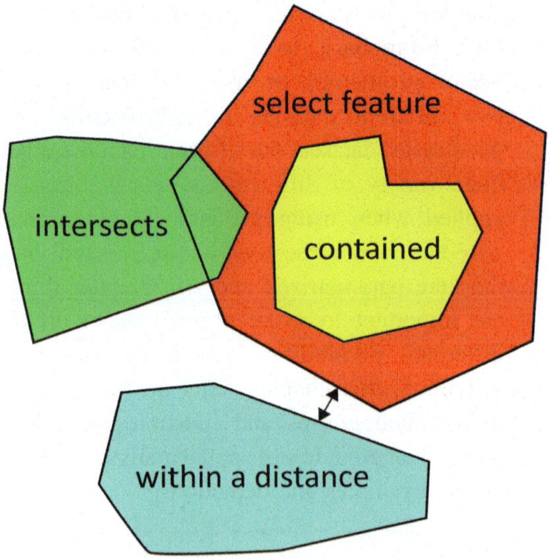

Fig. 3.2 Location-based selection methods. (Source: Authors)

(if no attribute is specified, all features will be merged). With some dissolve operators at the same time, other attributes of the features merged get aggregated by previously specified statistical functions (mean, sum, min, max, count, etc.) (Fig. 3.4).

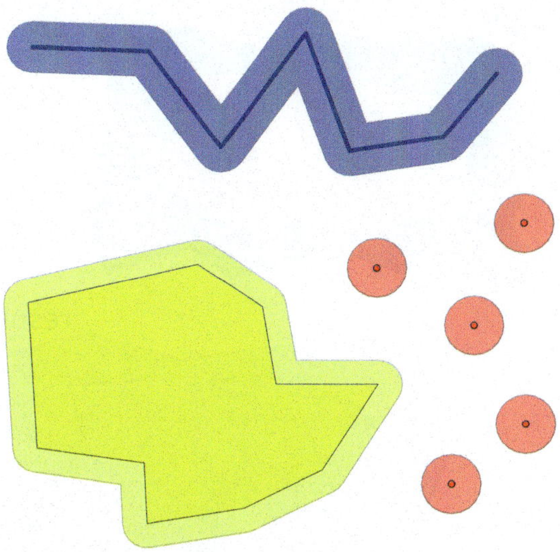

Fig. 3.3 Buffers with point-, line- and polygon-features. (Source: Authors)

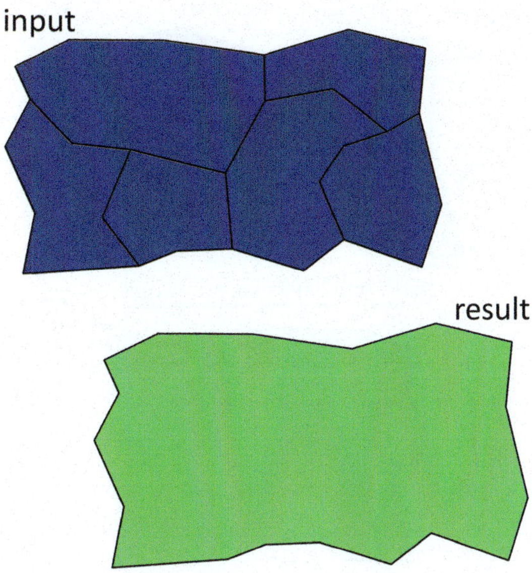

Fig. 3.4 Schematic example of a simple dissolve operation. (Source: Authors)

3.1.3 Overlay Operations

These operations combine two or more feature classes to gain new geo-datasets incorporating the extent of the features involved.

3.1.3.1 Clip

The clip-function creates a subset of features by cutting the features of one feature class by the polygon-features in another feature class. Only those parts of the features in the input layer that overlap with the polygons of the clipping layer will end up in the resulting feature class. It is often used to reduce the extent of a geo-dataset to that of the study area (area of interest, AOI) represented by a polygon feature. The attributes of the remaining features will not be changed (Fig. 3.5).

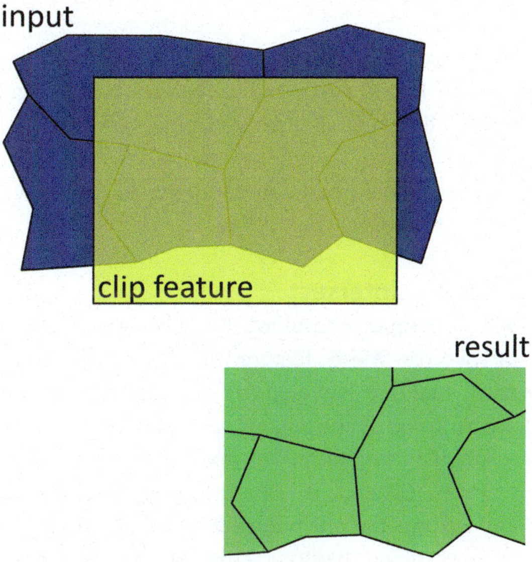

Fig. 3.5 Schematic example of a clip operation. (Source: Authors)

3.1.3.2 Difference

The difference-function also cuts the features of one feature class by the polygon features in another feature class. Only those parts of the features in the input layer that do not overlap with the polygons of the cut feature will end up in the resulting feature class. The attributes of the remaining features will not be changed (Fig. 3.6).

3.1.3.3 Union

This operator combines the polygon features of two or more feature classes. It does not create

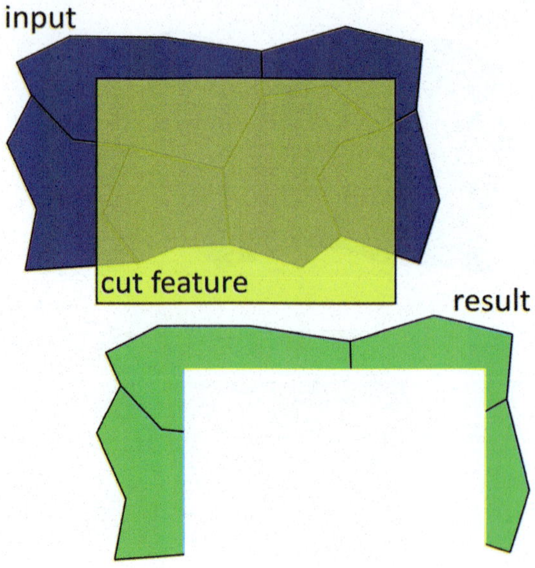

Fig. 3.6 Schematic example of a difference operation. (Source: Authors)

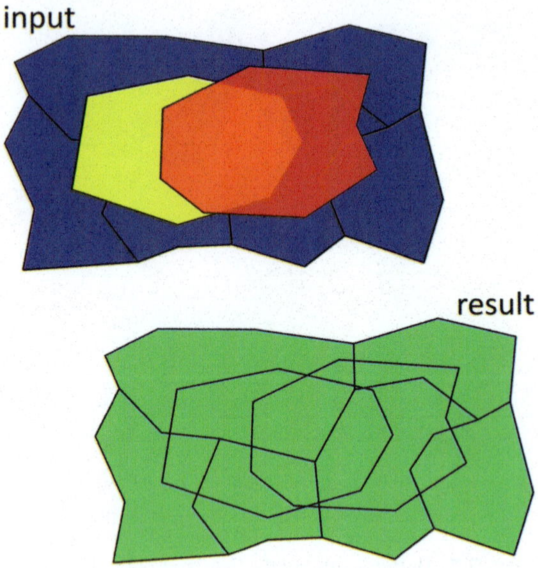

Fig. 3.7 Schematic example of a union operation. (Source: Authors)

overlapping features. Instead, it splits overlapping parts of features to subarea features and assigns the attributes of all involved objects to the new feature.

This spatial operation compares to the logical disjunction (OR) (Fig. 3.7).

3.1.3.4 Intersect

This operator combines the polygon features of two or more feature classes. Only those parts that are covered by a feature in every contributing feature class will be written to the result. The function does not create overlapping features. Instead, it clips the overlapping areas and assigns the attributes of all involved objects to the new feature. This spatial operation compares to the logical conjunction (AND) (Fig. 3.8).

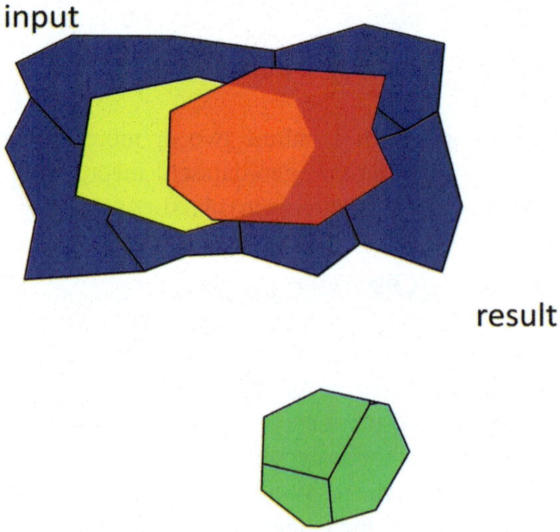

Fig. 3.8 Schematic example of an intersect operation. (Source: Authors)

3.1.3.5 Symmetrical Difference

The result of this function only contains those areas of the input features, that do not overlap. Hence it gives the same result as a union operation minus the result of an intersect.

This spatial operation compares to the logical exclusive disjunction (XOR) (Fig. 3.9).

3.2 Raster Analysis (by Jaroslav Burian)

Raster analysis (as part of spatial analysis) refers to the analytical operations with raster data. Map algebra (mathematical operations with rasters) is

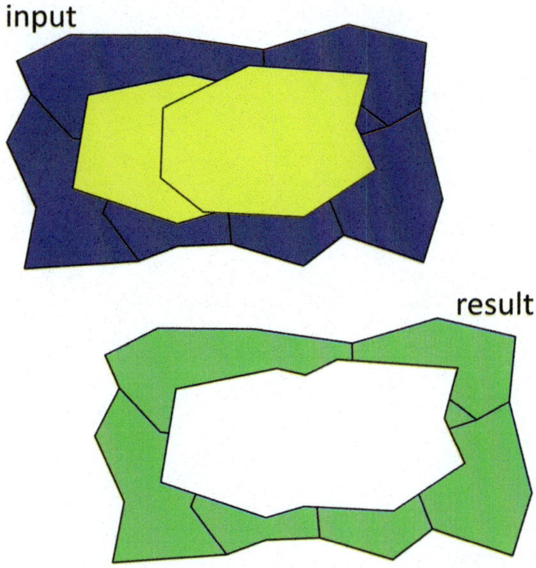

Fig. 3.9 Schematic example of a symmetrical difference operation. (Source: Authors)

used to processing this data. There exist many raster analysis options in GIS like hydrologic analysis, multi-criteria analysis, terrain analysis, surface modelling, surface interpolation, suitability modelling, statistical analysis, and image classification (processing of remote sensing data). Most of the application fields cover environmental issues (e.g. climatic change, weather forecasting, flood modelling) but there are also some focused on economic aspects (e.g. modelling of renewable energy potential, land suitability modelling, cost-distance analysis and many others).

3.2.1 Raster Data

As mentioned in Chap. 1, vector and raster data models are two main ways for geographic data representation. In a raster representation space is divided into an array of rectangular (usually square) cells (pixels). All geographic variation is then expressed by assigning properties or attributes to these pixels (Longley et al. 1999). The most significant characteristics of the raster is its spatial resolution that can be expressed as the length of a cell side as measured on the ground. As shown in Fig. 3.10, cell size can vary from centimetres (some aerial images) to kilometres (satellite images). The spatial resolution is a key characteristic that influences the quality and detail of any raster analysis. Higher spatial resolution leads to higher detail but also increases needed storage capacity and computational time.

3.2.2 Map Algebra

Mathematical operations that can be performed with rasters are referred to as raster or map algebra. Map algebra (also known as cartographic modelling) was defined by Dana Tomlin (Tomlin and Berry 1979; Tomlin 1994) as the informal computational language, that is the basis for raster data processing. Simply said, map algebra is the math applied to raster data. To formalise that, Tomlin defined raster operators and raster functions. Map algebra can be represented by arithmetic or simple analytical operations that are performed with one or more input raster layers (grids). In most software packages, the set of these features is referred to as a map or raster calculator (sometimes grid analysis) (Fig. 3.11).

3.2.3 Raster Operators

As part of the map algebra, operators and functions of mathematical language are used for data processing. Operators perform mathematical calculations with one or more raster layers. The basic type of operators are arithmetic operators (+, −, ∗, /). It is possible to add, subtract, multiply, divide, or perform the same single layer operations. In addition to arithmetic operators, there are Boolean operators (true, false), relational (greater than, smaller than or equal to), statistical (minimum, maximum, average and median), trigonometric (sine, cosine, tangent, arcsine), exponential and logarithmic.

3.2.4 Raster Functions

Tomlin (1994) classifies all GIS transformations of rasters into four basic classes, and it is used in several raster-centric GISs as the basis for their

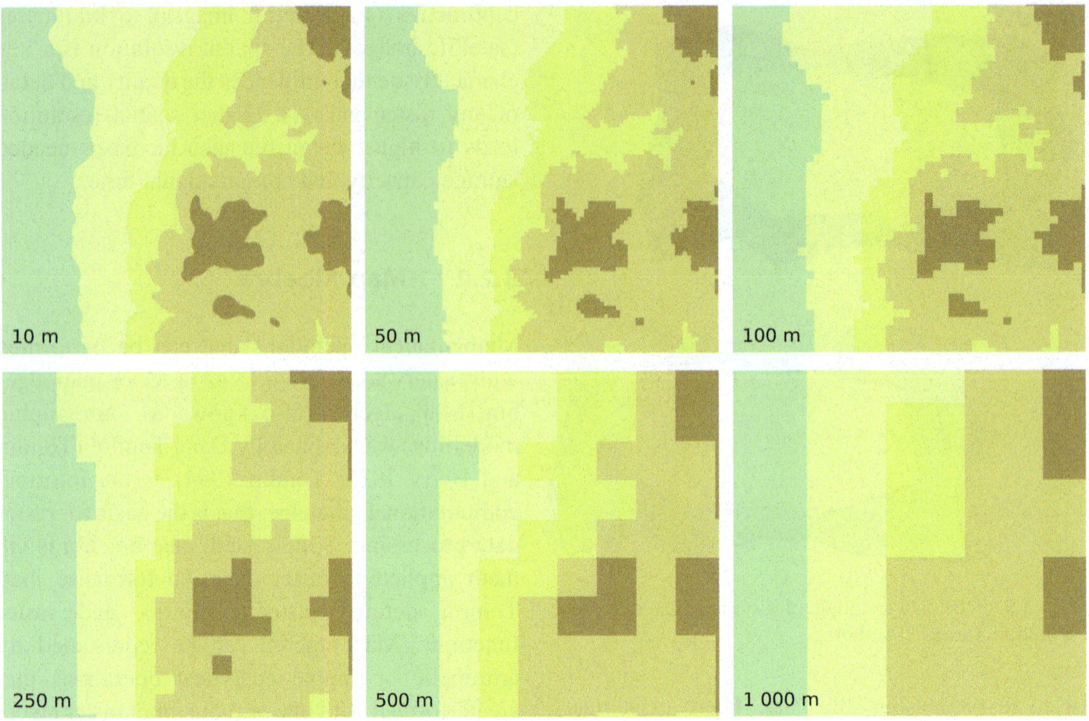

Fig. 3.10 Different cell size (10, 50, 100, 250, 500, 1000 m). (Source: Authors)

analysis languages. Depending on whether the functions work with only one raster cell or more, they are divided into local, focal, zonal, and global. Map algebra functions follow some rules (spatial resolution, the same coordinate system, mathematical operators) to combine all of its components.

3.2.4.1 Local

Local functions are always performed with one specific raster cell but in the entire grid. Using these functions, a new raster cell value is calculated from the values in one or more information layers. An example of a local function may be a simple combination of two raster layers (e.g. the combination of flood risk and earthquake risk) or multiplication of one raster layer by a specific value (e.g. prediction of the average temperature) (Fig. 3.12).

3.2.4.2 Focal

For focal functions, as with local functions, a new value is determined for each cell separately. However, it is calculated from the values in the defined area (neighbouring cells). The most common is the closest cell (3x3), but it can also be a larger area (square, triangle, circle, 4x4 matrix, etc.). On the principle of focal functions, the basic method of slope calculation works. For each cell in the defined area, the altitude difference is calculated from which the resulting gradient slope is calculated. The similar procedure is also used for aspect calculation and many hydrological modelling. Another example from the economic field is modelling of the city growth that uses cellular automata based on focal functions (Fig. 3.13).

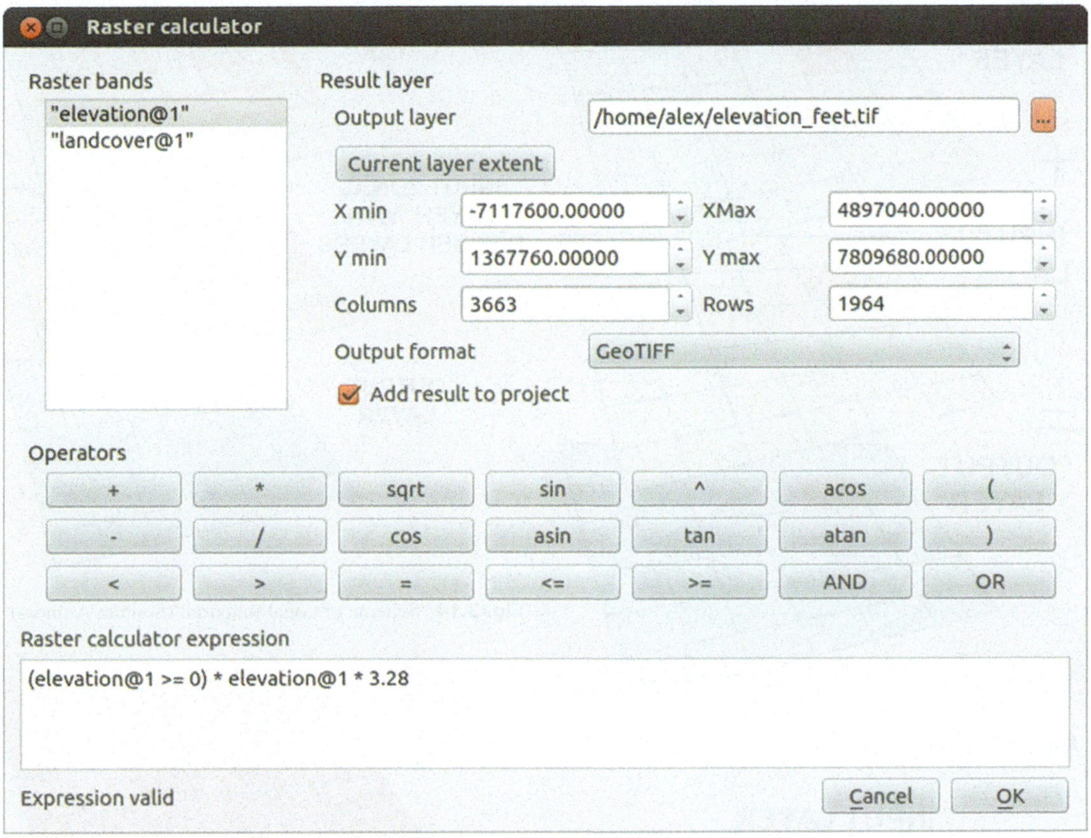

Fig. 3.11 Raster calculator in QGIS software. (Source: Authors)

3.2.4.3 Zonal

Calculation of zonal function is similar to the focal functions. The main difference is that the neighbourhood is defined by another raster (zonal) layer. The new value is calculated for each cell from the values that belong to the zone defined by another layer. The focal function is often applied to the calculation of a certain statistical indicator (average, median, etc.) for irregular areas defined in another layer (e.g. average altitude for individual forest areas or average highway accessibility for city districts) (Fig. 3.14).

3.2.4.4 Global

Global functions are performed from all grid cells. The result of global functions is usually several selected cells that meet the set conditions. They are mostly focused on distance analysis in the form of friction surfaces. An example of a global function may be to find the optimal route in a raster from A to B. For example, each cell in the entire raster represents the value of the friction (water, rock, forest – higher value, meadow, field – smaller value). The entire raster is then analysed to find the lowest cost path when moving from A to B (Fig. 3.15).

3.2.5 Selected Raster Analysis

Raster operators and raster functions can be applied to many different raster datasets to perform a wide range of raster analysis. For the purpose of this book, only a few selected analysis are described.

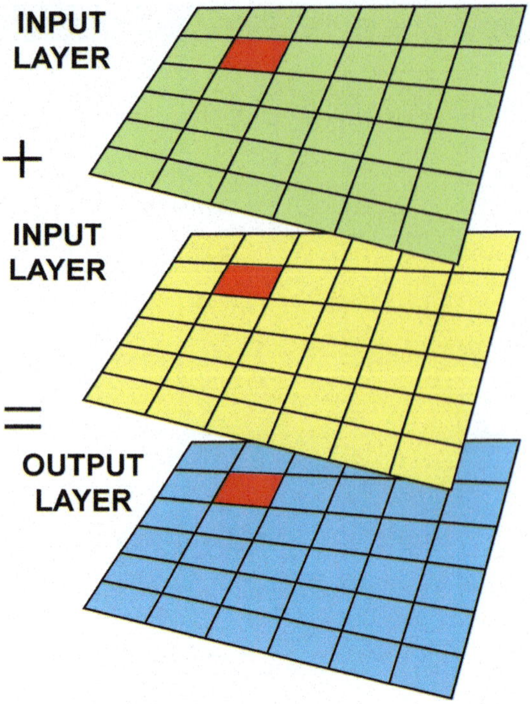

Fig. 3.12 Scheme of local function. (Source: Authors)

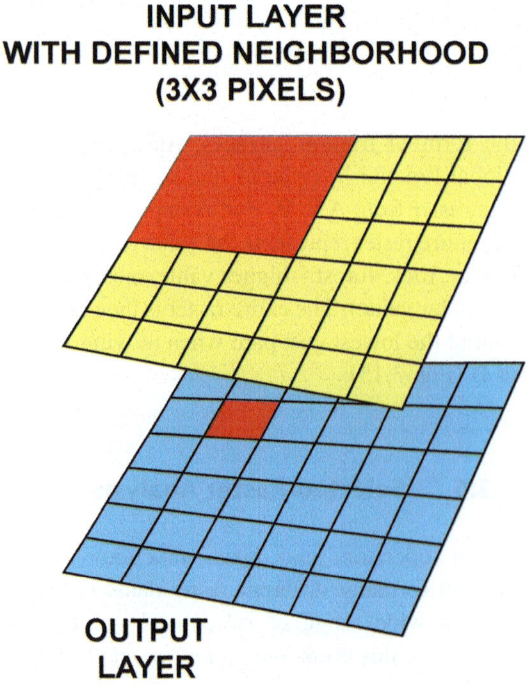

Fig. 3.13 Scheme of focal function. (Source: Authors)

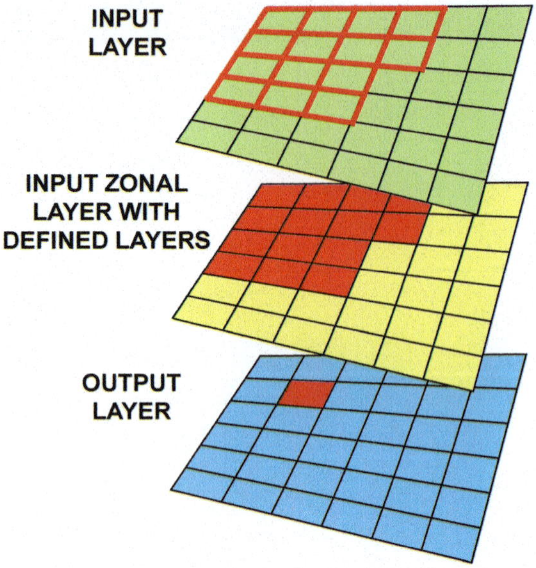

Fig. 3.14 Scheme of zonal function. (Source: Authors)

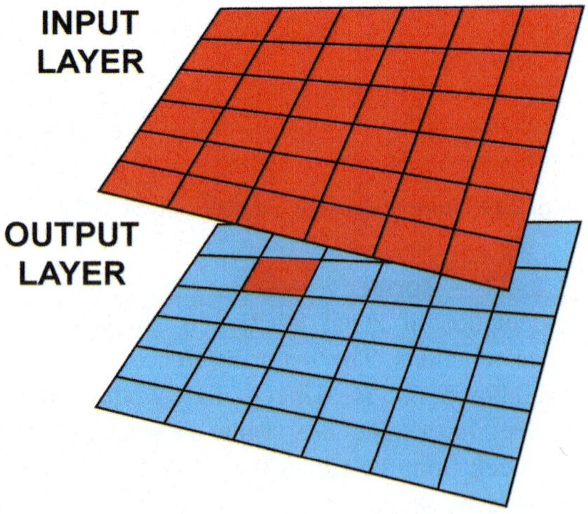

Fig. 3.15 Scheme of global function. (Source: Authors)

3.2.5.1 Resampling

To perform any raster analysis, input raster layers must have the same spatial resolution and coordinate system. Simply said, pixels (cell centres) have to match each other. To manage that several resampling methods are used. It means that one of the input rasters is resampled to the same resolution as another input layer. Original raster values are recalculated to the new ones based on nearest

Fig. 3.16 Example of resampling from 100 to 1000 m. (Source: Authors)

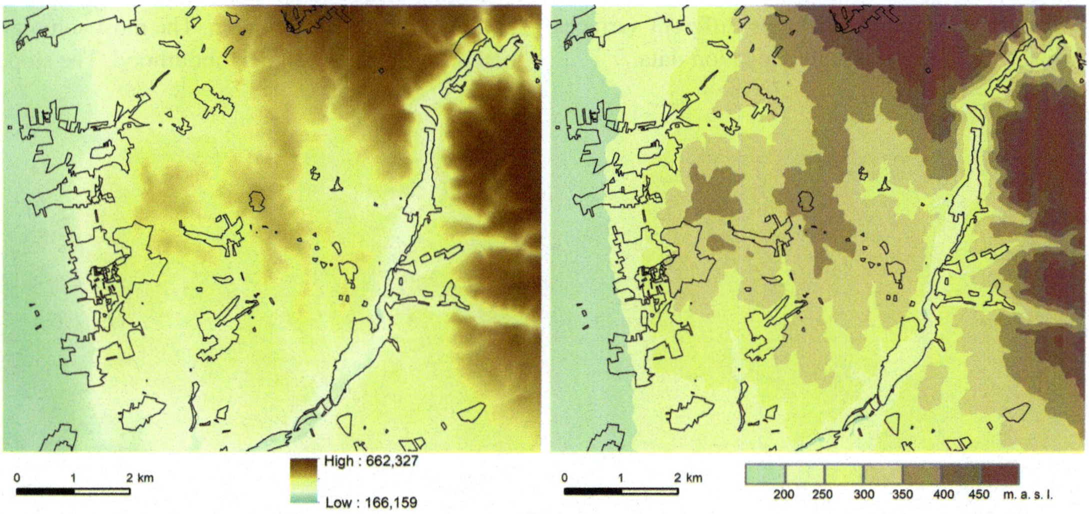

Fig. 3.17 Example of reclassification of elevation. (Source: Authors)

neighbourhood method (or other methods like bilinear, cubic convolution or majority) (Fig. 3.16).

3.2.5.2 Reclassification

One of the most common simple analysis is the reclassification. Reclassification is the process of reassigning a value, a range of values, or a list of values in a raster to new output values. In the case of continuous data (e.g. elevation, temperatures) reclassification creates a new raster with discrete values (several elevation zones – lowland, highland, etc.; or temperature zones). In the case of categorical data (e.g. 20 categories of land-use), reclassification creates a new raster with new discrete values (e.g. only 5 categories of land-use). Reclassification is a key process when you need to combine different data using a common value scale (Fig. 3.17).

3.2.5.3 Surface Analysis

Many raster analysis deal with the surfaces. Surfaces represent phenomena that have values at every point across their extent. Surfaces are derived from a limited set of sample values (e.g. elevation points, meteorological stations). A typical surface represents elevation, temperature, precipitation and many other continuous phenomena's. Surfaces can be represented by contour lines, points or TINs (triangulated irregular networks); however most surface analysis in GIS is done on raster data.

Spatial Interpolation

There exist several ways to create surfaces. Spatial interpolation is the most common way to do it. Interpolation creates a continuous surface from discrete samples with measured values (point layer mostly). There exist several interpolation methods with a variety of parameters that influence the resulting surface. Each method is suitable for different data set (different phenomena with different spatial distribution). The most common interpolation methods are kriging, natural neighbours, spline and IDW (inverse distance weighting). Figure 3.18 shows different surfaces using the same input point elevation data.

The surface analysis involves several kinds of processing, including extracting new surfaces from existing surfaces, reclassifying surfaces, and combining surfaces (ESRI 2018a). The most common surface analysis (slope, aspect, hillshade, viewshed and watershed) are applied to the elevation data (terrain surfaces – digital elevation models).

3.2.5.4 Slope

The slope represents the rate of maximum change in z-value (elevation) from each cell. The slope is calculated as the maximum rate of change in values between each cell and its neighbours. The neighbourhood can be defined by 4 or 9 neighbouring cells, and there exist several methods for slope calculation. The most common method uses 3 × 3 cell neighbourhood. The slope

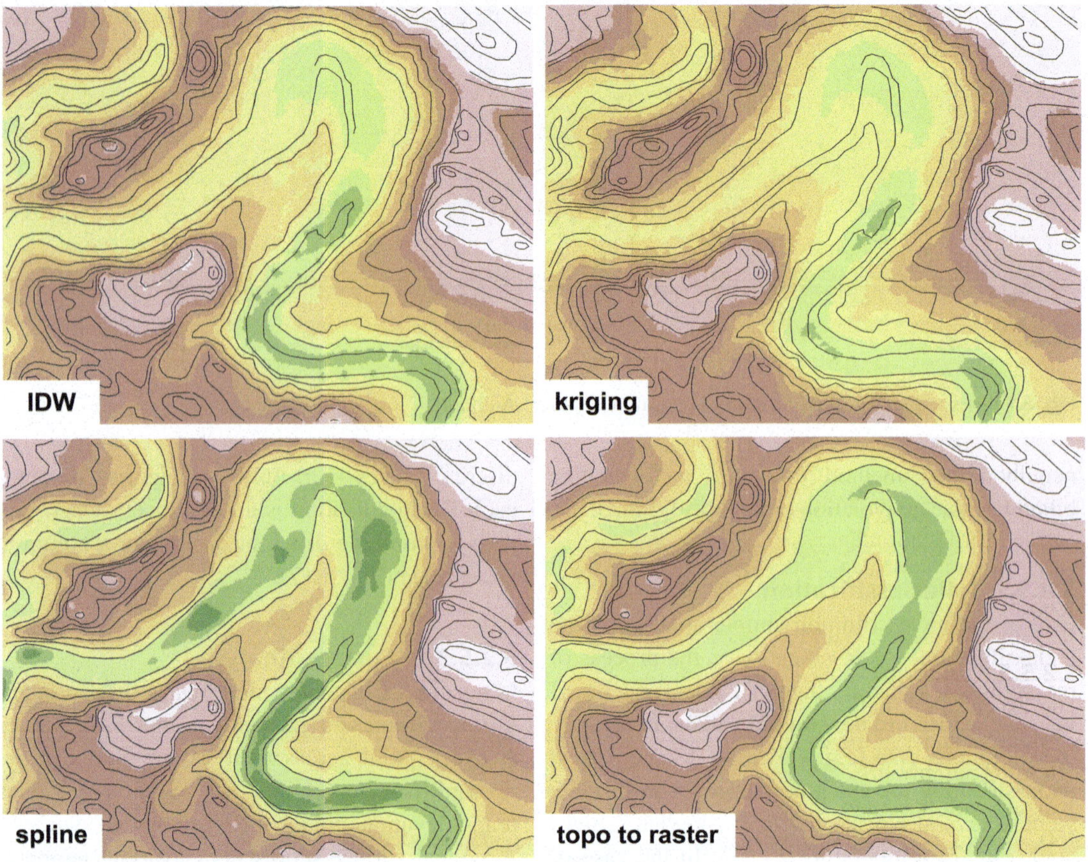

Fig. 3.18 Example of different interpolation methods. (Source: Authors)

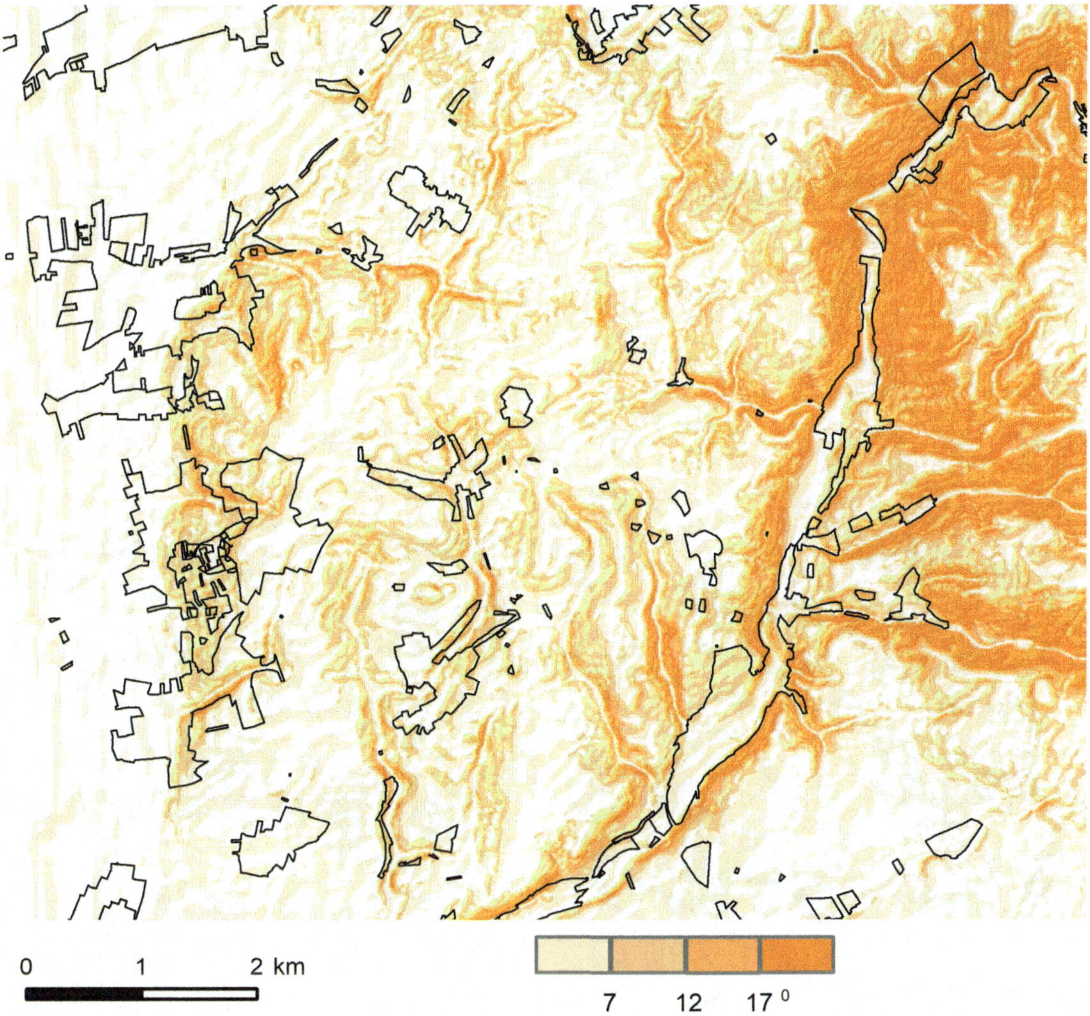

Fig. 3.19 Example of slope. (Source: Authors)

may be expressed as either degree (e.g., 45°) or percent (e.g., 50%). Information about slope can be used for location analysis in urban planning to find suitable places for new development (Fig. 3.19).

3.2.5.5 Aspect

The aspect identifies the orientation or direction of slope. Aspect is the down-slope direction of a cell to its neighbours. The cell values in an aspect grid are compass directions ranging from 0 to 360. North is 0, and in a clockwise direction, 90 is east, 180 is south, and 270 is west. Input grid cells that have 0 slope (flat areas) are assigned an aspect value of −1 (Albrecht 2005). Similarly, to slope analysis, aspect can be used for suitability and location analysis too (Fig. 3.20).

3.2.5.6 Hillshade (Illumination)

Hillshading is a technique used to create a realistic view (shades) of terrain by creating a three-dimensional surface from a two-dimensional display of it. Hillshading creates a hypothetical illumination of a surface by setting a position for a light source and calculating an illumination value for each cell based on the cell's relative orientation to the light or based on the slope and aspect of the cell (Albrecht 2005). Hillshades are often

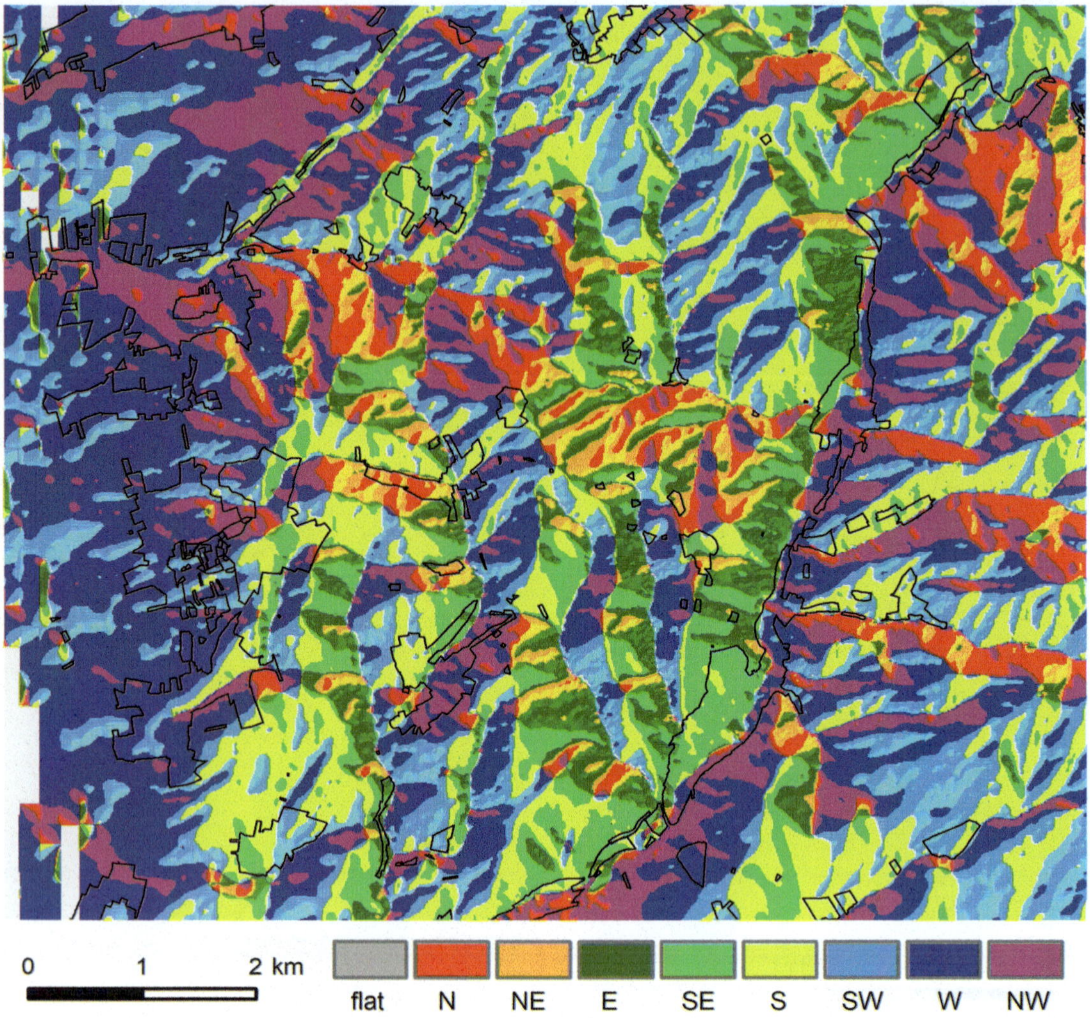

Fig. 3.20 Example of aspect. (Source: Authors)

used to increase the quality and readability of maps (Fig. 3.21).

3.2.5.7 Viewshed (Visibility Analysis)

The viewshed analysis identifies the cells in an input raster that can be seen from one or more observation points or lines. Each cell in the output raster receives a value that indicates how many observer points can see the location (Albrecht 2005). This raster analysis has a wide range of usage and applications. It can be used to determine the aesthetic impact of new city development (e.g. new houses), or for the placement of communications towers (if the direct visibility is needed) or optimal placement of a new lookout tower (Fig. 3.22).

3.2.5.8 Cost Distance Analysis (Least-Cost Path)

The cost distance analysis elaborates movement over continuous space, in which the cost of moving through any location is variable. Cost surface represents some factor or combination of factors that affect travel across an area (e.g. high values for steep terrain, low values for flat areas). In the second step, the least-cost path analysis

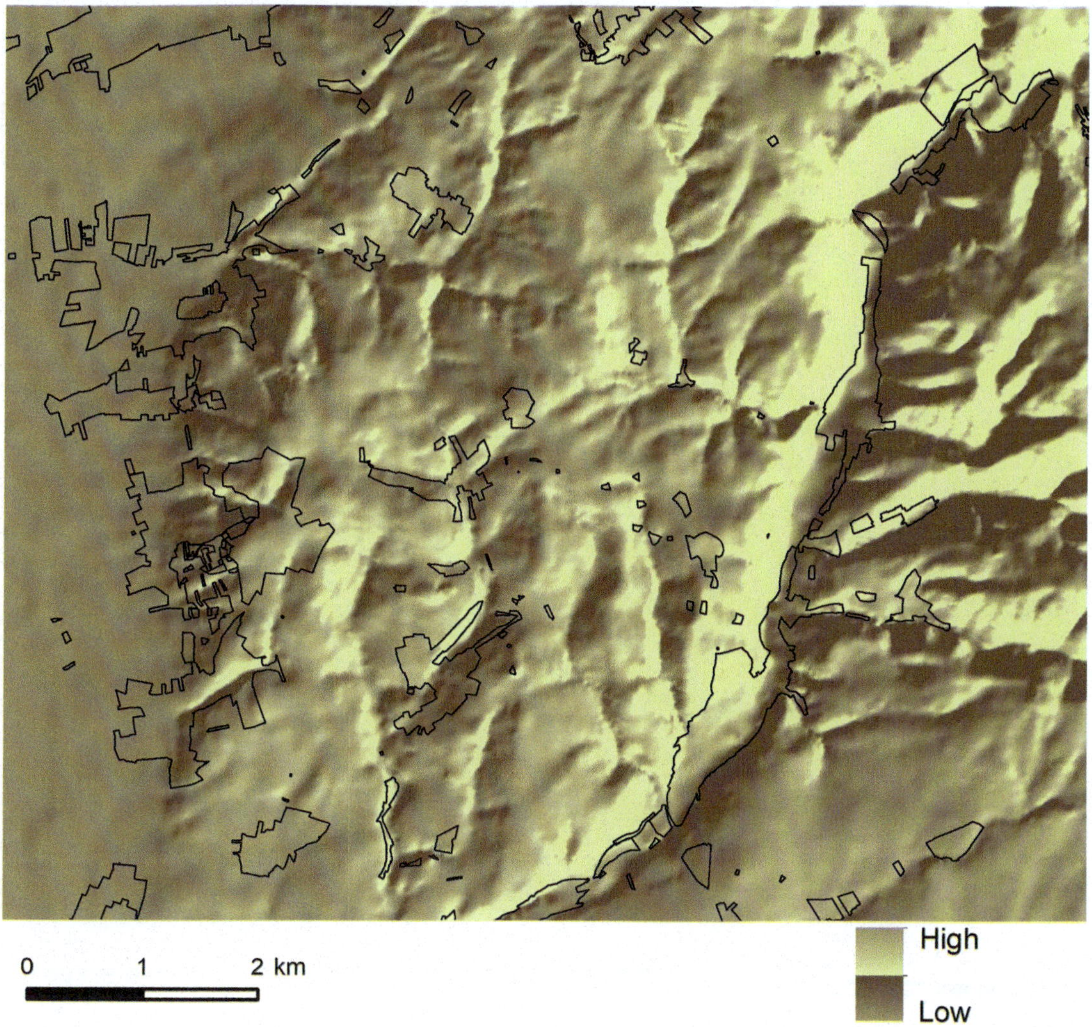

Fig. 3.21 Example of hillshade. (Source: Authors)

uses the cost-weighted distance and direction surfaces for an area to determine a cost-effective route between a source and a destination. This can be used e.g. for the planning of new highways to find the cheapest solution (Fig. 3.23).

3.2.5.9 Solar Radiation (Insolation) Analysis

The solar radiation analysis enables to calculate the amount of the solar energy over a geographic area for specific periods. It accounts for atmospheric effects, site latitude and elevation, steepness (slope) and compass direction (aspect), daily and seasonal shifts of the sun angle, and effects of shadows cast by surrounding topography (ESRI 2018a). Information about the amount of insolation is helpful for application in many fields, such as civil engineering, economy or agriculture research. It may be useful in localisation of a new site for a ski resort, wine yard or solar panels (Fig. 3.24).

3.2.5.10 Multi-Criteria Analysis

Multi-criteria analysis (MCA) is a method used to consider many different criteria when making a decision. In GIS, MCA is represented by overlay

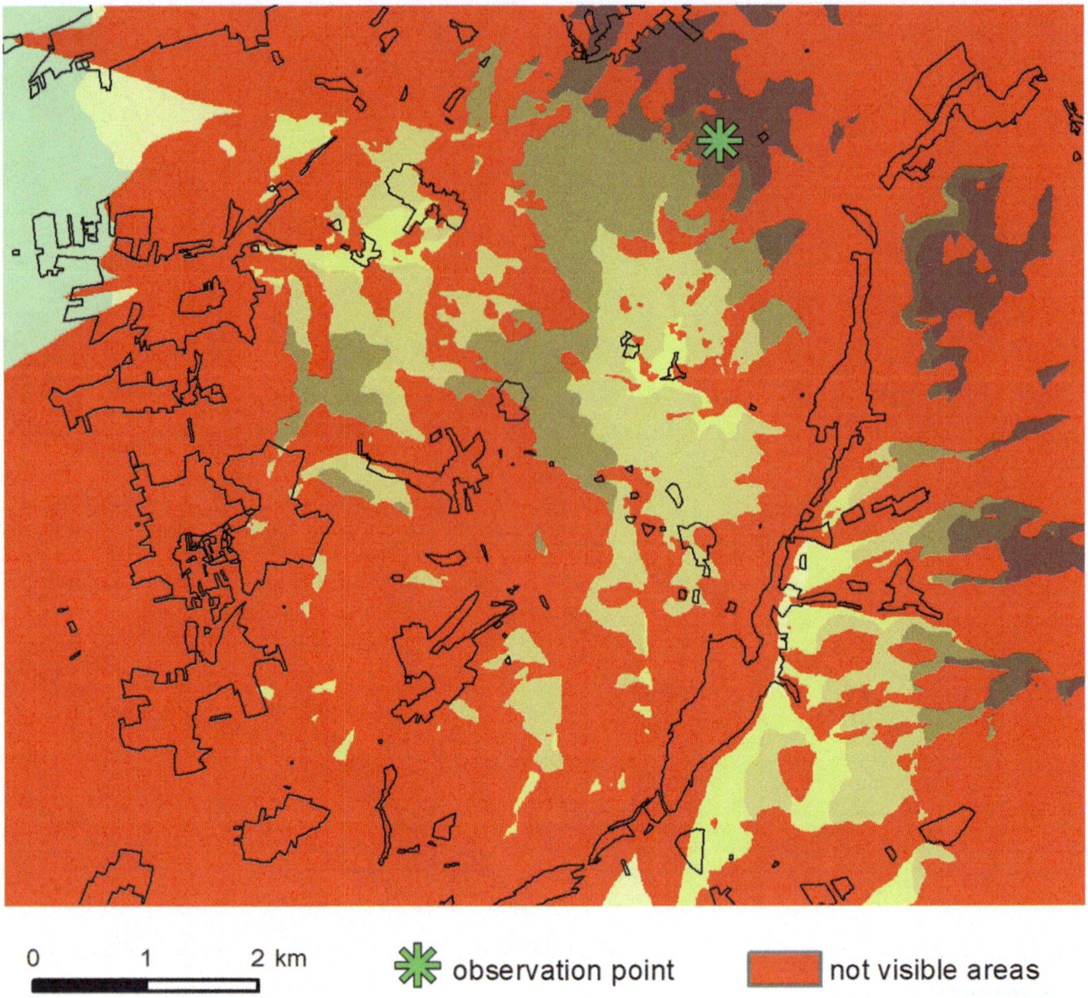

Fig. 3.22 Example of viewshed analysis. (Source: Authors)

analysis (weighted overlay) that overlays several rasters using a common measurement scale and weights each according to its importance. In this case, each criterion (or map layer) is brought to a common scale (reclassified) to simplify the process of combining the layers. Spatial MCA is used for decisions with a geographical factor (suitability analysis, location analysis), where multiple factors need to be considered (e.g. land-use, distances to public transportation, shops accessibility, park accessibility, etc.). In Fig. 3.25, you can see an example of multi-criteria analysis, that combines environmental (green), social (red), and economic suitability (blue) to obtain total land suitability for new housing development (dark green).

3.3 Network Analysis (by Nicolai Moos)

3.3.1 Introduction

Most people are familiar with using a navigation system, which means that they have at least once processed a basic network analysis by looking for

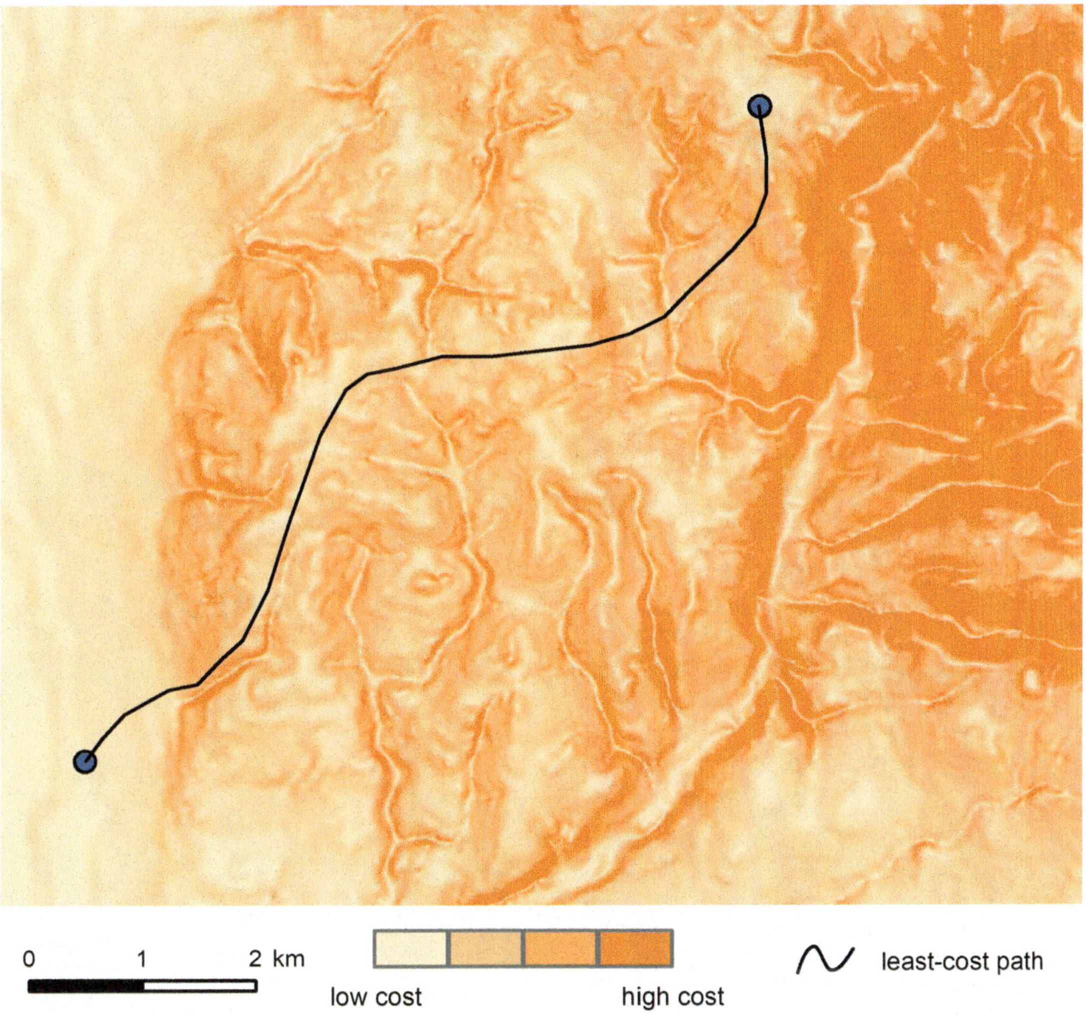

Fig. 3.23 Example of cost distance analysis. (Source: Authors)

the shortest path or fastest route to a different location from their own. Since GIS-Software is a lot more sophisticated than a common navigation system and consequently offers many more possibilities, this chapter will characterize the fundamental functions and approaches of network analyses in GIS.

When dealing with network analysis tools and functions it is necessary to prepare a suitable spatial network in form of a network dataset which is able to perform all functions that are included in a network analysis (DeMers 2008). Network datasets typically consist of line features that stand for the routes of motion in the network, enhanced with further features and premises to ensure proper usage (ESRI 2018b). Regular line features are generally not related to each other and have no or only a few connectivity rules. This means for instance if two different lines are intersecting each other, none of them is aware of it, what makes the dataset restrictive as you cannot turn from one to another. To make sure that the network recognizes these crossings as such, it is necessary to transfer the network into a new one that has nodes which allow a turn from one edge to another, except for over- or underpass lines (e.g. tunnels, bridges, etc.), where this intentionally should not be possible. Basically, it is necessary to decide whether the network dataset should have an end point or any vertex connectivity (see Fig. 3.26).

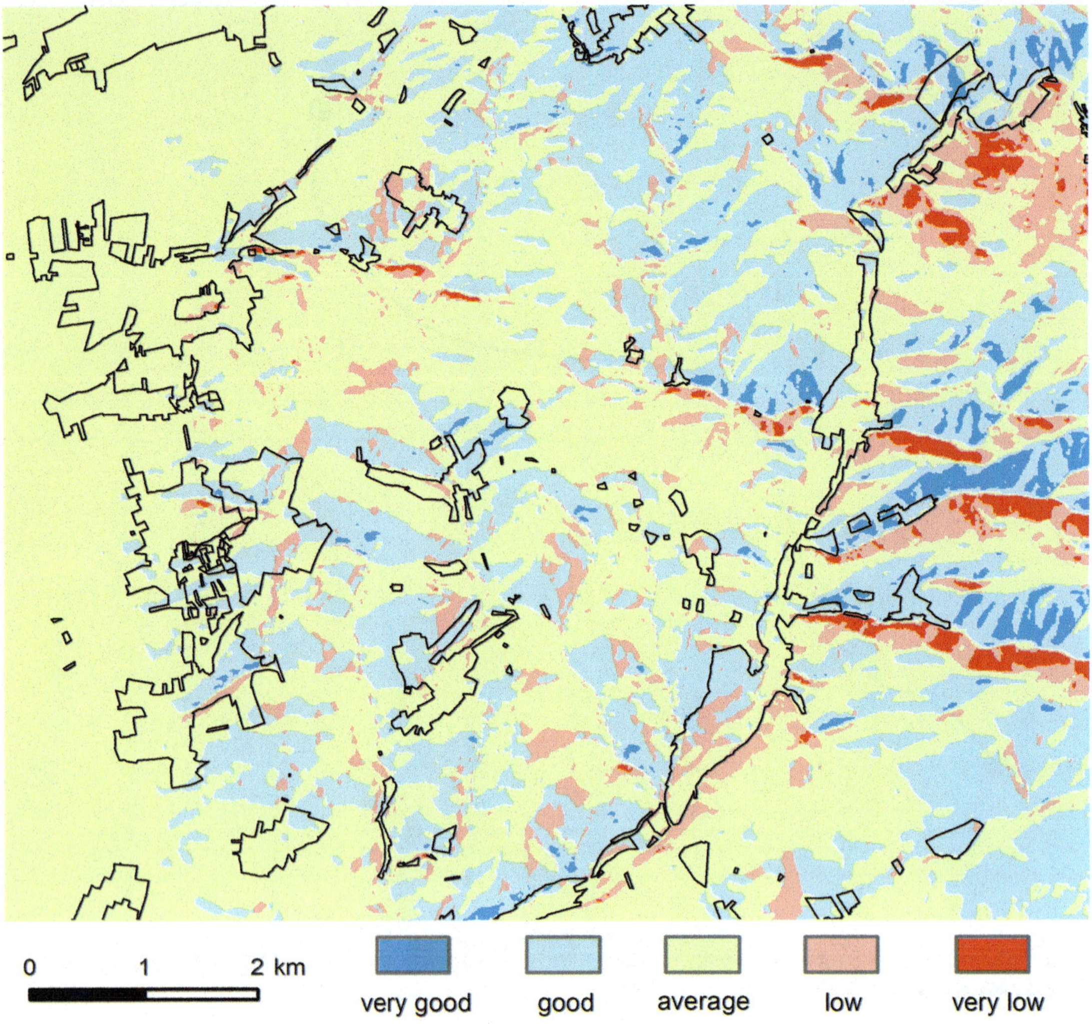

Fig. 3.24 Example of solar radiation intensity. (Source: Authors)

Furthermore, a working network dataset needs information on directions linked to each street segment as there are one-way streets as well as streets that can be driven on in both directions. If an outcome of the analysis should deal with a time or street capacity component these are also figures that have to be included as impedances constructing a network dataset (Chang 2010).

Once the network dataset is built up, there are several different opportunities of calculations in a network analysis.

3.3.2 Optimal Routes

The most basic function is the calculation of an optimal route from point A to point B and any number of intermediate stops, while the order of these stops is determined by the user and not the tool. This route can be either focusing on the shortest distance or the fastest route, depending on the needs of the respective user (Fig. 3.27). Relevant factors for this calculation can be the existence of one-way streets, barriers like construction sites

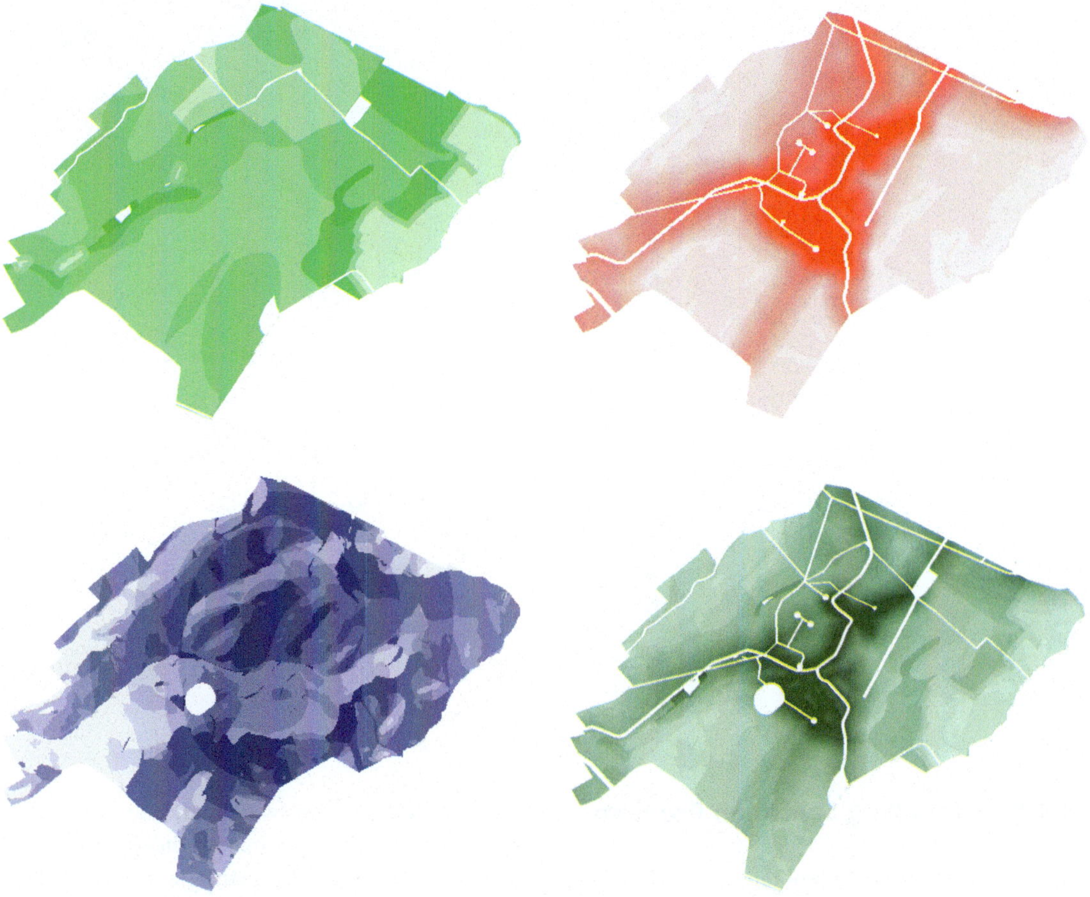

Fig. 3.25 Example of multi-criteria analysis. (Source: Authors)

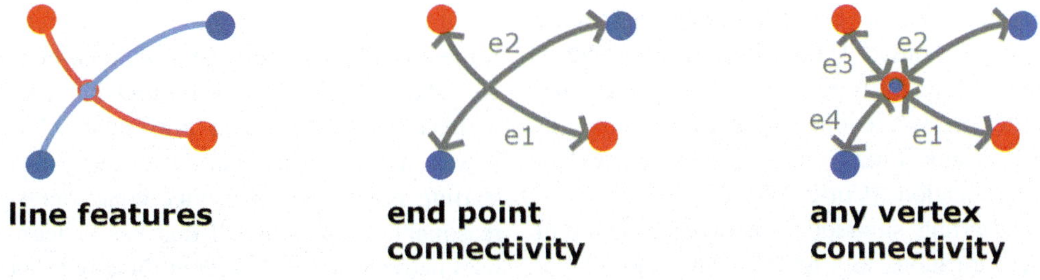

Fig. 3.26 Line Feature input with nodes and edges and different output connectivities for a network dataset. (Source: Authors)

or other obstacles, prohibited turns or other restrictions that influence the way the route will be computed (ESRI 2018b). All the different factors can be parametrized before the calculation and adjusted afterwards, as the result is only a virtual and therefore temporary layer in the map that can be permanently exported if necessary. Additionally to the line feature of the route, the user can generate driving directions containing specific information about the route, e.g. how long to stay on a certain street and when turn to another.

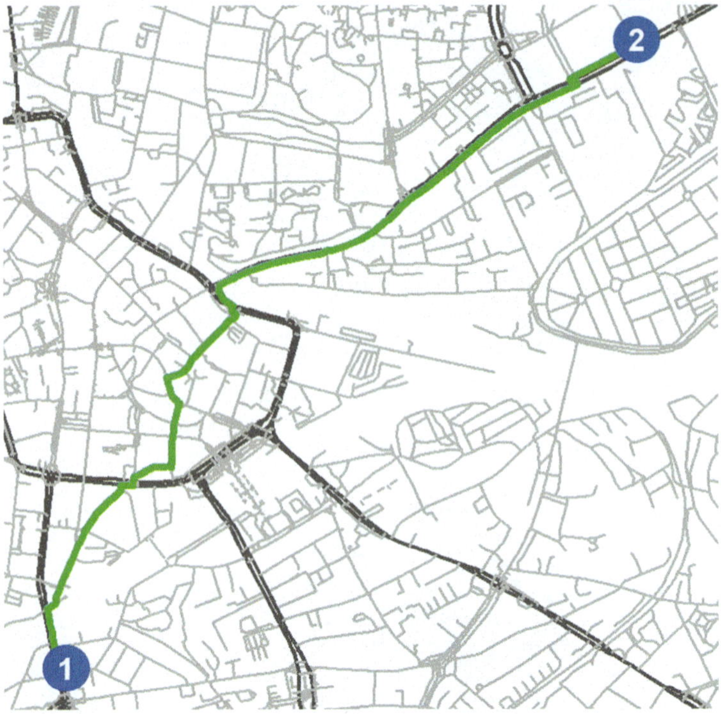

Fig. 3.27 Optimal Route from location 1 to location 2. (Source: Authors)

3.3.3 Traveling Salesman Problem

The traveling salesman problem is an issue that is not only present in classical spatial analyses, but also in various other fields, like e.g. logistics or designing products that contain several different spots which need to be connected in a specific chronological order. The most relevant aspect in this network analysis is the efficiency of a route no matter if it is a real person that is travelling or any other subject that is moving between several different locations (Curtin 2007).

The starting situation is a given amount of different locations in a network and a person or subject that has to visit each of these locations in a certain order. That order has to be the most efficient one regarding distance, required time or costs (see Fig. 3.28). Depending on how these different factors are weighted, the routes can vary significantly what makes choosing the right parameters essential for getting a proper result.

3.3.4 Service Areas

If there is a location in a network dataset and it is desirable to know for instance how far one can get with a car in a determined time range or how long it will last to reach a certain distance from that starting point, then a service area analysis is a reasonable approach. All that has at least to be parametrized are the different breaks in terms of certain time ranges or distances that define the borders between the different areas (see Fig. 3.29). These analyses could for example help to find a suitable location for a new hospital

Fig. 3.28 Fastest Route to stop once at every location, order calculated by the tool. (Source: Authors)

Fig. 3.29 Locations of hotels in blue, service areas for accessibility from five minutes (darkest blue) to 15 minutes (lightest blue). (Source: Authors)

as their output is the information about the size of an area (and the number of inhabitants) that is covered by e.g. an ambulance within two, four or 10 min during a specific time of the day.

3.3.5 Location-Allocation Analysis

How can we save money for transportation? Where should we place a new facility? How big is the potential area that is covered by the store? Are stores reachable for all customers in a certain amount of time? The location-allocation analysis has several approaches like minimizing impedances, number of facilities or maximize area coverage, accessibility or market shares (Chang 2010). It therefore combines the different methods of a network analysis. Each of these tasks implies the preparation of an analysis layer that can calculate the optimal location for the particular case of application.

Necessary inputs for this layer are a network dataset as well as facility locations and demand points. The facilities are split up into candidate facilities that represent the potential location of a new facility, competitor facilities that mark the existing sites of present competitors and required facilities that represent existing sites of say one's own organization. Demand points are locations that represent the different factors that determine the grade of suitability for a new candidate facility. These can be centroids of districts or other administrative units as well as different kinds of demand profiles like accumulation of students, families or workers of a certain business. The demand points contain information like income, age, social status, etc.

As there are too many possible cases of application, the maximize attendance approach as the most frequently used one is presented in this chapter. In this example the target is to detect the locations that would generate the most efficient business (maximized attendance) for a retail chain, assuming that the customers rather frequent stores that are close to dense population centers and don't want to travel for more than 5 min. Once everything is set up, the generated output layer shows the detected site(s) connected with lines to the most valuable demand points (number of population) that determine the chosen target classes (see Fig. 3.30).

3.3.6 Origin-Destination Matrices

For the creation of an origin-destination (OD) matrix it is essential to set a certain amount of starting point features as well as a certain amount of target point features that are all located within the network dataset. The analysis settings can vary between different impedances, barriers in the network, a certain point of time and other parameters that influence the result concerning the properties of the network dataset (Curtin 2007).

The result layer shows the shortest routes and directions from all starting point features to all target point features that are within a determined range of distance or time (see Fig. 3.31). This can be used e. g. for creating a new model of pedestrian movements or checking the suitability of new sites within a network dataset.

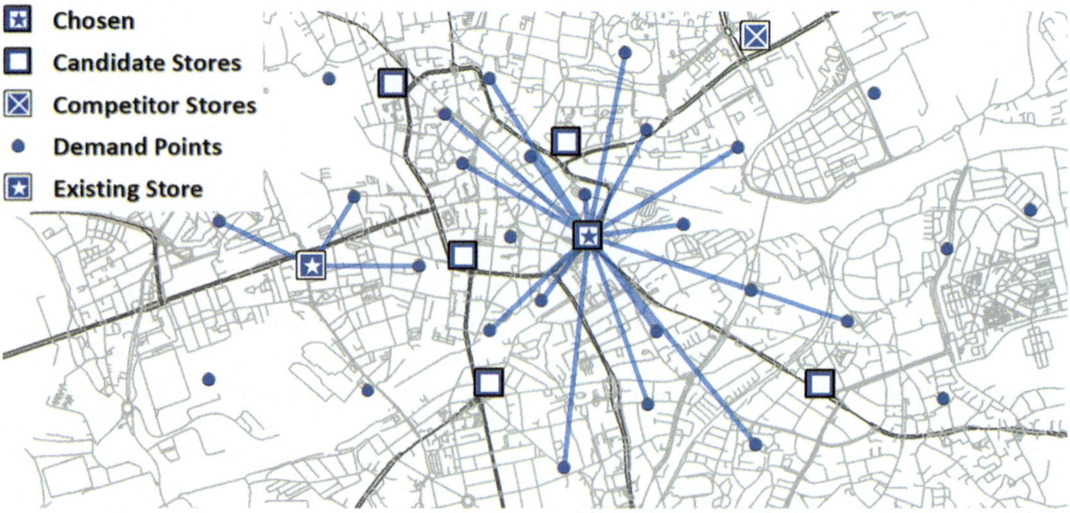

Fig. 3.30 Location-Allocation Analysis result with lines showing the connection to valuable and affecting demand points. (Source: Authors)

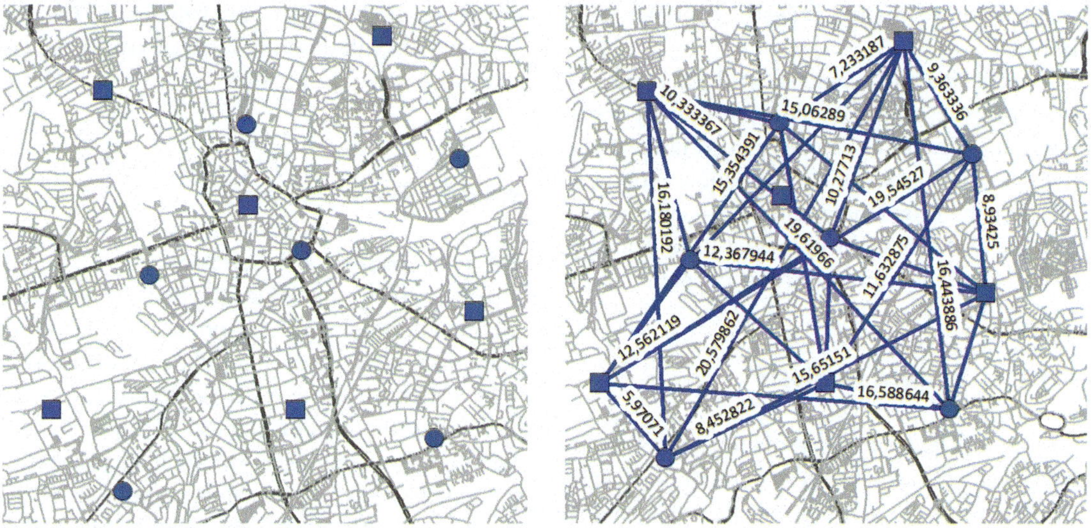

Fig. 3.31 OD Matrix for accessibility from certain locations (blue circle) to hotels (blue square) with needed amount of travel time (in minutes). (Source: Authors)

3.4 Spatial Statistics (by Karel Macků)

In the context of the Tobler's first law of geography saying "Everything is related to everything else, but near things are more related than distant things", spatial statistics is a set of exploratory techniques for describing and modelling spatial distributions, patterns, processes and relationships. This group of analyses is necessary for a deeper understanding of spatial data, which is provided with the use of statistical methods. In this chapter, the most frequently used methods of spatial statistics are briefly introduced.

Spatial statistics is a subcategory of spatial data analysis which is closely linked to mathematical statistics. Spatial statistics is a set of exploratory techniques for describing and modelling spatial distributions, patterns, processes and relationships (Bennett et al. 2017). According to Haining (2003) some of the spatial analyses include mathematical modelling where model outcomes are dependent on the spatial interaction between objects in the model, or spatial relationships of the geographical positioning of objects within the model. This statement represents the difference between simple spatial analyses and more advanced methods that approach the tasks using mathematical and statistical apparatus. Question is why any events happen on their location and not elsewhere? Is there any association with the environment? Are the events spread or clustered in any area? With proper data, these types of questions can be answered with spatial statistics.

Spatial statistics methods are based on the assumption that elements that are close to each other are also more closely related. A direct link to Tobler's first law of geography can be observed here: "Everything is related to everything else, but near things are more related than distant things" (Tobler 1970, p. 236). Spatial statistics can also be viewed as a complementary tool to spatial data analysis – it offers a mathematical apparatus and methods for evaluating spatial information, on the other hand, stands geography or other spatial science, which formulates a hypothesis or identifies the key parameters of these spatial data (Getis 2005). In the search for a high degree of certainty, the statistical approach is always recommended.

There should be no confusion between the terms spatial statistics and geostatistics – geostatistics is one of the spatial statistics sub-disciplines and has emerged as a tool for a probability prediction of the distribution of ore deposits in the mining industry (Longley et al. 2010).

Spatial statistics include methods based on stochastic (i.e. random) nature and pattern of phenomena. These tasks can be divided into descriptive (producing essential information about a set of elements) and interference – analysis of patterns and behaviour of spatial data. This type of analyses is the subject of this chapter.

3.4.1 Pattern Analysis

One of the frequently asked questions in the advanced spatial analysis regarding the spatial distribution is the question whether elements are in a random structure or there is a pattern in their behaviour. In particular, it is essential whether the presence of one value causes an increase or decrease in the probability of occurrence of another value in its vicinity (Longley et al. 2010). By their deployment in space, the elements can create one of the following structure:

- Random structure
- Cluster structure
- Regular/scattered structure

At the beginning, the term ‚cluster‘should be clarified. Clustering is a global property of the spatial pattern in a dataset, measured by a single statistics (Anselin 2005). Then cluster is a group of features, whose value and/or its locations are closer together than they would be by random. The purpose of pattern analyses is to determine whether the spatial behaviour of the geographic elements follows one of the above-mentioned options and if this behaviour is somehow statistically demonstrable. Actual spatial distribution is therefore tested against one of these options. Confirming the existence of significant clusters of similar values/clusters of points near one another is one of the most common tasks. Such a task could be based only on a visual analysis of spatially visualised data; however, the use of spatial statistics underlies this estimate by numerical tests and makes it more reliable. The resulting finding helps to understand the behaviour of the observed phenomenon and to support the hypotheses that explain this behaviour. The following lines will describe selected spatial pattern analysis.

3.4.2 Point Patterns

3.4.2.1 Ripley's K Function

The *K function* is one of the methods for assessing the randomness of the distribution of the set of point data. It allows seeing if the elements appear to be dispersed, clustered, or randomly distributed throughout the area of interest. The basis of this method is to monitor the occurrence frequency in a defined space – for example, the area in the distance *d* from each point. The K function is defined as the ratio of the number of occurrence points in the defined area (grid or defined distance *d*) and the expected density of points per area unit, how would it be within the random distribution of the elements (most often represented by the homogeneous Poisson process, also known as complete spatial randomness). This principle allows identification of deviations from spatially evenly distributed data (Dixon et al. 2002). If the number of observed points within a given area is higher than for a random distribution, the distribution is clustered. If the number is smaller, the distribution is dispersed (Gillan and Gonzalez).

For an example, data of position of small and medium enterprises in Olomouc region has been analyses with K function. In such a data, it is expected that companies are located in the sites that means they will be clustered within the city.

The result of point pattern analysis can be presented as a graph – see Fig. 3.32. The vertical axis is the K-function value; the horizontal axis is the searching distance *d*. The blue line represents the k value of the random distribution of the points, and the red line represents the *K* function value – the real observed distribution of the points (the position of companies). If the observed value is above the random, it means that points are clustered. If the observed were under the random, the data would be dispersed. In this case, result points to strongly clustered data. This supports the original hypothesis about the location of enterprises.

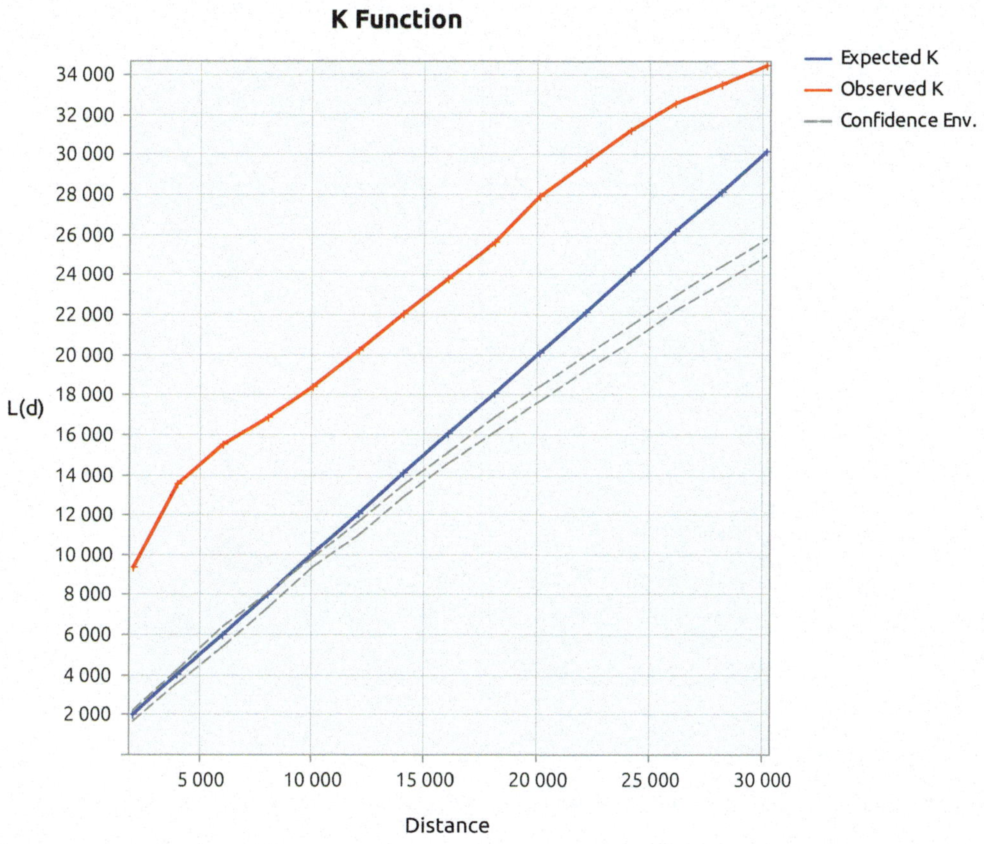

Fig. 3.32 A graphical output of K function. (Source: Authors)

3.4.2.2 Kernel Density

Kernel smoothing methods are used to transform data from a discrete representation (geolocated points) into a continuous array. This process is particularly useful for better interpretation of spatial distinction of variables behaviour. The kernel density estimate works with localised data, which are used for the expression of the spatially smoothed estimate of the local intensity of the occurrence of objects/events. This local smoothed intensity can also be understood as the surface of the risk of occurrence of these objects/events (INSEE Eurostat 2018). The application on spatial data is based on density estimation, a function of estimating the values occurrence based on observed data (Silverman 1986).

Conceptually, a smoothly curved surface is fitted over each point. The surface value is highest at the location of the point and diminishes with increasing distance from the point (ESRI 2018c). The final surface is created by estimating the intensity at any point using the appropriate probability density function (K – kernel function). It is necessary to determine the area in which the algorithm will assess the density of the phenomenon. This sphere – so-called bandwidth, might be calculated on all input points and median distance between its centre and all input points. The bandwidth parameter essentially determines the degree of smoothing of the resulting surface. The different kernel functions can be used to make the result of density estimation different. The application on the spatial data implemented in ArcMap software uses the quartic function, which approximates to the normal distribution.

The resulting surface is represented in the form of a raster, which can be conveniently visualised for the purpose of overview of the phenomenon

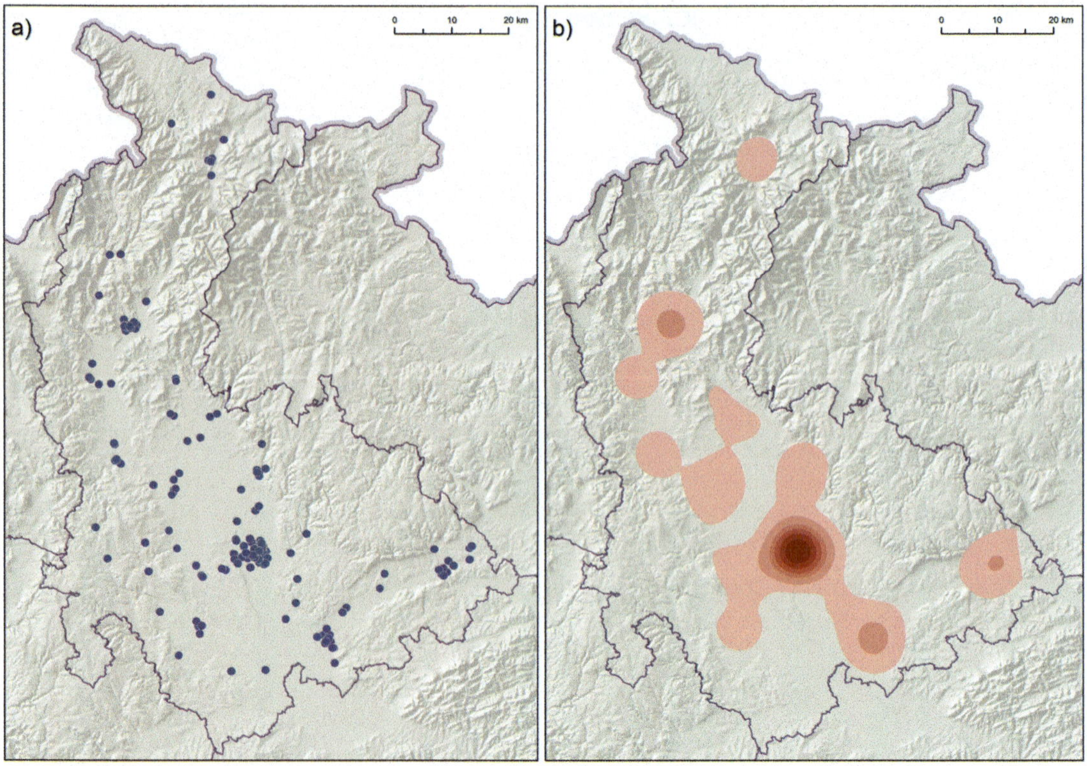

Fig. 3.33 (**a**) input point data, (**b**) result of Kernel density function. (Source: Authors)

and revealing point patterns. The same data describing location of enterprises in Olomouc region (as for the K function in previous chapter) was used for demonstration of Kernel density function. Figure 3.33 shows the visualisation of input points and output surface. The bandwidth was set to five kilometres, with the aim to produce more smoothen output. The output presents the probability of enterprise occurrence, this type of visualisation brings a generalized overview of the spatial distribution of points in the area of interest. The aim is not to estimate the correct probability of occurrence, but rather to get overall impression about spatial distribution of points. For that reason, a legend for interpretation of result is not included.

3.4.3 Spatial Autocorrelation

The previous chapter has described how clusters of point phenomena can be identified based on their location. A following task can be to identify clusters based on the location combined with the value of the observed phenomenon at the same time. Such an analysis makes possible to evaluate whether there are spatially closer elements that have similar values of the observed phenomenon and form together high or low-value clusters, or whether the elements in the space are located at random. In the natural world, we expect some influence of environment on the monitored phenomenon. For example, analysing a strong economic region concerning the GDP per capita, we naturally assume that the regions in its immediate neighbourhood will be similar, as the whole area is characterised by similar conditions. Similarly, we expect these regions to differ from other, more remote areas. To support such a claim, an analysis of spatial autocorrelation can be used.

Spatial autocorrelation is a correlation between the values of one variable, and it allows to evaluate the degree of similarity of one object with objects in its neighbourhood and comparison with more remote objects (Cliff and Ord 1973). First, it is necessary to define relations of the object with its surrounding objects, which is provided by the matrices of spatial weights. Here, the distance of objects enters as a weight for defining spatial relationships – the autocorrelation of neighbouring objects will have more importance than the autocorrelation of distant objects. If positive autocorrelation occurs, we conclude that objects with similar values are spatially located near each other, forming spatial clusters of similar values. Negative autocorrelation indicates the proximity of different values, autocorrelation around zero indicate randomity in the spatial distribution of values.

Autocorrelation can be measured by several measures – an example of them is the **Moran's I** or **Geary's** criterion. Positive index value indicates a positive autocorrelation, and negative values represent negative autocorrelation. These indicators, however, measure autocorrelation only at the *global* level, that is the whole area of interest. If the result of these tests come out positively, it makes sense to ask how the autocorrelation varies in the space. A *local* test – **LISA** (Local Indicators of Spatial Association) serves for this task. Since the method of identifying spatial autocorrelation is based on traditional statistical methods, the calculation is complemented by the statistical significance, represented by p-value. This makes it possible to assess whether the result obtained is statistically significant or not.

The initial analysis of autocorrelation reveals spatial dependence, so it is known that clusters of high and low values occur in the area of interest. Local Moran's I can be visualised to identify these areas. However, it is still unknown whether the high value of autocorrelation means clustering of high or low values. For a deeper understanding of the phenomenon, it is possible to visualise the observed variable depending on the average value in its surroundings – this is presented by Moran's plot (Anselin 1996).

Using LISA and Moran's plot as supporting tools, all objects can be classified into four groups corresponding to the quadrants in Moran's plot. Spatial clusters showing above-average or below-average values of a variable in a particular unit consistent with its surroundings are found in the graph in the top right (hot spots, high-high) and left-low (cold spots, low-low) quadrants. This is evidence of high autocorrelation. On the contrary, the areas identified in the left upper (LH) or right lower (HL) quadrants are characterized by the existence of a low value surrounded by high and vice versa (Anselin 1995) (Fig. 3.34).

Similar output as provided by LISA is available also with **Getis – Ord G∗.** The main difference is that for LISA, the value of the feature being analysed is not included in that analysis, only neighbouring values are. Alternatively, when the local analysis is being done with Getis-Ord Gi∗, the value of each feature is included in its analysis (Getis and Ord 2010). The local sum for a feature and its neighbours is compared proportionally to the sum of all features; when the local sum is very different from the expected local sum, and when that difference is too large to be the result of random chance, a statistically significant z-score results (ESRI 2018d).

The output of this indicator is the so-called z-score for each analysed object. The higher (positive) the z-score value, the higher the intensity of clustering of high values in the area (so-called hotspot), and vice versa – the smaller (negative) the z-score is, the higher the intensity of clustering with a cold spot.

An example demonstrating the use of spatial autocorrelation methods is described in the analysis of the economically strongest and also the weakest regions in Europe. The monitored variable is GDP, the spatial unit is NUTS 3 regions.

The GDP is expressed in purchasing power standard per inhabitant in the year 2015. In Fig. 3.35a, a choropleth map is used to display the GDP in regions. By this visualisation, areas with the highest or lowest values can be defined, especially the big difference between east and west are visible. But can be said with certainty which regions are the strongest and which are the weakest? In many cases, when the data doesn't have a clear pattern, or inappropriate visualisation is used, it might be a difficult task. For that reason, spatial autocorrelation is calculated. Figure 3.35b

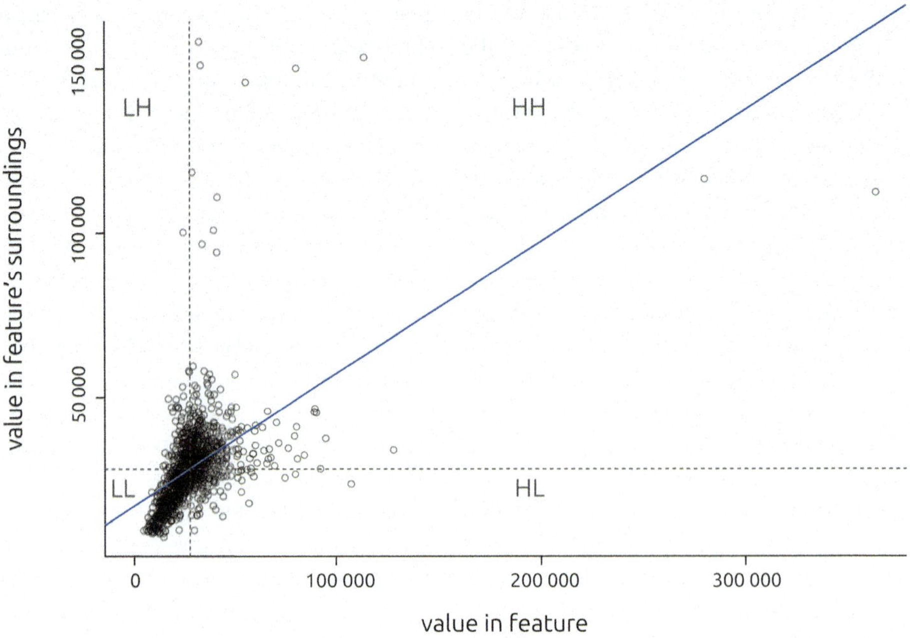

Fig. 3.34 Moran's plot. (Source: Authors)

shows the distribution of spatial autocorrelation calculated for the GDP data. Only the shades of green are statistically significant; darker shade stands for higher autocorrelations = clusters of low or high values are present. The final step is the derivation of cluster type based on the value of autocorrelation in every region and its neighbourhood. This can be done by LISA analysis (Fig. 3.35c) or Getis-Ord G∗ (Fig. 3.35d). See the difference between this methods caused by the different approach, how they calculate the membership to any clusters.

Now user can state that regarding the spatial distribution of GDP, there is a great cluster of low values in the eastern European and several small clusters of high value in the central Europe, Sweden and UK. In the rest of the area of interest, the GDP value has a random distribution without statistically significant patterns.

3.4.4 Geostatistics

As mentioned in the introduction to the chapter, the term spatial statistics is often confused with the term geostatistics. In the narrower sense, geostatistics is used only to define a set of interpolation algorithms – algorithms used to estimate the values of the continuous phenomenon or its intensity in any location of the controlled area where no measurements have been made. The continuous character is typical of environmental phenomena such as temperature, air pressure or soil concentration. In the context of economic

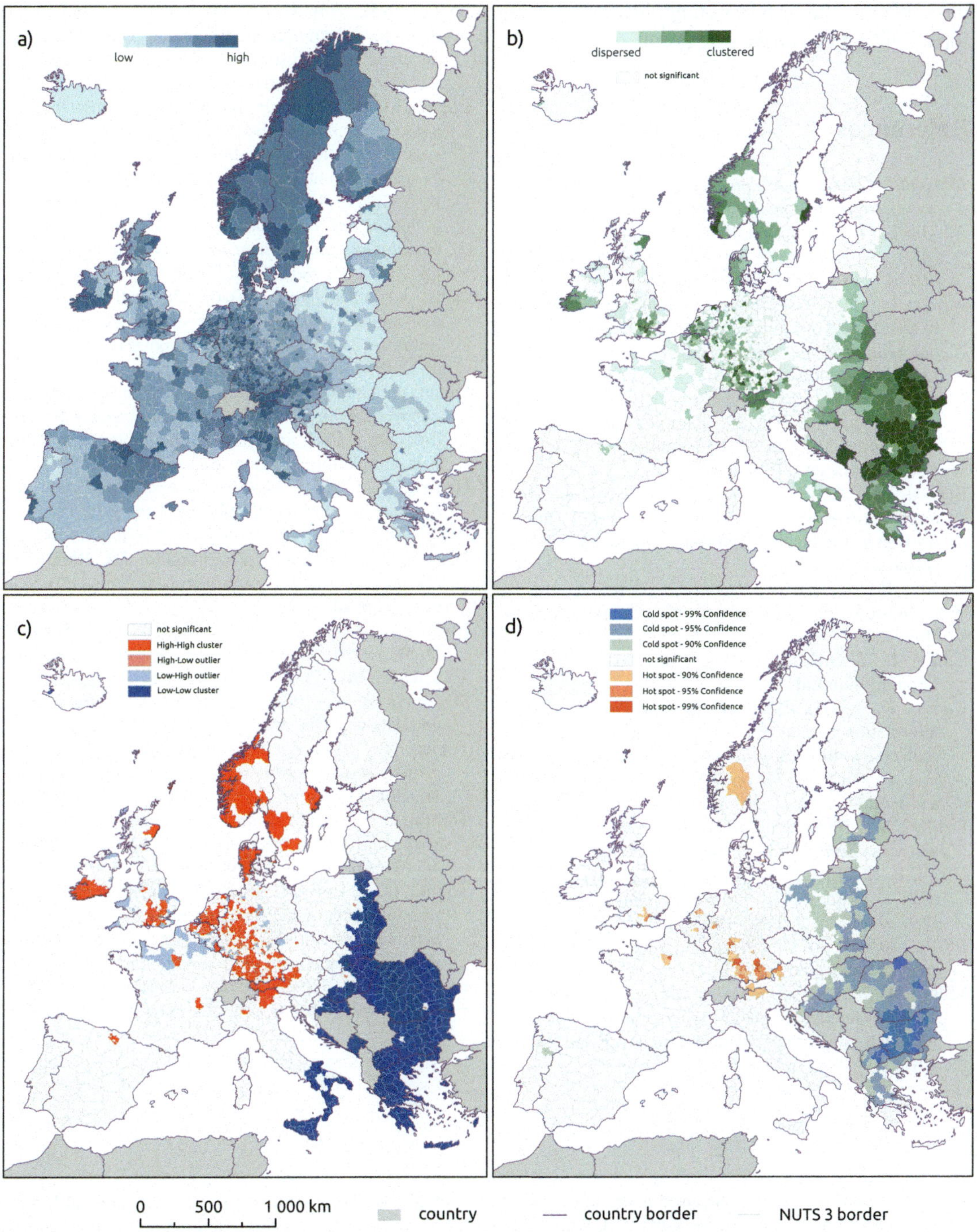

Fig. 3.35 Analysis of spatial autocorrelation of GDP in Europe. (Source: Authors)

data, there would be a lack of applications, so this topic will not be further discussed.

References

Albrecht, J. (2005). *Geographic information science*. http://www.geography.hunter.cuny.edu/~jochen/gtech361/.

Anselin, L. (1995). Local indicators of spatial association – LISA. *Geographical Analysis, 27*, 93–115. https://doi.org/10.1111/j.1538-4632.1995.tb00338.x.

Anselin, L. (1996). The Moran scatterplot as an ESDA tool to assess local instability in spatial association. In *Spatial analytical perspectives on GIS* (pp. 111–125). London: Taylor and Francis.

Anselin, L. (2005). Spatial statistical modeling in a GIS environment. In D. J. Maguire, M. Batty, & M. Goodchild (Eds.), *GIS, spatial analysis, and modeling* (1st ed.). ESRI Press.

Bennett, L., Vale, F., D'Acosta, J. (2017). *Spatial statistics: Simple ways to do more with your data*. https://www.esri.com/arcgis-blog/products/product/analytics/spatial-statistics-resources/

Chang, K. T. (2010). *Introduction to geographic information systems* (5th ed.). New York: McGraw-Hill.

Cliff A. D., & Ord J. K. (1973). *Spatial autocorrelation*. London: Pion Ltd.

Curtin, K. (2007). Network analysis in geographic information science: Review, assessment, and projections. In *Cartography and geographic information science* (Vol. 34, pp. 103–111). London: Taylor and Francis.

DeMers, M. (2008). *Fundamentals of geographic information systems* (4th ed.). Hoboken: Wiley.

Dixon, P. M., El-shaarawi, A. H., & Piegorsch, W. W. (2002). Ripley's K function. *Encyclopedia of Environmetrics, 3*, 1796–1803.

ESRI. (2018a). *ArcGIS Desktop Help*. http://desktop.arcgis.com/en/arcmap/10.3

ESRI. (2018b). *What is a network dataset?* https://desktop.arcgis.com/en/arcmap/latest/extensions/network-analyst/what-is-a-network-dataset.htm. Accessed 28 Dec 2018.

ESRI. (2018c). *How Kernel density works—Help | ArcGIS desktop*. http://pro.arcgis.com/en/pro-app/tool-reference/spatial-analyst/how-kernel-density-works.htm. Accessed 22 Nov 2018.

ESRI. (2018d). *How hot spot analysis (Getis-Ord Gi*) works*. http://pro.arcgis.com/en/pro-app/tool-reference/spatial-statistics/h-how-hot-spot-analysis-getis-ord-gi-spatial-stati.htm. Accessed 23 Oct 2018.

Getis, A. (2005). Spatial statistics. In *New developments in geographical information systems: Principles, techniques, management and applications* (2nd ed., pp. 239–252). New York: Wiley.

Getis, A., & Ord, J. K. (2010). The analysis of spatial association by use of distance statistics. *Geographical Analysis, 24*, 189–206. https://doi.org/10.1111/j.1538-4632.1992.tb00261.x .

Gillan, J., Gonzalez, L. *Ripley's K Function and pair correlation function*. http://wiki.landscapetoolbox.org/doku.php/spatial_analysis_methods:ripley_s_k_and_pair_correlation_function. Accessed 22 Oct 2018.

Haining, R. (2003). *Spatial data analysis: Theory and practice*. Cambridge: Cambridge University Press.

INSEE Eurostat. (2018). *Handbook of spatial analysis*.

Longley PA, Goodchild MF, Maguire DJ, Rhind DW (1999) Geographical information systems: Principles and technical issues.

Longley, P. A., Goodchild, M., Maguire, D. J., & Rhind, D. W. (2010). *Geographic information: Systems and science* (3rd ed.). Wiley Publishing.

Silverman, B. W. (1986). Density estimation for statistics and data analysis. *Monogr Stat Appl Probab*.

Tobler, W. (1970). A computer movie simulating urban growth in the Detroit region. *Economic Geography, 46*, 234–240.

Tomlin, D. (1994). Map algebra: One perspective. *Landscape and Urban Planning, 30*, 3–12. https://doi.org/10.1016/0169-2046(94)90063-9.

Tomlin, D., & Berry, J. K. (1979). A mathematical structure for cartographic modeling in environmental analysis. In *Proceedings of the 39th symposium of the American conference on surveying and mapping*.

w3schools.com. (2018) Retrieved December 30, 2018, from, https://www.w3schools.com/sql/sql_where.asp

Business Informatics Principles

4

Simona Sternad Zabukovšek, Polona Tominc, and Samo Bobek

Abstract

Business informatics consists of two major areas which are developing separately but are closely related and also to some extent integrated within information systems architecture in organisations. The first area is business solutions as core information systems for support of business operations – they consist of Enterprise resources planning solutions (ERP), Customer relationship management solutions (CRM) and other specialised solutions. The second area is Business Intelligence (BI). After describing the basic concepts of these information solutions/business information systems and their functionality, the chapter explains the emerging integration of business informatics and geo-informatics. Developments provided by solution providers are analysed and discussed. The chapter concludes with a bibliometric analysis of research which shows areas and dynamics of business informatics and GIS integration.

Keywords

Enterprise resource planning - ERP · Business intelligence - BI · Geographical information systems - GIS

S. S. Zabukovšek (✉) · P. Tominc · S. Bobek
Faculty of Economics and Business, University of Maribor, Maribor, Slovenia
e-mail: simona.sternad@um.si; polona.tominc@um.si; samo.bobek@um.si

4.1 Introduction

Modern organisation is viewed as a group of people with a common goal, which has certain resources at its disposal to achieve goals. In the traditional approach, the organisation is divided into different units based on the business functions, such as manufacturing or production department, production planning department, purchasing department, sales and distribution department, finance department, research and development (R&D) department etc. (Anderegg 2000; Sneller 2014). These departments have not been integrated through business processes, and each department has been a »silos« and has its own goal and objectives (Magal and Word 2011). They had their departmental information systems with their own databases where they collected data and performed the analysis. Because of that the information was created or generated by the various departments, in most cases is available only to the top management and not to other departments (Anderegg 2000). Departments didn't know what others do, and sometimes the departments' objectives could be conflicting. When all departments don't know what other departments are doing and for what purpose, different kinds of conflicts often arise.

Global competition requires companies to behave as an integrated organisation - the entire organisation is considered an integrated enterprise system and is supported by enterprise information systems. Information about all business

V. Pászto et al. (eds.), *Spationomy*, https://doi.org/10.1007/978-3-030-26626-4_4

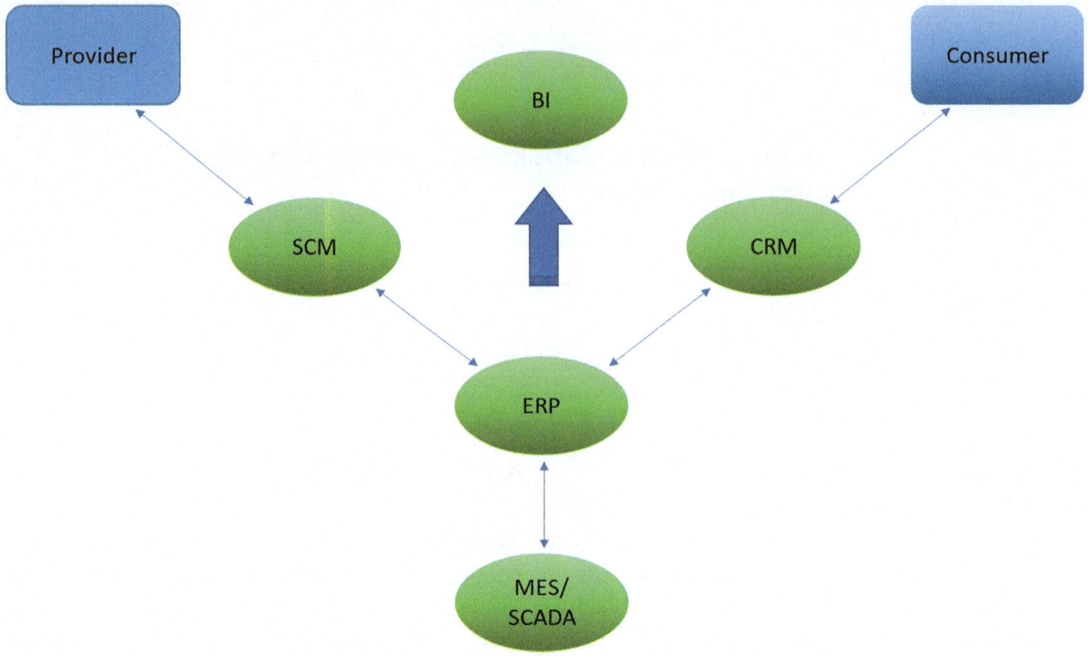

Fig. 4.1 Business information systems

functions is stored centrally and is available to all departments (Bradford 2016). This transparency and information access ensure that the departments are no longer working in isolation pursuing their own departmental goals. Each part of the enterprise knows what others are doing, why they are doing it and what should be done to move the company towards the common goal (Magal and Word 2011). Therefore, the prerequisite for successful modern organisations is enterprise information systems.

Enterprise information systems can be divided into two categories: enterprise information systems for operational support which are focused in business transactions (online transaction processing – OLTP) and enterprise information systems for management support (online analytical processing – OLAP) (Fig. 4.1). The first category – enterprise information systems for operational support – consists of the following types: manufacturing resource planning (MRP); enterprise resources planning (ERP) – systems used to register most of the events in the enterprise concerning economics and finance issues as well as business processes (supply, production and sale); customer relationship management (CRM) and some other specialised enterprise information systems needed in certain industries. The second category – enterprise information systems for management support – is also referred as business intelligence (BI) – systems for the analysis of the company's condition, they support higher information functions; there are also other specialised information systems which provide specialised functionality. In this category one of the most important is Geographic(al) information systems (GIS) which enable spatial location of existing infrastructure etc. (Stepniak and Turek 2014). Companies can use various combinations of these systems, where the mixture depends on the industry. This usually leads to the integration of information systems in the enterprise.

In the next section, the major categories of enterprise information systems, namely Enterprise resource planning systems (ERP) and Customer relationship management systems (CRM) will be explained, and concepts of their integration with GIS announced and implemented by vendors will be discussed. In the section that follows the Business intelligence systems will be explained and their integration with Spatial analytics, sometimes also referred to as Geospatial Business Intelligence foreseen and implemented by vendors will be summarised and discussed. In the last section, the findings of a bibliometric analysis of scientific publishing in these areas will be presented what will show the state of the art of research.

4.2 Enterprise Information Systems for Operational Support

4.2.1 Enterprise Resource Planning (ERP) Information Systems and Customer Relationship Management (CRM) Information Systems

Enterprise resource planning (ERP) solutions usually refer to the business-management support software. Typically, this is an integrated application which an organisation can use to collect, store, manage and interpret data from their daily business activities (Bradford 2016). ERP solutions provide an integrated and continuously updated view of core business processes using a common database. ERP solutions track business resources—cash, raw materials, production capacity—and the status of business commitments: orders, purchase orders, and payroll. The applications that form the system share data across various departments (manufacturing, purchasing, sales, accounting, etc.) that provide the data (Almajali et al. 2016). ERP facilitates information flow between all business functions and manages connections to outside stakeholders (Bidgoli 2004). ERP predicts and balances demand and supply. It is an enterprise-wide set of forecasting, planning and scheduling tools, which links customers and suppliers into the supply chain, employs proven processes for decision-making and coordinates business areas such as sales, marketing, operations, logistics, purchasing, finance etc. Most ERP systems incorporate best practices which means the software reflects the vendor's interpretation of the most effective way to perform each business process (Monk and Wagner 2009; Sneller 2014). The most widely used integrated solutions for business in companies from almost all industries worldwide are Enterprise Resource Planning (ERP) solutions. About 90% of the Fortune 500 companies use ERP solutions (HubPages 2018). A number of ERP implementations and because of that also a number of ERP users within organisations is growing very fast as well; employees are using ERP solutions daily at their work.

The organisation Gartner Group first defined ERP as a concept more than 25 years ago (Montgomery et al. 2018). ERP systems initially focused on automating back-office functions (functions which did not directly affect customers), while front office functions (functions which directly dealt with customers), e-business or supplier relationship management (SRM) became integrated later when the Internet enabled the simplified communication with external parties.

An ERP system covers the following common functional areas. In many ERP systems these are called and grouped as ERP modules (Anderegg 2000; Bradford 2016; see Fig. 4.2):

- Finance & Accounting: general ledger, fixed assets, payables including vouchering, matching and payment, receivables, cash management and collections, cash management, and financial consolidation.
- Management Accounting: budgeting, costing, cost management, and activity-based costing.
- Human resources: recruiting, training, rostering, payroll, benefits, retirement and pension plans, diversity management, retirement, and separation.
- Manufacturing: engineering, bill of materials, work orders, scheduling, capacity, workflow management, quality control, manufacturing process, manufacturing projects, manufacturing flow, and product life cycle management.
- Order Processing: order to cash, order entry, credit checking, pricing, available to promise,

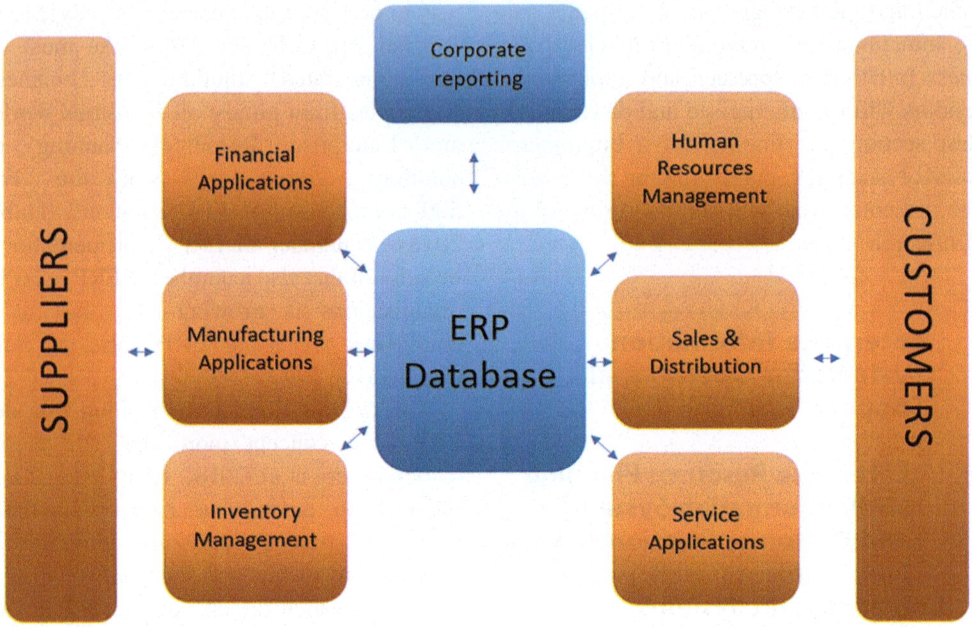

Fig. 4.2 Modules of enterprise resource planning (ERP) information systems

inventory, shipping, sales analysis and reporting, and sales commissioning.

- Supply chain management: supply chain planning, supplier scheduling, product configurator, order to cash, purchasing, inventory, claim processing, and warehousing (receiving, put away, picking and packing).
- Project management: project planning, resource planning, project costing, work breakdown structure, billing, time and expense, performance units and activity management.
- Customer Relationship Management (CRM): sales and marketing, commissions, service and customer contact.
- Data services: various "self–service" interfaces for customers, suppliers and/or employees.

In 2013 the organisation Gartner Group (Ganly et al. 2013) introduced the term "postmodern ERP" (also called as the eXtended ERP – xERP). According to Gartner's definition of the postmodern ERP strategy, legacy systems of monolithic and highly customised ERP suites, in which all parts are heavily inter-dependent, should be replaced by a mixture of both cloud-based and on-premises applications, which are more loosely coupled and can be easily exchanged if needed. The organisation Gartner Group has evolved its definition over time and now defines ERP as an application strategy focused on several distinct enterprise applications suites. They segment ERP into four major business process support areas: financial management systems, human capital management (HCM), enterprise assets management (EAM), and manufacturing and operations (Montgomery et al. 2018). Key characteristics of postmodern ERP are (ACC Software Solutions 2018):

- it delivers business agility and flexibility;
- flexibly coupled applications;
- not a single product suite;
- plays well with others through APIs and pre-built integrations;
- modern, web and mobile-friendly user interface (UI);
- flexible, deployment options (on-premises, private/public cloud, or hybrid);

- business-specific integrations;
- reduced complexity of the core ERP;
- SaaS deployment means near "painless" upgrades; improved business outcomes;
- and flexible licensing options.

Early ERP providers focused on large enterprises, but smaller enterprises are increasingly using ERP systems as well (Phillips and Ryan 2013). Main reasons for the growth of the ERP market are (Bradford 2016): it enables improved business performance (i.e. cycle time reduction, increased business agility, inventory reduction), supports business growth requirements (i.e. new product or product lines, new customers, global requirements including multiple language and currencies), provides flexible, integrated, real-time decision support (i.e. improve responsiveness across the organization), eliminates limitation in legacy systems (i.e. century dating issues, fragmentation of data and processing, inflexibility to change, insupportable technologies), takes advantage of small and medium-size organizations (i.e. increased functionality at a reasonable cost, cloud computing compatibilities, vertical solutions). These are just some of the reasons for the growth rate of the ERP market. Company SAP is a market leader, followed by Oracle, Sage, Infor and Microsoft (SMRC 2017). It is expected that ERP will remain the important basic software in the organisations (Pelphrey 2015).

Some big organisations require more advanced support for customer relationship management which is beyond basic functionalities of ERP. Such organisations are using customer relationship management (CRM) solutions to obtain advanced functionality (Fig. 4.3). CRM solutions compile data from a range of different communication channels, including a company's website, telephone, email, live chat, marketing materials, and more recently, social media (Starzyczna et al. 2017; Yerpude and Kumar Singhal 2018). Through the CRM approach and the systems used to facilitate it, businesses learn more about their target audiences and how to best cater to their needs.

The primary goal of CRM systems is to integrate and automate sales, marketing, and customer support (Lizzote 2017). Therefore, these systems typically have a dashboard that gives an overall view of the three functions on a single customer view, a single page for each customer that a company may have. The dashboard may provide client information, past sales, previous marketing efforts, and more, summarising all the relationships between the customer and the firm. Operational CRM is made up of 3 main components: sales force automation, marketing automation, and service automation (Buttle and Maklan 2015).

- Salesforce automation works with all stages in the sales cycle, from initially entering contact information to converting a prospective client

Fig. 4.3 Features of ERP and CRM systems

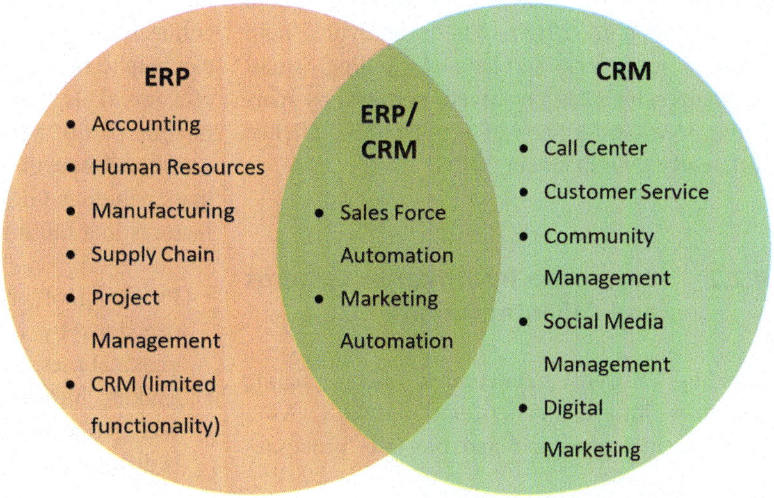

into an actual client (Yerpude and Kumar Singhal 2018). It implements sales promotion analysis, automates the tracking of a client's account history for repeated sales or future sales and coordinates sales, marketing, call centres, and retail outlets. It prevents duplication of efforts between a salesperson and a customer and automatically tracks all contacts and follow-ups between both parties.

- Marketing automation focuses on easing the overall marketing process to make it more effective and efficient. CRM tools with marketing automation capabilities can automate repeated tasks, for example, sending out automated marketing emails at certain times to customers, or posting marketing information on social media. The goal of marketing automation is to turn marketing leads to a full customer. CRM systems today also work on customer engagement through social media.
- Service automation is part of the CRM system that focuses on direct customer service technology. Through service automation, customers are supported through multiple channels such as phone, email, knowledge bases, ticketing portals, FAQs, and more.

The role of analytical CRM systems is to analyse customer data collected through multiple sources and present it so that business managers can make more informed decisions (Wan and Xie 2018). Analytical CRM systems use techniques such as data mining, correlation, and pattern recognition to analyse customer data (Chorianopoulus 2016). These analytics help improve customer service by finding small problems which can be solved, perhaps, by marketing to different parts of a consumer audience differently (Nussbauman 2015).

4.2.2 Enterprise Information Systems and GIS Integration

Existing software architectures often separate enterprise information systems and GIS. As a result, spatial processes and business processes are also perceived as distinct processes (Treiblmayer et al. 2011). As explained previously, ERP solutions usually refer to the business-management support software. Typically, this is an integrated application which an organisation can use to collect, store, manage and interpret data from their daily business activities. On the other hand, GIS are information systems for capturing, storing, checking, and displaying data related to positions. GIS is often isolated in information system landscapes. But decision support requires integrated workflows that cover business processes and spatial processes. Since most business data has a geographic or spatial component that can be geo-referenced on a GIS map to visualise, understanding and interpretation of data through a spreadsheet or table is not possible (Abou-Ghanem and Arfaj 2008). Although the GIS can include data about daily business activities, for example, actual and potential customers on the map for market analysis etc., they usually are not part of ERP, because of the complexity involved in handling each system. By integrating ERP systems for workflow management and GIS for location-based information management, this integration can bring many advantages in both fields. Abou-Ghanem and Arfaj (2008) pointed out that this could result in a loss of opportunities to leverage spatial analysis capabilities of GIS and business transaction management tools of ERP systems. Therefore, in recent years the level of interest in integrating GIS with ERP and legacy systems has been growing significantly. By visualising relationships, connections and patterns in business data, GIS can help to make informed decisions and increase efficacy. This makes ERP and GIS systems an integral part of a powerful IT strategy.

Abou-Ghanem and Arfaj (2008) pointed out that including GIS into business process offers features that fall into the following categories:

- Provision of geographic context to business decisions by integrating business data with geodatabases and feature tables.
- Linking business functions in an ERP system with geo-processes combined functionality that is distributable across the software architecture.

- Fusion of business and geographic information and functionality in the common operational picture on both high and low levels.

With integration, a user can visualise ERP system data within the GIS and can get direct access to the GIS within the ERP system. A user can accomplish more since he/she has the ability to make decisions by visualising the output of both systems on a screen in a simple visible manner, without the need to switch between systems and correlate the data between several systems. Specific ERP and GIS capabilities include functions such as (Horwitt 2009):

- Map relationships between customers and markets, and plant and warehouse locations. This enables companies to manage transport costs and product distribution better, as well as to optimise the location of new manufacturing facilities (for example, putting a site in a low-rent district near a highly populated area).
- Analyse supply chain, distribution and service vulnerabilities from a geographic perspective. Some automakers have been using GIS demographic data to help figure out which dealerships to close. Companies can also analyse service response time by region, identifying areas where it tends to be slow. They can then institute training programs or hire more people in those regions.
- Sort customers by attributes such as a preference for a product or attribute (low cost, low calorie, luxury), then map those preferences by region, state or neighbourhood and use the data, in combination with demographic data, to focus marketing, sales or distribution. Example of sales by customer visualised on the map is shown in Fig. 4.4.

Organizations that integrate GIS with ERP systems include (ESRI 2007): utilities (water, electric, gas, waste, recycling), local government, oil and gas production, defence and public security, service providers (routing and logistics), real estate, forestry and forest products, waterways, airports, ports etc. With systems linked, the user can do an array of functions that could impact corporate running cost by accomplishing the following (Patel and Doctor 2013):

- Improve resource utilisation, analysis, safety and asset integrity through an ability to represent work orders and notifications at their exact location on a GIS map.

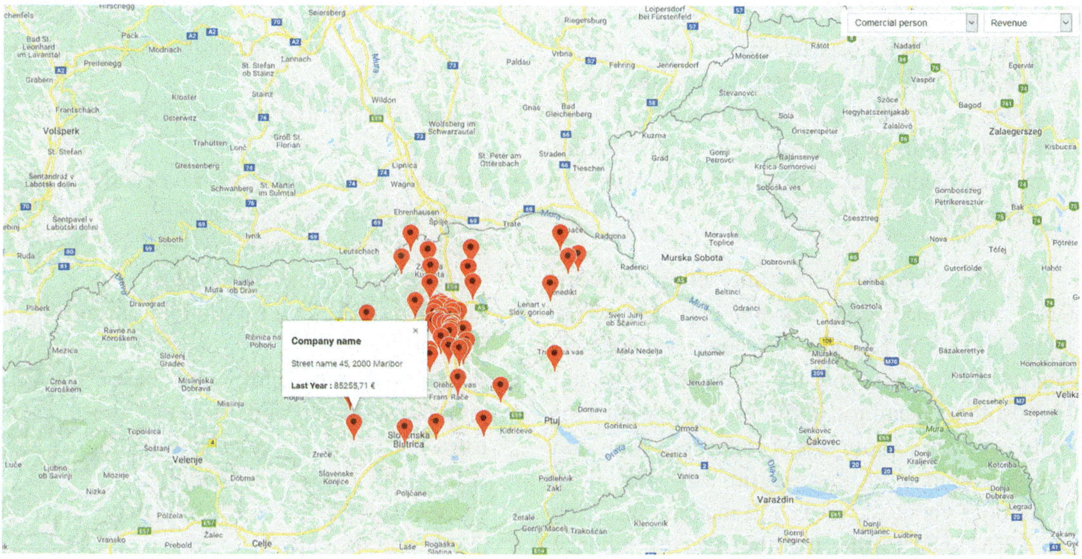

Fig. 4.4 Sales by customers visualised on a map

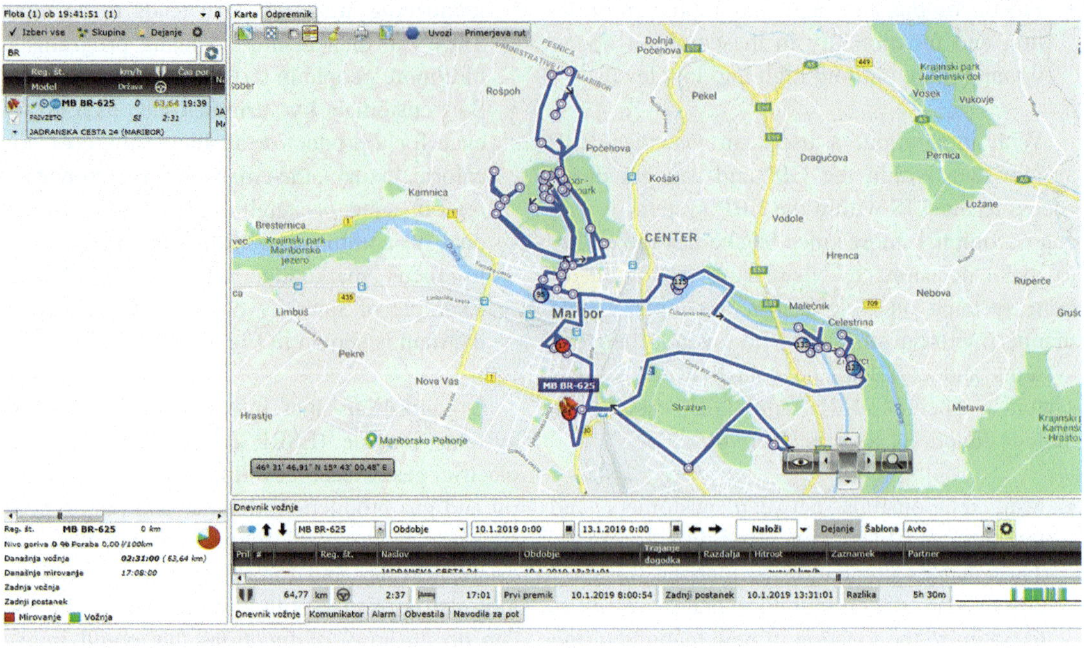

Fig. 4.5 Visualisation of the daily route of the employee

- Improve operational efficiency (e.g. safety, dispatching) by representing work activities that impact an area as polygons on the map and identification of shortest routes.
- Improve workforce productivity by providing broader browser-based access to ERP and GIS-based assets, maintenance, and customer information.
- Reduce service costs by expanding a no-line self-service portal with essential business processes (e.g. report fault equipment)
- Elevate customer service and decision-making with expanded access to ERP data from the GIS maps.

Another example is the interaction between CRM and GIS. GIS system Esri Maps for Dynamics CRM provides fast and easy integration of Esri maps and the Esri platform into Microsoft's Dynamics CRM product. With Esri Maps we can see our CRM data on a map and visualise it in more meaningful way (Fig. 4.5), learn more about the areas and markets in which our CRM data reside, use the map to enhance analysis, and share all of this with others in our organisation. It includes the following features (ESRI 2018) – the user can:

- map CRM data to see easily where concentrations of customers or constituents exist;
- customise maps and add them to Dynamics CRM dashboards, access rich collection of base maps, demographic maps, image services, and data;
- visualise neighbourhood demographics using pop-ups;
- calculate routes and build travel activity based on customer/prospect location;
- geocode, geo-enabled, and geo-enrich CRM data using a Locate workflow;
- visualise CRM data by grouping according to attribute;
- save CRM data into views based on geographic selection and then use those views for other CRM activities like adding to marketing lists and sending emails;
- share maps with the rest of the organisation using ArcGIS Online;

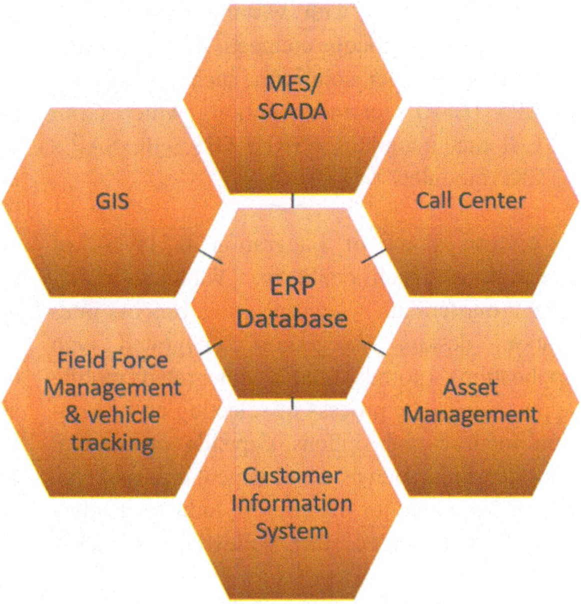

Fig. 4.6 ERP/CRM and GIS-based asset/resources management

- single sign-on capability (with CRM on-premises, organisations can use the same login for their organisation, for CRM, and for ArcGIS Online or with CRM Online, organisations can use the same login for their organisation and ArcGIS Online).

Both ERP/CRM and GIS industry leaders have identified several options and methods for integration of GIS and ERP systems (Fig. 4.6). The options and methods are derived based on the system functions and ERP objects. They can build or purchase software connectors that directly connect a given ERP and GIS package. They can use passive middleware, which is fine if they can stick with generic ERP and GIS and don't need to customise their processes. Or they can deploy frameworks for comprehensive integration with a given GIS package from an ERP vendor (Horwitt 2009).

For example, the most successful integration GIS – ERP was done by Esri and SAP, the industry leaders of GIS and ERP systems respectively, has resulted in the identification of five main technical interfaces available for integration (Abou-Ghanem and Arfaj 2008; Patel and Doctor 2013):

- SAP RFC (Remote Function Calls) connectors, which allow for remote calls between SAP ERP and ESRI GIS software.
- Third-party connectors include the Way Control Broker (CB) from Information Builders which is component-based and allows director connection to packages and legacy back-end systems. They provide an application level approach for greater control, performance and scalability.
- SAP generic GBC (GIS Business Connector) is a passive middleware that mediates between SAP solution and ESRI GIS. The method is best used when there is flexible processing modelling with existing methods in both software.
- Third-party Enterprise Application Integrators (EAI) is generally defined as the combination of platforms, business processes, standards and applications that results in the seamless integration of two or more enterprise systems. It can be used when there are non-standard processes that require custom development.
- SAP EAI called SAP Exchange Infrastructure is used to facilitate the exchange of information between both and can be used to mediate

between entities with varying requirements in terms of protocols, connectivity and format.
- Vendors partner solutions, such as SICAD-APX (ESRI's partner solution), which is an EAI that integrates ESRI's GIS with SAP's ERP modules.

Another point of integration is layers of integrations which could be (Yaptenco et al. 2005): master data synchronisation, process integration, desktop integration, integrated web applications (portals) and/or integrated mobile applications. The process of selecting an integration method and options depends on several considerations such as development cost and corporate directions for integration. Industry leaders have recommended the following selection criteria:

- Corporate standards in exposing business data either on real-time bases or using previously prepared reports from a business warehouse.
- Data exchange technology adopted on the corporate level such as using Service Oriented Architecture (SOA) and web services, application integrators or third-party solutions.
- Business process requirements such as data visualization or data editing and updating.
- Technology availability based on ERP objects to be integrated.
- Vendors recommendation methods.

In addition to these criteria, also development cost plays an important role in the selection process. Some solutions could provide rapid deployment, but the cost of development, consulting and maintenance could be high, and this could have a stronger impact on the technology and methods of integration.

But the most important part of ERP integration is to define the business processes and the corresponding business requirements. Yaptenco et al. (2005) define the user cases and business processes, the user interface requirements, the data model requirements, select technical connector approach based on business processes and requirements and design business logic in conjunction with the requirements and technical connector approach.

Starting from software development patterns and three-tier applications development, there are three layers of software integration: data integration, service integration and process integration (Litan et al. 2011). They also added that the level of complexity rises from data to process integration, and the level of abstraction as well. Service and process integration lead to the completeness and coherence of integrated systems. The most commonly used approaches in service integration are SOA (Service Oriented Architecture) and ESB (Enterprise Service Bus). The concept of SOA provides an approach to integrate heterogeneous software applications (Treiblmayer et al. 2011). SOA allows interacting software components to be loosely-coupled. Web service interfaces can facilitate the exchange of data and services between GIS and enterprise information systems. Thus, a workflow can be established that integrates the business process as it is covered by the enterprise information system and the spatial process as the GIS covers it. ESB generally provides an abstraction layer on top of an implementation of an enterprise messaging system, which allows developers to exploit the value of messaging without allowing writing code (Litan et al. 2011). Unlike the classical EAI approaches, ESB cuts the number of interfaces for interconnection between different systems, being capable of translating interfaces. Therefore, technological progress in Internet technology and the development of the SOA concept made it possible to embed GIS applications into common activities as well as integrating them with different systems such as ERP and CRM. Benefits of integration are efficient to use of resources, reduced costs, enhancement of customer satisfaction, increased interoperability of different departments, quick and accurate analysis and better business process management.

As we mentioned before the synergy between GIS and ERP/CRM systems offers competitive advantages to any enterprise (Fig. 4.7) in supply chain management and marketing areas. In supply chain management offer shorter order cycle, more reliable deliveries, better warehouse

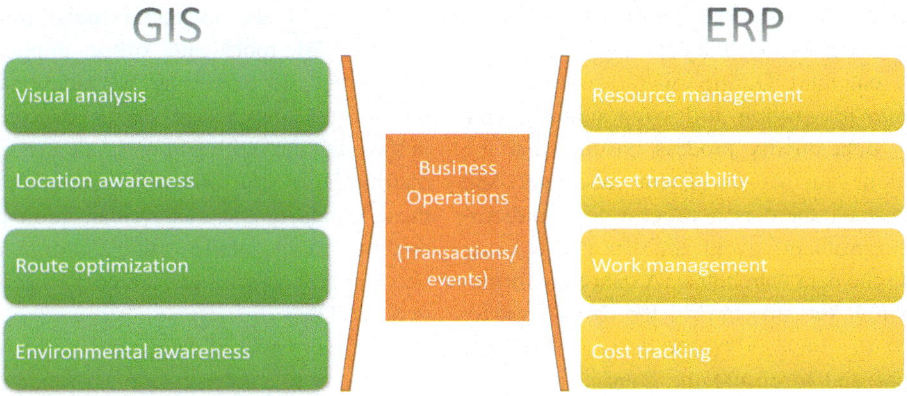

Fig. 4.7 ERP systems and GIS integration

management and lower transportation costs (Aydin and Sarman 2006). In marketing, offer segmentation of customer by lifestyle and product category, implementation of pricing policy depending on location, site selection and delivery routing, development of target promotions and campaigns, geocoding of customers, understanding of customer spending (Hessa et al. 2004).

4.3 Information Systems for Management Support

4.3.1 Business Intelligence Systems

The term *business intelligence (BI)* emerged in the mid-1990s to describe the concepts of transforming business data from an information system in which operational data of business transactions are captured and stored into a database which is being used for management support (Sherman 2015). BI comprises the solutions and technologies used by organisations for the business data analysis used in management reporting. Business intelligence can be used by organisations to support a wide range of business decisions - ranging from operational to strategic. BI is a priority for organisations interested in gaining a competitive advantage. BI leverages corporate data and empowers managers with insights needed for sound business decisions. BI technologies provide historical, current and predictive views of business. BI technologies can handle large amounts of structured and sometimes unstructured data to prepare business reports for managers (Howson 2014). BI is most effective when it combines data derived from external data sources (external data) with data from company internal data sources such as financial and operations data (internal data). When combined, external and internal data can provide a holistic picture which, in effect, creates an "intelligence" that cannot be derived from any partial set of data. BI tools empower the organisation to gain insight into new markets, to assess demand and suitability of products and services.

Business intelligence (BI) and business analytics (BA) are sometimes used interchangeably, but there are distinctions between them. The term business intelligence usually refers to collecting business data for business reporting, and online analytical processing (Trieu 2017). Business analytics, on the other hand, refers to statistical and quantitative tools for explanatory and predictive modelling.

BI/BA systems use the extraction, transformation, and loading (ETL) processes which are used to retrieve data from information systems on an operational level. The BI processes collect data directly from the point where it's generated. Data can be originated from many types of systems and applications, including business software such as enterprise resource management (ERP) or customer relationship management (CRM) applications, plain text files, or office application files such as spreadsheets. The data is moved or forwarded from its source location to a data warehouse or data mart. During this process, which is called *data integration*, following subtasks take place. A data quality process ensures that the information remains consistent, accurate, and "clean"— i.e., there is a process to avoid/correct/detect problems within the data that is being moved to the data warehouse. A data transformation modifies the structure of the data to satisfy the conditions imposed by the design of the data warehouse and to ensure the consistency of all information. The load process allocates the information into an information repository (such as a data warehouse or data mart) (Trieu 2017).

Users interact with an easy-to-use interface to use of tools for querying, reporting, online analytical processing tool (OLAP) etc. The same interface is also the gateway into a structured reporting environment that distributes operational reports and business decision results throughout the organisation (Fig. 4.8).

Reporting, a main task of BI, has become more graphics intensive. Business graphics, typically charts, are now a common component of reports. Access to BI data became more timely, because of that graphic dashboards were developed to monitor key business processes. Dashboards, named for their similarity to automobile dashboards, convey operational information at a glance. Dashboards and scorecards nowadays comprise only part of the available tools. Interactivity between BI tools and office applications is increasing and extending BI functionality, and mobile technologies are taking their own place in the equation, with BI providers now capable of distributing BI information to mobile devices. All these trends in the data visualisation phase are enabling more people within the organisation to become BI software consumers or users (Chung et al. 2002).

A primary goal of data visualisation used in BI systems is to communicate information clearly and efficiently via statistical graphics, plots and information graphics. Numerical data may be presented using dots, lines, or bars, to communicate a quantitative message visually. Effective visualisation helps users analyse and reason about data and evidence. It makes complex data more accessible, understandable and usable. Users may perform analytical tasks, such as comparisons or understanding causality, and the design principle of the graphic (i.e., showing comparisons or showing causality) follows the task. Tables are generally used where users will look up for a specific measurement, while charts of various types are used to show patterns or relationships in the data for one or more variables.

Dashboards are using principles described above, and they often provide at-a-glance views of Key Performance Indicators (KPIs) relevant to an objective or business process also referred to

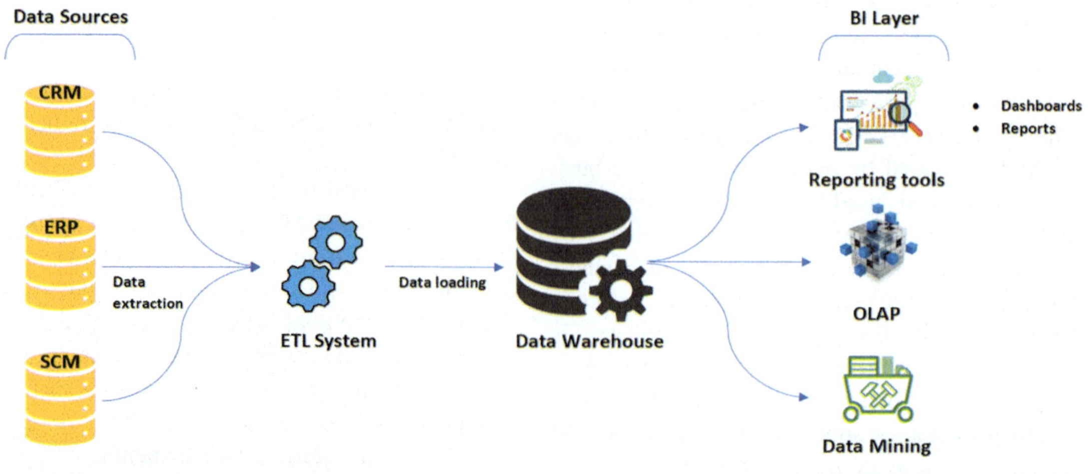

Fig. 4.8 Business intelligence system

as Critical Success Factors (CSFs). The "dashboard" is often displayed on a web page which is linked to a database that allows the report to be constantly updated. For example, a manufacturing dashboard may show numbers related to productivity such as the number of parts manufactured, or a number of failed quality inspections per hour. Similarly, a human resources dashboard may show numbers related to staff recruitment, retention and composition, for example, number of open positions, or average days or cost per recruitment. The term dashboard originates from the car dashboard where drivers monitor the major functions at a glance via the instrument cluster. Digital dashboards allow managers to monitor the contribution of the various departments in their organisation. To gauge exactly how well an organization is performing overall, digital dashboards allow to capture and report specific data points from each department within the organisation, thus providing a "snapshot" of performance.

Dashboards can be divided according to the role and are either strategic, analytical, operational, or informational (Few 2006). Strategic dashboards support managers at any level in an organisation and provide a quick overview that decision makers need to monitor the "health" and opportunities of the business. Dashboards of this type focus on high-level measures of performance, and forecasts. Strategic dashboards benefit from static snapshots of data (daily, weekly, monthly, and quarterly) that are not constantly changing from one moment to the next. Dashboards for analytical purposes often include more context, comparisons, and history, along with subtler performance evaluators. Analytical dashboards typically support interactions with the data, such as drilling down into the underlying details. Dashboards for monitoring operations are often designed differently from those that support strategic decision making or data analysis and often require monitoring of activities and events that are constantly changing and might require attention and response at a moment's notice.

Digital dashboards may be laid out to track the flow inherent in the business processes that they monitor. Graphically, users may see the high-level processes and then drill down into low-level data. This level of detail is often buried deep within the corporate enterprise and otherwise unavailable to the senior executives (Chen et al. 2012).

Balanced Scoreboards and Dashboards have been linked together as if they were interchangeable. However, although both visually display critical information, the difference is in the format: Scoreboards can open the quality of operation while dashboards provide calculated direction. A balanced scoreboard has what they called a "prescriptive" format. It should always contain these components:

- Perspectives – groupings of high-level strategic areas
- Objectives – verb-noun phrases pulled from a strategy plan
- Measures – also called Metric or Key Performance Indicators (KPIs)
- Spotlight Indicators – red, yellow, or green symbols that provide an at-a-glance view of a measure's performance.

Each of these sections ensures that a Balanced Scorecard is essentially connected to the business's critical strategic needs.

With the dynamic economic landscape, businesses are increasingly looking for ways to do more with less and maximise their existing assets to extract the most value. To achieve this, BI has been a significant component in many organisations' technology portfolios. Business intelligence can provide a pro-active approach, such as alert functionality that immediately notifies the end-user if certain conditions are met. For example, if some business metric exceeds a pre-defined threshold, the metric will be highlighted in standard reports, and the business analyst may be alerted via e-mail or another monitoring service. This end-to-end process requires data governance, which should be handled by the expert. Well-implemented BI allows organisations to focus on what's important and make business decisions to drive performance.

4.3.2 Business Intelligence and Spatial Analytics

Historically, business intelligence (BI) and geographic information system (GIS) technology have followed separate development and implementation paths. Customer requests for a complete operational picture and the ability to be more proactive have led to the combination of these two technologies. Regulatory requirements have also raised the visibility of both technologies within many organisations. In response to BI and GIS users, leading BI providers have been integrating the two technologies and providing innovative solutions to a growing number of end users. The users are responding with new applications that leverage the synergy of the combined technologies (Bimonte et al. 2010).

Organisations today are collecting data at every level of their business and in volumes that in the past were unimaginable. On the other hand, it is predicted that nearly 80% of all data has a kind of spatial component. Traditionally such data would be presented to users in long reports, either with graphs and pie charts or in spreadsheet format. Today the complex interrelationships of multidimensional data, integrating spatial data and visualisation are offering high impact insight to business intelligence users.

As BI has matured, the reach of GIS has expanded significantly as well (Posthumus 2008). In addition to speciality IT groups, GIS provides agility to a multitude of departments in many industries. It allows users to visualise and intelligently analyse historically underutilised data in ways not typically seen in traditional BI implementations (Gideon Adewale et al. 2016). Given the complementary natures of BI and GIS, the adoption of geographic analysis to enhance business intelligence is growing rapidly. Through the fusion of these two enterprise technologies, organisations can visualise and analyse key business data through "smart" maps to discover patterns and trends that would have been easily overlooked with traditional BI tables and charts (Fig. 4.9).

Humans think visually, therefore spatially. While traditional methods used to represent information and gain insight have been helpful, they have been limited in capabilities when it comes to

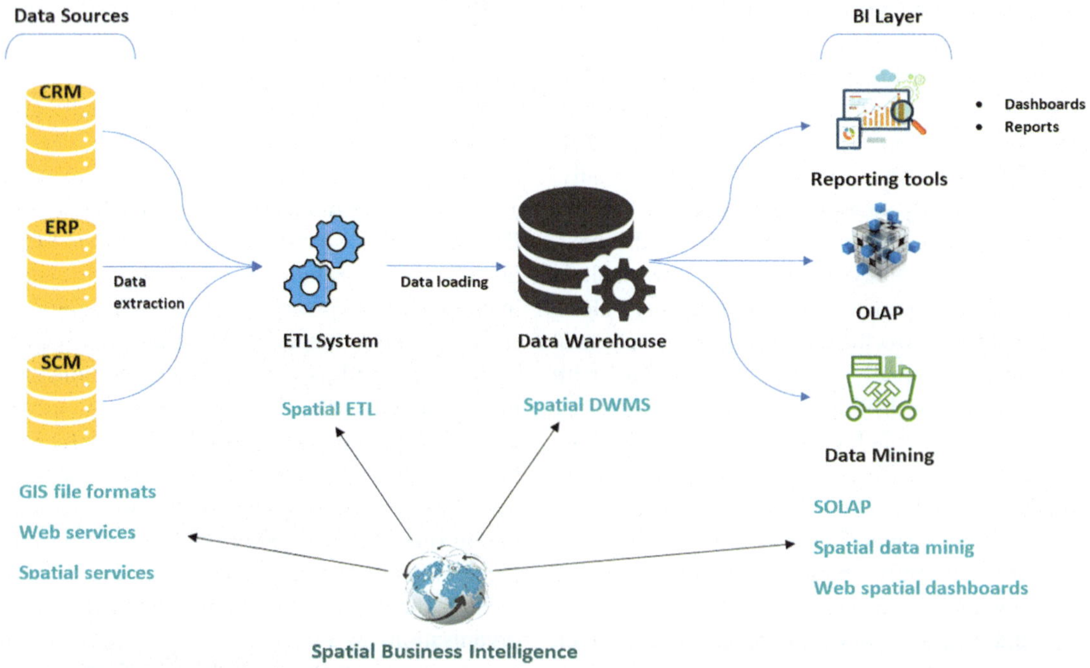

Fig. 4.9 From BI to spatial BI

performing quick visual decoding and comparison of data. Data gains immediate visual impact with the help of maps, more emphatically true for data with a spatial dimension (Rivest et al. 2005). Maps best represent spatial phenomena or relationships such as the flow of proximity, while also facilitating visualisation of statistical measures for an area or region. In addition, maps allow multi-measure displays.

Today's GIS recognises the location component of data and associates data with geographic features maintained in a GIS. Features in a GIS are graphic representations of actual features, such as roads, rivers, and forests, and conceptual features such as political boundaries or service areas (Fig. 4.10). Associating data with features lets users organise data based on the geographic location of each record in the data. This geographic organisation, presented as a map, reveals spatial relationships and influences that cannot be identified in traditional tabular views of data.

Geographically organising data allows the utilisation of new data that may not have anything in common with existing data other than location. For instance, GIS analysts for insurance companies can map the addresses of insured structures and overlay floodplain boundaries to identify all structures within the floodplain. With this information, they can calculate the total financial impact on reserves from a potentially catastrophic flood. Other organisations, private and public, can perform this same analysis to determine the potential impact on facilities, supply chain, and employees. By carrying out spatial analysis using varied BI tools, decision-makers are able to better understand the historical, current and future aspects of business operations, derive useful insights and make the most effective decisions for their business.

GIS and BI were being implemented as the IT landscape was evolving to embrace common ways of compiling, storing, using, and distributing data. Knowing how BI and GIS were deployed in organisations presented opportunities for the proliferation of these technologies. If BI and GIS applications could work together, the benefits of these respective technologies could be realized by operational units not currently using both technologies. This would result in integrated applications expanding throughout the enterprise. Innovators in the public sector who wanted to extract more actionable information from existing data came to the same conclusion. Exposure to "new" technologies in the context of homeland security raised interesting possibilities for improving processes not

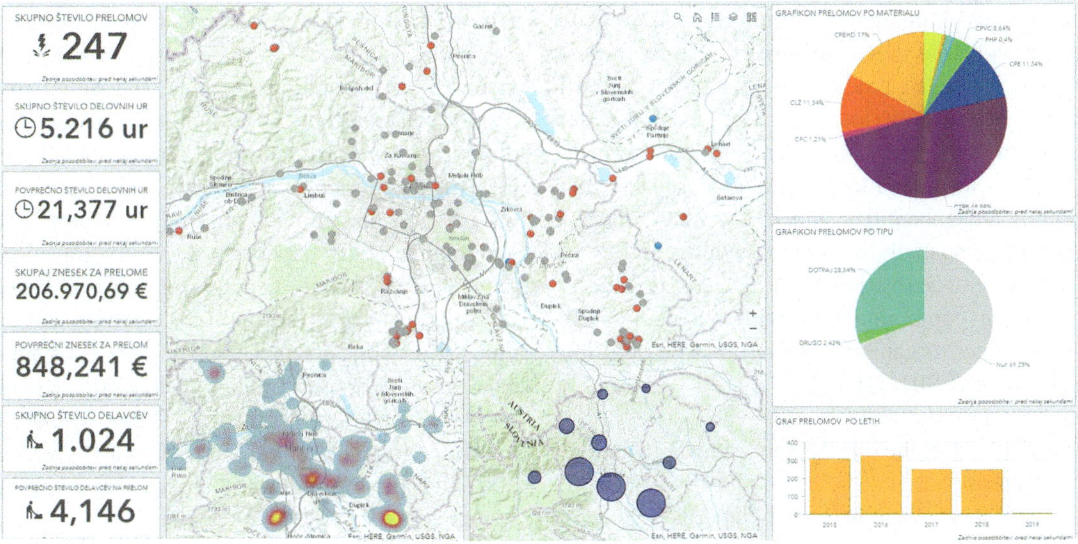

Fig. 4.10 GIS-based dashboard visualising information for managers

directly related to homeland security. The fact that public agencies were looking at BI with the idea of integrating it with GIS was not lost on the BI providers whose success in the private sector had not been matched in the public sector. Business charting abilities of BI applications, conversely, GIS brings unique charting capabilities to BI in the form of spatial relationship and distribution charts. The portrayal of BI data as maps addresses a recognised shortcoming in BI graphics—the lack of context needed for informed decisions. For example, node-to-node supply chain performance data presented as bar charts or dashboards does not supply the location information needed for planning improvements. The same performance report presented as a map immediately shows spatial relationships between nodes that could explain variations in performance. Many organisations, both public and private, have come to understand the business cases for integrating BI and GIS and are actively exploring integration strategies (Wickramasuriya et al. 2013).

Most BI users are not accustomed to using maps as analytical tools. They typically analyse business data for patterns and trends using tables, charts, and graphs. They also benefit from OLAP data, which involves users analysing the major dimensions of business by drilling up and down through business data to uncover trends and anomalies. Although traditional BI tools are powerful and have delivered proven results, they do not incorporate a crucial component of most business information: location. Most business data contain some sort of location information: office locales, customer addresses, sales territories, marketing areas, facilities, and so on. When this data is viewed spatially on a map, patterns and trends that were once overlooked are clearly revealed.

When combining GIS with business intelligence data, organisations can answer questions like these (ESRI 2012):

- Who are my best customers and where are they located?
- Are there location-based patterns related to customers' purchase decisions?
- Where can I find potential customers like existing customers? Where have our marketing efforts been the most or least successful and why?
- Do we have customers at risk from physical phenomena, such as severe weather events, and what impact might that have on our business?

Answers to these and other critical questions are delivered through the successful integration of BI and GIS that provides the following (ESRI 2012):

- Powerful visual analysis capabilities for key BI data such as customers, retail outlets, assets and infrastructure, partner locations, supply chains, operational events, and utility grids
- Easy correlation of BI data with demographics, occupational classifications, lifestyle and socioeconomic factors, and consumer information to analyse and optimise product and service sales across multiple geographies

Immediate insight to enable rapid and informed decision making, including clear visualisation of what matters and where it matters, complemented with supporting business analytics, allowing knowledge workers to prioritise efforts and immediately become more productive.

Spatial analytics is increasingly becoming essential for obtaining accurate and actionable insight because there is a significant geographic dimension to every business transaction. There are two categories of industries. The first set is industries somehow naturally rooted in geographies like transport, telecommunication, or real estate, who depend on location-based information. The second set comprises of industries who not necessarily use geospatial data on an everyday basis, but they still depend on it for better performance. These would be retailers, insurance, banking (Devillers et al. 2007). Following industries are actively bringing spatial analytics to BI:

- Retail industry: Along with physical store level detail, spatial analysis is helping the industry to gain invaluable insights in stock delivery, store management, inventory management, marketing and sales
- Healthcare: healthcare providers use spatial analytics to see issues from a location perspective, leading to more efficient practices and higher standards of patient care
- Banking: location data and analytics of customers help banks to carry out effective customer segmentation and profiling, develop more successful marketing and sales campaigns, tailor products and services according to the customer base within each region or division and identify and actively retain and pursue profitable customers,
- Insurance: Spatial analytics of location-based data, such as crime rates, weather patterns etc. is enabling insurance companies to identify high or low-risk cases and develop marketing strategies, policies and pricing models accordingly.
- Telecommunications: Spatial analytics enable the industry to assess strengths of the current infrastructure, analyse competitor network coverage etc. Understanding which locations will provide the best network coverage at the lowest costs possible helps the industry to function more effectively.
- Airports and airlines: Airlines use spatial analytics to track flight operations more closely and accurately. Airport, meteorological, and fleet data are monitored in real time, and operations crew reroute flight paths to optimise fuel and staff costs.
- Education: Adding a geographical component to students data enable higher education institutions to develop more effective marketing campaigns, understand better where to put facilities, and work on research grants.
- Energy: Spatial analytics is enabling the energy industry to discover patterns and uncover hidden insights in seconds rather than minutes and hours.
- Transport and logistics: Spatial analytics is helping to determine fastest transportation routes, enabling effective forecasting, optimising warehousing processes and stock flows based on consumption rates of particular products by locality.

More and more industries are integrating spatial analytics in BI as such a system provides more comprehensive information. Spatial analytics can be used to gain operational, transactional and competitive advantage (Devillers et al. 2007).

More and more vendors of BI platforms and tools like Tableau (2018) are embedding spatial analytics functionality in their solutions. Tableau is enabling instant geocoding and automatically turns the location data into interactive maps with 16 levels of zoom or alternatively enables the use of custom geocodes to map what matters for the business. Tableau supports Choropleth maps, Proportional symbol maps, Point distribution maps, Flow maps, origin-destination spider maps, Heath maps, etc.

Open Geospatial Consortium (OCG) develops standard protocol Web Map Service (WMS) which is a standard protocol for serving georeferenced map images over the internet that are generated by a map server using data that is typically sourced from GIS database. Recently it becomes well recognised and deployed standard. The Map intelligence WMS capability provides a standard generic method of exchanging data between Business Intelligence tools and map servers capable of handling WMS. These removes two major concerns for companies which want to utilise their organisation's BI data views. BI vendors' platforms for which there is a Map Intelligence (MI) Client, can use map servers that can provide WMS – these are: Esri ArcGIS Server, Spectrum spatial, GE Smallworld, GeoWebPublisher, GeognoSIS, GeoMedia, Oracle MapViewer, ObjectFX Web Mapping Tools, LizardTech Express Server and SuperMap and also open source GeoServer and MapServer.

4.4 Bibliometric Analysis of Research Publishing on Spatial Data Issues in Business Information Systems

4.4.1 Background, Aims and Scope of the Bibliometric Analysis

In previous sections, we focused mainly on platform/solution/tools vendors viewpoints connected with functionality embedded in their software. In this section, we will discuss the research on spatial data issues in business information systems. Using the bibliometric study, we aim to identify the stage of integration of the two fields, namely:

- Enterprise Resource Planning (ERP) and Geographical Information Systems (GIS),
- Business Intelligence (BI) and Geographical Information Systems (GIS).

We analysed the development in the past period, and we want to identify the future development trends within these fields, as well. Therefore, the main objectives of the bibliometric study answered the questions, what are the dynamics of research literature production in the area of ERP and GIS integration on one side, and BI and GIS integration on the other side, and which are the most productive research topics in this field.

There are some the most widely known definitions of the bibliometric research: Hawkins (2001) defined bibliometrics as "the quantitative analysis of the bibliographic features of a body of literature", consist of bibliographic units - books, monographs, reports, theses, and papers in serials and periodicals are analysed. For analysing research literature production (to identify patterns in the literature), the bibliometric analysis uses quantitative methods (De Bellis 2009). Moreover, Garfield (2009) is convinced that with bibliometric analysis, we can also examine "the history and structure of a field, the flow of information into a field, the growth of the literature, the patterns of collaboration amongst scientists, the impact of journals, and the long-term citation impact of a work".

We performed the bibliometric analysis by using the Scopus database (on December 20, 2018). The Scopus database was selected, because it is easy to use, and it is also easy to transfer data into the program VOSviewer (Leiden University, the Netherlands) for further data analysis (van Eck and Waltman 2013). Namely, the VOSviewer was used in the second step of the bibliometric analysis, to obtain the bibliometric maps.

Bibliometric mapping is used with the purpose to represent scientific publications based on bibliographic data visually. With bibliometric mapping, we can produce different bibliometric maps which provide an overview of the structure of the scientific publications in a specific research field. One of the most popular ways to use bibliometric mapping is to identify specific research areas within a selected science field, with the purpose of getting a view of the size of the field and relevant subfields, and how they relate to each other (van Eck 2011). The VOS mapping technique has been implemented in a computer program, called VOSviewer (Leiden University, Netherlands) (van Eck and Waltman 2013), that is available at www.vosviewer.com. The VOSviewer software has visualisation capabilities. Therefore bibliometric maps can be displayed in various ways and consequently emphasise different aspects of a map. Additionally, VOSviewer allows for the use of different colours to indicate clusters of objects. Moreover, the VOSviewer software also merges terms that may be closely related to term clusters denoted by the same cluster colour (van Eck 2011). According to van Eck, the proximity of the terms can be interpreted as an indication of their relatedness.

4.4.2 Research Publishing on Enterprise Resource Planning (ERP) and Geographical Information Systems (GIS)

In the first part of the bibliographic analysis, the search results from the Scopus database were obtained by using the "enterprise resource planning" key-phrase at the first level. A search revealed 12,906 bibliographic units that included this key-phrase in the title, keywords or in abstract. In the second step, the "GIS" keyword was used to identify the bibliographic units identified within the first step, that included both, "enterprise resource planning" and "GIS" key-phrases. The search revealed 40 units published between 1998 and 2018.

The search in Scopus database revealed, that the first two identified bibliographic units, that combines Enterprise Resource planning - ERP and Geographic(al) Information System – GIS, date to 1998 (although Scopus identifies three units, as presented by Fig. 4.1, two are identical). In his article, author Wilson (1998) discussed that Enterprise Resource Planning (ERP) and GIS are offering new opportunities for utility applications; article presented the IS/GIS integration, cost benefits, organisational re-engineering, as well as the geospatial implications. In the other article (Anon 1998) different software solutions that were used to predict the hydraulic behaviour of water distribution networks, to manage all data related to network assets, customer location and billing, and to demand profiles, pump curves and schedules, were discussed. On the other hand, a new software package was designed to help organisations to collate, evaluate and report corporate environmental information.

While no bibliographic units, covering the enterprise resource planning and GIS, were identified by Scopus in 1999, in 2000 and 2001 one per year were identified. A particularly interesting paper was published in 2000 (Zipf 2000), where the technology-enhanced project management in the Port Authority of New York and New Jersey was discussed. Author has stressed the importance of integrated project management systems, including GIS, electronic project management systems and enterprise-wide database systems; he argued that these technologies made it possible for timely information to be provided to project managers so that they could manage the project more effectively.

In 2004, a paper, covering the importance of integration of the Enterprise Resource Planning and GIS, in a field of tourism industry was presented to the professional public (Yan et al. 2002). Authors presented the advantages of an integrated information system of the tourism industry including its construction and functional realisation. The fusion of spatial information in the ERP system is discussed, and a spatially integrated information scheme is proposed.

After 2004 the number of publication was on average increasing – a positive trend is identified, with the peak in the volume of publications in 2006 in 2010, with 5 and 4 bibliographic units published.

That the 2004–2006 was the period when the "basic" research results on the integration of ERP and GIS were developed and publish, is also suggested by the analysis of the citations. The two most frequently cited bibliographic units, both with over 40 citations, are from 2005 (Li et al. 2005) with 61 citations, and from 2004, with 44 citations (Gayialis and Tatsiopoulos 2004). Li et al. (2005) presented a study on applying an integrated Global Position System (GPS) and Geographical Information System - GIS technology to the reduction of construction waste, where the integrated GPS and GIS technology is combined to the Enterprise Resource Planning system. Authors presented a case study with the purpose to demonstrate the deployment of the system that resulted in the minimisation of the amount of onsite material wastage.

The second highly cited publication (Gayialis and Tatsiopoulos 2004), presents the development of a decision support system used by an oil downstream company for routing and scheduling purposes. The delivery process of oil products from a number of distribution centres to all customers is very complex. The development of the operations research enabled the development of applications of the advanced planning and scheduling systems, that can be applied in practice if they are embodied in packaged information technology solutions. The second important condition is that the interface problems to mainstream ERP software applications are solved. In this study, the utilisation of advanced IT systems supports the planning and management of distribution operations effectively. This study shows that the combination of a supply chain management application with a geographical information system (GIS) integrated with an enterprise resource planning (ERP) software resulted in the innovative decision support tool; its' use may have many benefits: optimal use of the distribution network resources, transportation cost reduction and customer service improvement.

In the second step of the bibliometric analysis, the identified set of 40 bibliographic units were used in the mapping of clusters by using the VOSviewer.

In VOSviewer, based on the title and abstract the relevant terms were identified. The minimum number of occurrence of a term was set to 3. Out of 1186 terms, identified by VOSviewer, 76 met the threshold. For each of the 76 terms, the relevance score was calculated, and the most relevant terms were selected (60% the most relevant terms). This process resulted in the identification of 46 terms. After deleting terms that are general and not associated with the topic investigated (article, research, country etc.), the process of mapping of terms was performed. Three clusters were identified, that is presented in Fig. 4.11.

Results reveal that three clusters are observed. The common characteristics of bibliographic units in a green cluster are defined by terms of resource planning, organisation, management efficiency, implementation. Thus, the bibliographic units in this cluster relate with terms associated in particular to management. The bibliographic units in the red cluster contain terms like an information system, support, function and optimisation, while in the blue cluster, in particular terms GIS, GIS

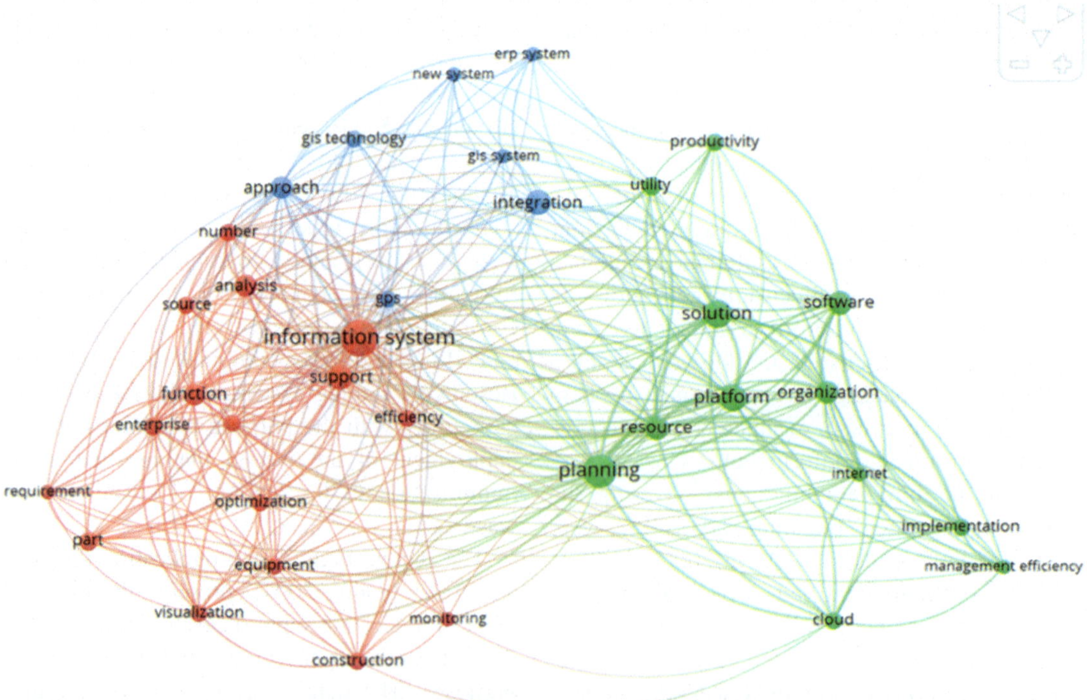

Fig. 4.11 Clusters – mapping of terms for "enterprise resource planning" and "GIS" key-phrases

technology, GIS system, ERP and integration are the most emphasised.

The three clusters are interrelated, and the dense net of connections is visible among the terms within each cluster, and among terms between clusters, as well. If the term integration is put into the central place, the most emphasised connections are presented in Fig. 4.12.

Figure 4.12 shows that the integration of GIS, ERP, information system and software solution is very topical for research as well as for the implementation in organisations.

4.4.3 Research Publishing on Business Intelligence (BI) and Geographical Information Systems (GIS) Integration

The second part of the bibliometric analysis was also performed based on the Scopus database; we used the "business intelligence" key-phrase at the first level. A search revealed 6631 bibliographic units that included this key-phrase in the title, keywords or in abstract. In the second step, the "GIS" keyword was used to identify the bibliographic units identified within the first step, that included both, "business intelligence" and "GIS" key-phrases. The search revealed 112 units.

The first article that was identified in Scopus and is combining the Business Intelligence and the Geographical Information Systems was published in 2002. In their conference paper authors (Osianlis and Arnott 2002) presented the use of data warehousing and business intelligence technologies in an Australian water business for the management of water and waste-water services provision.

In 2004 two bibliographic units were identified; Weigang et al. (2004) presented a dynamic information system for urban bus passengers in Brasilia, using business intelligence, that was developed to optimise bus operations and increase the satisfaction of urban transportation users. To achieve these objectives,

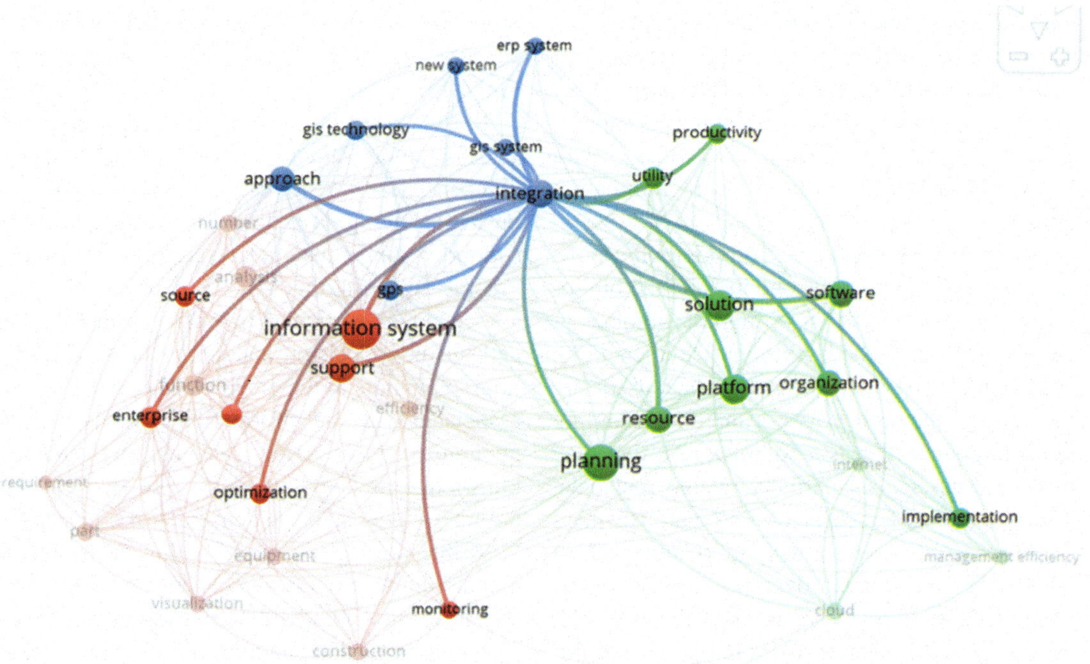

Fig. 4.12 The most emphasised connections with the term integration – "enterprise resource planning" and "GIS" key-phrases

the system involves the convergence of a number of different technologies, including Global Positioning System (GPS), Geographic Information System, database, data mining, Internet and telecommunications. Thrall (2005) presented the software solution that provides the essential tools and functionality needed for deriving geospatial business intelligence, while Barnes (2005) emphasized many "layers" of GIS in today's business and everyday life: GIS plays a key role in natural resource extraction, infrastructure management, intelligence and military defence, homeland security, business intelligence, navigation, etc.

In 2005, a groundbreaking bibliographic unit was published that links BI and GIS, which to date is the most widely cited reference (108 citations) in the Scopus database in this integrated area. The authors (Rivest et al. 2005) emphasise the importance of on-line analytical processes in which companies combine data warehouses and analytical tools to access, visualise and analyse their integrated, aggregated and summarised data. The authors emphasise that a large part of this data has a spatial component, so they emphasise the importance of spatial online analytical processing, which allows interactive spatial-time data exploration. The purpose of their paper is to show how these concepts support spatial-temporal research of data with geo-visualisation, interactivity and animation options.

The number of published bibliographic units was constantly increasing, and in 2007 the bibliographic unit with the next highest number of citations was published. Authors (Devillers et al. 2007) emphasise that geospatial data users are often facing the need to assess and understanding the data quality, that is a complex task that may involve thousands of partially related metadata. The combining concepts of GIS and Business Intelligence represent such a complex case where heterogeneous datasets have to be integrated. Authors, therefore, describe and present the approach, that provides interactive, multi-granularity and context-sensitive spatial data quality indicators that help experts to build and justify their opinions and business decisions.

The highest number of publications was in 2012 (14 bibliographic units); in the years to 2018, the number of published bibliographic units covered by Scopus ranges around 10 per year. The latest publications in 2018 show, how the integrated approach of GIS and BI may be beneficial for different aspects of quality of life of different social groups (Szewrański et al. 2018a), for detecting and predicting the flood risks for improving the water management infrastructure modelling (Szewrański et al. 2018b), for market segmentation and visualization, based on user behavior geographical distributions (Kamthania et al. 2018) etc.

In the second step of the bibliometric analysis, the identified set of 112 bibliographic units was used in the mapping of clusters by using the VOSviewer.

In VOSviewer, based on the title and abstract the relevant terms were identified. The relevance score was calculated, and the most relevant terms were selected; the process resulted in the identification of 51 terms. After deleting terms that are general and not associated with the topic investigated (article, research, country etc.), the process of mapping of terms was performed. Two clusters were identified, that is presented in Fig. 4.13.

Results reveal that two clusters are observed. The common characteristics of bibliographic units in a red cluster are defined by terms of business intelligence, decision making, decision support systems, data, analysis. Thus, the bibliographic units in this cluster relate with terms associated in particular to business intelligence and business decision-making process. The bibliographic units in the green cluster contain terms associated with the information systems and the geographic information systems, performance, knowledge and effectiveness. The two clusters are interrelated, and the dense net of connections is visible among the terms within each cluster, and among terms between clusters, as well.

The results of both bibliometric analyses therefore confirm, that integration of both fields, namely, the Enterprise Resource Planning and the Business Intelligence, with Geographic Information Systems, is very topical and important

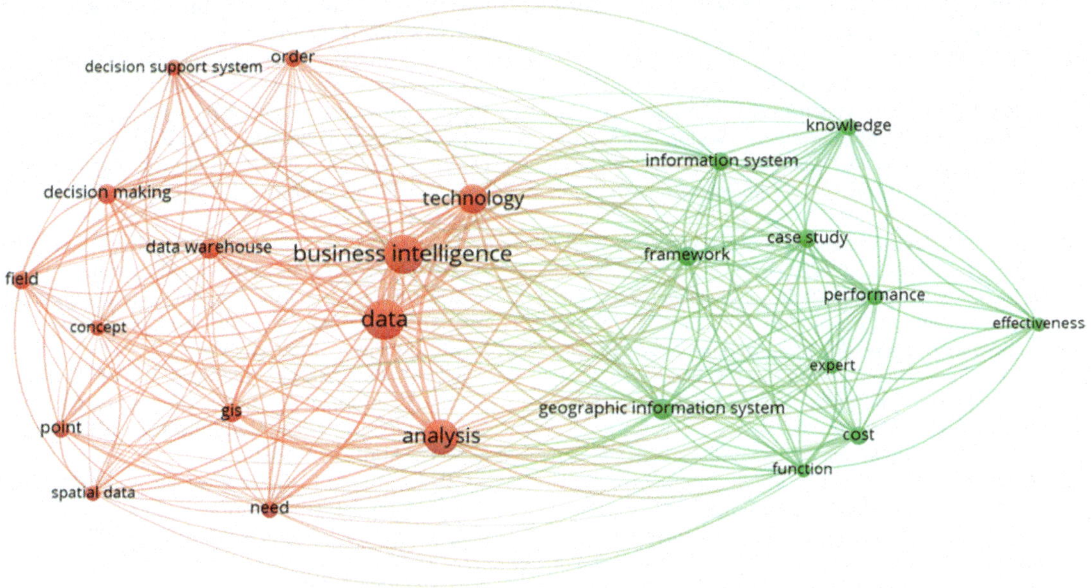

Fig. 4.13 Clusters – mapping of terms – "business intelligence" and "GIS" key-phrases

field in academic research as well as in the applied field, especially in the field of business decision making, by providing information, that could not be gathered without such an integration (or, at least, could not be captured on such a quality level). Linking ERP and BI with GIS thus represents an important quality information basis for business decision making.

4.5 Conclusion

Organisation-wide information systems like ERP, CRM and BI, are core information systems of organisations because they provide core and crucial support for daily operations/business activities. They are used in nearly every company/organisation. They are also already mature technologies – solutions available on the market offered by providers belong to third wave/generation. They are not only technological sophisticated, but they are also following the demands and expectations of organisations. Solutions providers are expanding their functionality all the time by adding new modules and features. A new generation of solutions can be easily integrated with other information systems which are more specialised in a certain area. One such category of information systems is GIS. Their use is expanding more and more to business processes, and therefore more and more organisations are starting to use GIS. Years ago they were used as stand-alone systems, but recently they are in many companies integrated with other information systems. In such a way they become an important part of information support to the operational level and also to management level. At the operational level, they are primarily used in companies with location-based business events, with location-based resources and with location-based workflows (i.e. route planning). At management level companies are using GIS-enabled reporting with GIS-based dashboards.

References

Abou-Ghanem, M. & Arfaj, K. A. (2008). *SAP/GIS integration case studies & techniques*. Retrieved from https://pdfs.semanticscholar.org/6a84/21e2698462aa8e47a8cfc537639a8825788b.pdf (20.12.2018).

ACC Software Solutions. (2018). *Future-proof your business with postmodern ERP (e-book)*. NJ: ACC

Software Solutions. Retrieved from https://4acc.com/future-proof-with-postmodern-erp-ebook/ (18.10.2018).

Almajali, D. A., Masa'deh, R., & Tarhini, A. (2016). Antecedents of ERP systems implementation success: A study on Jordanian healthcare sector. *Journal of Enterprise Information Management, 29*(4), 549–565.

Anderegg, T. (2000). *ERP: A-Z Implementer's guide for success*. Resource Pub.

Anon, J. (1998). Information technology and GIS. *Water Services, 102*(1225), 20–22.

Aydin, A. O. & Sarman, S. (2006). A management information system: Integrating GIS, GPS and ERP. In *The Proceedings of the 4th International Logistics and Supply Chain Management Congress*, Turkey.

Barnes, S. (2005). The many layers of GIS. *GeoSpatial Solutions, 15*(7), 26–29.

Bidgoli, H. (2004). *The internet encyclopaedia* (Vol. 1). Hoboken: Wiley.

Bimonte, S., Tchounikine, A., Miquel, M., & Pinet, F. (2010). When spatial analysis meets OLAP: Multidimensional model and operators. *International Journal of Data Warehousing and Mining, 6*(4), 33–60.

Bradford, M. (2016). *Modern ERP: Select, implement and use Today's advanced business systems*. Raleigh, NC: North Carolina State University (lulu.com.

Buttle, F., & Maklan, S. (2015). In London/New York: Routledge/Taylor & Francis Group (Ed.), *Customer relationship management: Concepts and technologies* (3rd ed.).

Chen, H., Chiang, R. H. L., & Storey, V. C. (2012). Business intelligence and analytics: From big data to big impact. *MIS Quarterly, 36*(4), 1165–1188.

Chorianopoulus, A. (2016). *Effective CRM using predictive analytics*. West Sussex: Wiley.

Chung, W., Chen, H. & Nunamaker Jr, J. F. (2002). Business intelligence explorer: A knowledge map framework for discovering business intelligence on the web. *HICSS'03*.

De Bellis, N. (2009). *Bibliometrics and citation analysis*. Lanham, MD/Toronto/Plymouth, UK: The Scarecrow Press, Inc.

Devillers, R., Bédard, Y., Jeansoulin, R., & Moulin, B. (2007). Towards spatial data quality information analysis tools for experts assessing the fitness for the use of spatial data. *International Journal of Geographical Information Science, 21*(3), 261–282.

ESRI (2007). *GIS best practices – Using GIS and SAP*. Retrieved from https://www.aeroterra.com/PDF/using-gis-and-sap.pdf (15.10.2018).

ESRI (2012). *ESRI location analytics for business intelligence*. Retrieved from https://www.esri.com/library/whitepapers/pdfs/esri-location-analytics-for-bi.pdf (15.10.2018).

ESRI (2018). *Esri maps for dynamics CRM*. Retrieved from http://doc.arcgis.com/en/maps-for-dynamics-crm/ (20.11.2018).

Ganly D., Kyte A., Rayner N., & Hardcastle C. (2013). *Predicts 2014: The Rise of the Postmodern ERP and Enterprise Applications World*. Retrieved from https://www.gartner.com/doc/2633315/predicts%2D%2Drise-postmodern-erp (10. 3. 2017).

Garfield, E. (2009). From the science of science to Scientometrics visualizing the history of science with HistCite software. *Journal of Informetrics, 3*, 173–179.

Gayialis, S. P., & Tatsiopoulos, I. P. (2004). Design of an IT-driven decision support system for vehicle routing and scheduling. *European Journal of Operational Research, 152*(2), 382–298.

Gideon Adewale, A., Gideon Olukunle, B. & Ede, A. (2016). Geographic information systems application in sustainable business intelligence analysis. *3rd International Conference on African Development Issues*. Covenant University Press.

Hawkins, D. T. (2001). Bibliometrics of electronic journals in information science. *Information Research, 7*, 1. Retrieved from http://InformationR.net/ir/7-1/paper120.html (20.12.2018).

Hessa, R. L., Rubin, R. S., & West, L. A. (2004). Geographic information systems as a marketing information system technology. *Decision Support Systems, 38* (2), 197–212.

Horwitt, E. (2009). *ERP and GIS integration brings spatial dimension to key business apps*. Search Manufacturing ERP. Retrieved from https://searcherp.techtarget.com/news/1373664/ERP-and-GIS-integration-brings-spatial-dimension-to-key-business-apps (20.12.2018).

Howson, C. (2014). *Successful business intelligence – Unlock the value of BI & Big Data* (2nd ed.). New York: Mc Graw Hill Education.

HubPages. (2018). *Enterprise Resource Planning (ERP) and Fortune 500 Companies (2017)*.Retrieved from https://hubpages.com/business/Enterprise-Resource-Planning-ERP-and-Fortune-500-Companies (18.7.2018).

Kamthania, D., Pahwa, A., & Madhavan, S. S. (2018). Market segmentation analysis and visualisation using K-mode clustering algorithm for E-commerce business. *Journal of Computing and Information Technology, 26*(1), 57–68.

Li, H., Chen, Z., Yong, L., & Kong, S. C. W. (2005). Application of integrated GPS and GIS technology for reducing construction waste and improving construction efficiency. *Automation in Construction, 14*(39), 323–331.

Litan, D., Mocanu, A.-M., Copcela, L., Surugiu, I., Teohari, M., & Raduta, O. (2011). Business' new requirement: Information systems integration – Methods and technologies. *International Journal of Computers and Communications, 3*(5), 133–145.

Lizzote, O. (2017). *Do you CRM me?: An analytical guide to customer relationship management*. North Charleston: CreateSpace Independent Publishing Platform.

Magal, S. R., & Word, J. (2011). *Integrated business processes with ERP systems*. Hoboken: Wiley.

Monk, E., & Wagner, B. (2009). In Course Technology Cengage Learning Boston (Ed.), *Concepts in Enterprise Resource Planning* (3rd ed.). Boston.

Montgomery N., Anderson, R. P., Kostoulas J., and Woodyer A. (2018). *High-tech Tuesday webinar:*

Best opportunities and bets for growth in Enterprise Resource Planning. Retrieved from https://www.gartner.com/doc/3640429?ref=SiteSearch&sthkw=postmodern%20erp&fnl=search&srcId=1-3478922254 (20.5.2018).

Nussbauman, K. C. (2015). *Storytelling with data: A data visualization guide for business professionals*. New Jersey: Wiley.

Osianlis, A., & Arnott, D. R. (2002). The application of business intelligence and data warehousing in an Australian water business. *Management Information Systems*, 369–378.

Patel, J. & Doctor, G. (2013). *Exploring Enterprise Resource Planning (ERP) and Geographic Information System (GIS) integration*. Retrieved from https://www.researchgate.net/publication/259462868_EXPLORING_ENTERPRISE_RESOURCE_PLANNING_ERP_AND_GEOGRAPHIC_INFORMATION_SYSTEM_GIS_INTEGRATION (13.8.2018).

Pelphrey, M. W. (2015). *Directing the ERP implementation: A best practice guide to avoiding program failure traps while tuning system performance*. London: CRC Press.

Phillips, S. S. & Ryan, E. (2013). *Control your ERP Destiny: Reduce project costs, mitigate risks, and design better business solutions*. US: Street Smart ERP Publications.

Posthumus, R. (2008). *GIS as a tool in business intelligence*. PositionIT. Retrieved from http://www.ee.co.za/wp-content/uploads/legacy/AppT-GIS%20as%20a%20tool.pdf

Rivest, S., Bedard, Y., Proulx, M.-J., Nadeau, M., Hubert, F., & Pastor, J. (2005). SOLAP technology: Merging business intelligence with geospatial technology for interactive spatio-temporal exploration and analysis of data. *ISPRS Journal of Photogrammetry & Remote Sensing, 60*, 17–33.

Sherman, R. (2015). *Business intelligence guidebook – From data integration to analytics*. Waltham: Elsevier.

SMRC (2017). *ERP software market report 2017, trends, analysis, share, estimates and forecasts to 2022*. Retrieved from http://www.mynewsdesk.com/us/pressreleases/erp-software-market-report-2017-trends-analysis-share-estimates-and-forecasts-to-2022-2287038. (5.6.2018).

Sneller, L. (2014). *A guide to ERP – Benefits, implementation and trends*. Bookboon.com.

Starzyczna, H., Pellešova, P., & Stoklasa, M. (2017). The comparison of customer relationship management (CRM) in Czech small and medium enterprises according to selected characteristics in the years 2015, 2010 and 2005. *Acta Universitatis Agriculturae et Silviculturae Mendelianae Brunensis, 65*, 1767–1777.

Few, S. (2006). *Information dashboard design: The effective visual communication of data*. Cambridge, MA: O'Reilly.

Stepniak, C., & Turek, T. (2014). Integration of spatial information resources on the example of utility companies in Czestochowa region. *Online Journal of Applied Knowledge Management, 2*(2), 97–108.

Szewrański, S., Chruściński, J., van Hoof, J., Kazak, J.K., Świader, M., Tokarczyk-Dorociak, K., Zmuda, R. (2018b). A location intelligence system for the assessment of pluvial flooding risk and the identification of stormwater pollutant sources from roads in suburbanised areas. Water (Switzerland) 10 (6), art. no. 746.

Szewrański, S., Świąder, M., Kazak, J. K., Tokarczyk-Dorociak, K., & van Hoof, J. (2018a). Socio-environmental vulnerability mapping for environmental and flood resilience assessment: The case of ageing and poverty in the City of Wrocław, Poland. *Integrated Environmental Assessment and Management, 14*(5), 592–597.

Tableau. (2018). *Solutions maps*. Retrieved from www.tableau.com/solutions/maps (8.10.2018).

Thrall, G. I. (2005). ESRI enables ArcGIS for business analyst. *GeoSpatial Solutions, 14*(6), 45–48.

Treiblmayer, M, Tso-Sutter, K. H. L., & Krüger, A. (2011). Interfacing business processes and spatial processes. In *Proceedings 2011 IEEE international conference on spatial data mining and geographical knowledge services*, Fuzhou, 174–180.

Trieu, V.-H. (2017). Getting value from business intelligence systems: A review and research agenda. *Decision Support Systems, 93*, 111–124.

van Eck, N. J., & Waltman, L. (2013). *VOSviewer Manual*. Leiden: Universiteit Leiden.

van Eck, N.J. (2011). Methodological advances in bibliometric mapping of science. PhD Series Research in Management, 247. Rotterdam: Erasmus Research Institute of Management.

Wan, J. & Xie, L. A. (2018). *Bibliometric review of research trends in Social CRM*. Association for Information Systems, 2018. *WHICEB 2018, 63*. Retrieved from http://aisel.aisnet.org (19.8.2018).

Weigang, L., Yamashita, Y., Koendjbiharie, M. W., De Moura Juca, R. C., & Maclver, A. (2004). The development and implementation of the operation system and data bank for the intelligent transportation system – Sitcuo. *Journal of Advanced Transportation, 38*(2), 163–186.

Wickramasuriya, R., Ma, J., Berryman, M., & Perez, P. (2013). Using geospatial business intelligence to support regional infrastructure governance. *Knowledge-Based Systems, 53*, 80–89.

Wilson, J. D. (1998). Enterprise resource planning creates new opportunities for GIS in utilities. *GIS World, 11* (5), 51–53.

Yan, L., Wang, J.-B., Ma, Y.-A., & Dou, J. (2002). The Enterprise Resource Planning (ERP) system and spatial information integration in tourism industry – Mount Emei for example. *Lecture Notes in Computer Science, 2331*(Part 3), 966–974.

Yaptenco, E., Piepel, C., Rowland, B. & Causley, M. (2005). *Integrating GIS within the enterprise-options, considerations and experiences*. ESRI International User Conference 2005. Retrieved from http://downloads.esri.com/support/ProjectCenter/tw_1111_Mark_Causley_PRES_1.pdf (20.10.2018).

Yerpude, S., & Kumar Singhal, T. (2018). Internet of Things based Customer Relationship Management – A Research Perspective. *International Journal of Engineering & Technology, 7*(2.7), 444–450.

Zipf, P. J. (2000). Technology-enhanced project management. *Journal of Management in Engineering, 16*(1), 34–39.

Methods in Microeconomic and Macroeconomic Issues

5

Jarmila Zimmermannová

Abstract

This chapter focuses on microeconomic and macroeconomic issues. Regarding microeconomics, the key microeconomic topics, as supply, demand, product markets and factor markets, will be presented. The main microeconomic variables connected with consumer and producer behaviour will be described, including the marginal variables. Decision-making of producer and the possibilities of market equilibrium will be discussed, depending on different types of competition. Specifics of factor markets will be explained, precisely the labour market, land market, and capital market. The macroeconomic part focuses on key macroeconomic issues connected with spatial aspects. Firstly, the key macroeconomic indicators and their features will be described. Then the spatial view will be included, a comparison of selected macroeconomic indicators in EU28 and NUTS2 regions will be presented. The question of economic growth will be discussed. The last part of the chapter deals with economic modelling, both in microeconomic and macroeconomic areas. The short overview of possibilities of modelling strategic behaviour of particular economic subjects and agent-based modelling, as well as options of macroeconomic modelling in a short period (I-O analysis), medium period (CGE models) and long-term models will be presented. The spatial view in the whole chapter is underlined.

Keywords

Supply · Demand · Market · Consumer · Producer · Decision-making · Macroeconomic indicators · Economic growth · Economic modelling

5.1 Methods in Microeconomics

5.1.1 Microeconomics and Relationships Between Variables

5.1.1.1 Microeconomic Issues

Microeconomics studies individual prices, quantities and markets, on the other hand macroeconomics studies the behaviour of the economy as a whole. It examines the forces that affect firms, consumers and workers in the aggregate.

Regarding variables, microeconomics can examine for example price, quantity, hours worked, acres of land, incomes in currency units, number of employees, etc. A functional dependency between two variables exists in the case when one variable depends on the other variable, that is, the value of the dependent variable is determined by the independent variable.

J. Zimmermannová (✉)
Department of Economics, Moravian Business College Olomouc, Olomouc, Czech Republic
e-mail: jarmila.zimmermannova@mvso.cz

V. Pászto et al. (eds.), *Spationomy*, https://doi.org/10.1007/978-3-030-26626-4_5

For example, the price (independent variable) determines the quantity of goods we buy (dependent variable).

In general, it can be stated that there are three possible types of relationships between variables:

- direct - a positive relationship expressing that the growth of one variable is accompanied by the growth of the other variable (e.g., the price and the offered quantity = the supply function),
- indirect - a negative relationship expressing that the growth of one variable is accompanied by the simultaneous decline of the other variable (e.g., the price and the demanded quantity = the demand function),
- neutral - mutual independence of variables, this means that the growth of one variable does not cause a change in the other variable.

Every society must answer three basic economic questions:

- What should we produce?
- How should we produce it?
- For whom should we produce it?

Although every society answers the three basic economic questions differently, in doing so, each confronts the same fundamental problems: resource allocation and scarcity. The classical microeconomic theory was developed by Adam Smith in 1776 and later economists, such as David Ricardo. The essential aspect of the classical microeconomic theory includes determination of market price and output and market-equilibrium. Adam Smith is well known for his 'invisible hand of the market.' In his approach, people act out of self-interest and markets tend to provide goods and services which are demanded by the population. Market forces respond to changes in demand and supply, e.g. a shortage pushes up the price and causes demand to fall.

Smith also investigated topics such as the division of labour, specialisation and economies of scale. The early classical economists emphasised the importance of costs to firms and consumers. The following sections will focus on these topics more in detail.

5.1.1.2 Economic Circle, Economic Entities, Different Kinds of Markets

One of the main basic models in economics is the circular-flow model, which describes the flow of money and products throughout the economy in a very simplified way. The model represents all of the actors in an economy as either households or firms (companies), and it divides markets into two categories:

- markets for goods and services (product markets)
- markets for factors of production (factor markets)

Product market represents the marketplace in which final goods or services are offered for purchase by consumers, businesses, and the public sector.

Factor market represents the marketplace for the services of a factor of production. A factor market facilitates the purchase and sale of services of factors of production, which are inputs like labour, capital, land and raw materials that are used by a firm to make a finished product.

The circular flow model is shown in Picture 5.1.

Consumers buy goods and sell factors of production, businesses sell goods and buy factors of production. Consumers use their income from the sale of labour and other inputs to buy goods from businesses, businesses base their prices of goods on the costs of labour and property. Prices in goods markets are set to balance consumer demand with business supply; prices in factor markets are set to balance household supply with business demand.

5.1.1.3 Resources, Scarcity

Production factors are natural, human, financial and other resources that enter into production and help to create the final economic output. They are scarce since the amount of these factors is limited. We can understand the production factors as

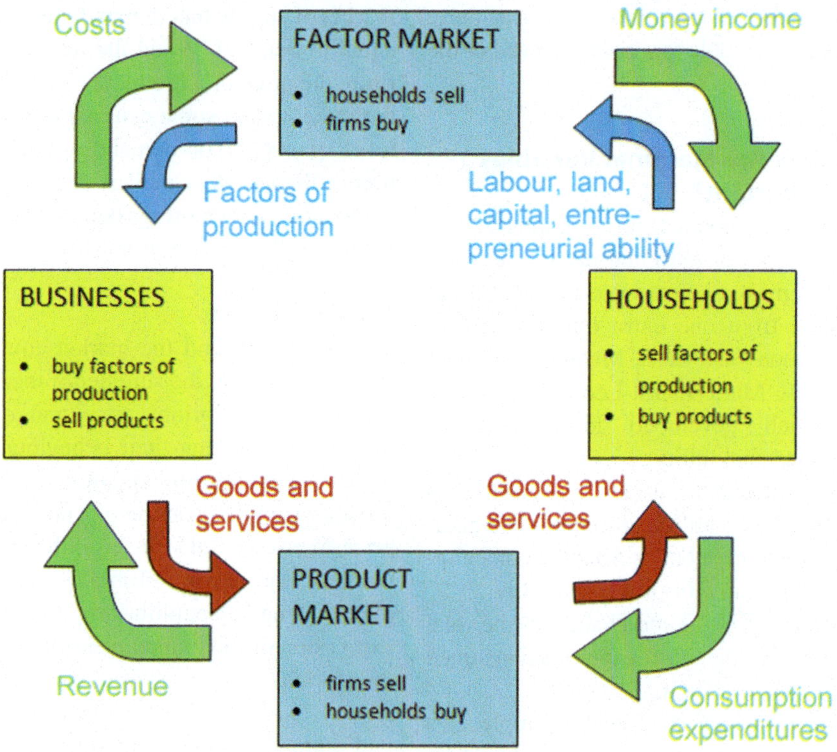

Picture 5.1 **The circular-flow model**. (Source: Own processing, based on Samuelson and Nordhaus 2010)

resources, which are transformed in the manufacturing process into the desired products and services. In other words, the elements of production are the inputs, which the firm uses to produce output desired by consumers and to deliver this output to the market.

The factors of production include:

- labour,
- land and natural resources (raw materials),
- capital (financial capital, machinery and tools, equipment),
- the level of technology (level of knowledge), or entrepreneurship.

The demand for a given input is the so-called derived demand, that means the demand for the goods, which is produced using the given input.

The market of production factors is the point at which the demand for factors of production meets with the supply of factors of production. The demanders in this market are the firms, that - via the use of production factors - offer on the final product market a specific product to achieve profit. The suppliers on the market of production factors are the households, i.e. the owners of the factors of production. These households then rent out these factors to get income.

Scarcity means that resources (production factors) are limited, and hence the number of manufactured goods cannot satisfy all human needs. The category of needs is very broad; however, economics does not address its scope, structure and classification. Within economic theory, only economic needs that are satisfied by using up produced goods and services are relevant. The essence is to satisfy the sense of lack of something that is desirable for the consumer and the fact that the overall satisfaction is never found, as the fulfilment of a need arises the additional one, as well as both the intensity of a need and the hierarchy of needs vary, etc. Satisfying needs is related to the concept of consumption. Consumption of economic goods leads to the satisfaction of the human economic needs, provided that these goods are useful, i.e. able to satisfy the need of the consumer.

5.1.1.4 Why Are Marginal Variables Important?

In neo-classical economics, more emphasis was placed on concepts of marginal utility and marginal cost. We make choices depending on satisfaction we get from one extra unit of a good. Economists such as Carl Menger, William Stanley Jevons, Marie-Esprit-Léon Walras and Alfred Marshall developed ideas such as diminishing marginal utility.

Focusing on consumer issues, we can find as an example marginal utility; it denotes the additional utility consumer obtains from the consumption of one additional unit of a commodity.

The marginal utility is the change in the total utility that is caused by the change of a consumed quantity of goods by a unit.

Regarding producer, a typical example is a marginal product, the extra output produced by one additional unit of one input while other inputs are held constant, or marginal cost of production, the additional cost incurred in producing one extra unit of output.

5.1.1.5 Market Equilibrium - Product Markets, Factor Markets

Market equilibrium, also known as the market clearing price, refers to a perfect balance in the market of supply and demand, i.e. when supply is equal to demand. When the market is at equilibrium, the price of a product or service will remain the same, unless some external factor changes the level of supply or demand. According to economic theory, in a market economy, there is a single price which brings demand and supply into balance – the equilibrium price.

Product Markets

We can understand the market equilibrium as a state where either the supplier or the buyer had, at the given equilibrium price and quantity, any interest in changing their behaviour. In the case of Market equilibrium, quantity demanded equals quantity supplied and there is no tendency for the price to rise or fall - the equilibrium price is the market-clearing price. Table 5.1 shows an example of a market equilibrium, using a numerical expression of the market clearing price setting on the market.

Figure 5.1 shows the same example of the market clearing price using the graphical expression. The quantity demanded connected with retail price is represented by the demand curve, the quantity supplied by the supply curve. The equilibrium price P_e (17,500 EUR) shows the market clearing price where the supply is equal to the demand (500 cars).

Price elasticity is a crucial methodological approach in economics. Generally, the price

Table 5.1 Example of market equilibrium using a numerical expression

Market Equilibrium – Car Market		
Retail Price (EUR)	Quantity demanded	Quantity Supplied
30 000	0	1000
27 500	100	900
25 000	200	800
22 500	300	700
20 000	400	600
17 500	500	500
15 000	600	400
12 500	700	300
10 000	800	200

Source: Own processing, based on picture on https://marketbusinessnews.com/financial-glossary/market-equilibrium/

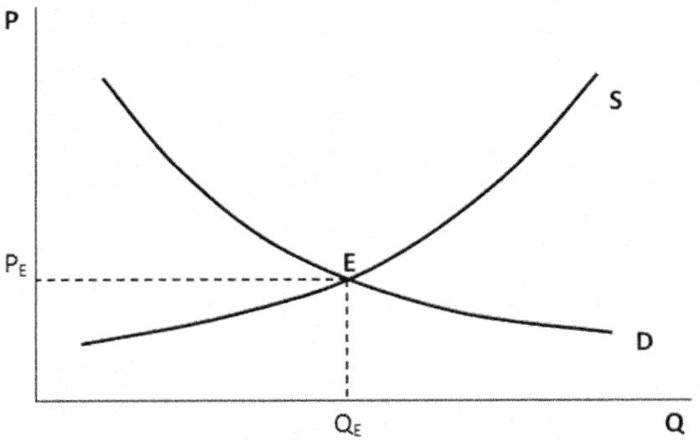

Fig. 5.1 Example of market equilibrium using graphical expression. (Source: Own processing based on Samuelson and Nordhaus 2010; *P* price, *Q* quantity, *S* supply, *D* demand, *E* equilibrium, *Pe* equilibrium price, *Qe* equilibrium quantity)

elasticity of demand for some product is a sensitivity of consumers to a price change. There are some products like meat or fruits for which small change in price causes a great change in the quantity purchased. So, the demand for such products is elastic. But, there are other products (electricity, medical devices) for which the large price change causes the small change in the quantity purchased. Thus, the demand for such products is inelastic. We use the coefficient of price elasticity of demand EDP, to determine whether the demand is elastic or inelastic:

$$\mathbf{EDP} = \%\mathbf{Change\ in\ Quantity\ Demanded}/\%\mathbf{Change\ in\ Price} \quad (5.1)$$

There can be three cases: if $|EDP| < 1$, then demand is inelastic; if $|EDP| > 1$, then demand is elastic and, finally, if $|EDP| = 1$ we can observe the unit elasticity. The slope of demand curves can be different from our basic case, presented in Fig. 5.1; therefore Fig. 5.2 shows the selected examples of particular demand curves, including the boarder elastic and inelastic demand curves.

Factor Markets

On factor markets, companies represent demand side and households represent supply side. The demand for factors of production depends on the demand for final products. Therefore, the demand for a given input is also called derived demand. This is derived from the demand for the goods, which is produced using the given input.

Particular factor markets are characteristic by specific market prices, as follows:

- Labour market: the price of labour is the wage rate (w), for example, the amount of money for one hour of work.
- Land market: the price of land is the rent (r), for example, the amount of money for one hectare of land.
- Capital market: the price of capital is the interest rate (i), for example, the amount of money for one machine hour. The firm can compare the price of capital to the interest rate it could gain from the money spent on the purchase of the capital goods if it had put it in a bank. The interest rate thus represents the alternative cost of the ownership of the capital goods.

Regarding spatial aspects of particular production factors, we will focus on more details of the land market and labour market.

Land Market

The land market is characteristic by a fixed supply of land and a rent, as a price of land. Generally, some goods or productive factors are entirely set in amount, regardless of price. Nature's original endowment of land can be taken as fixed in amount. Giving the quantity supplied is constant at every price, the payment for the use of such a factor of production is called rent or pure economic rent.

When supply is independent on price, the supply curve is vertical in the relevant region. Figure 5.3 shows the case of land, for which a higher price cannot coax out any increase in output. An increase in the demand for this fixed production factor will affect only the price of this production factor.

It is important to underline the spatial aspect of land price; there can be observed differences in land price in particular regions, depending on the level of demand for some particular sort of land. In areas with a better economic situation, the

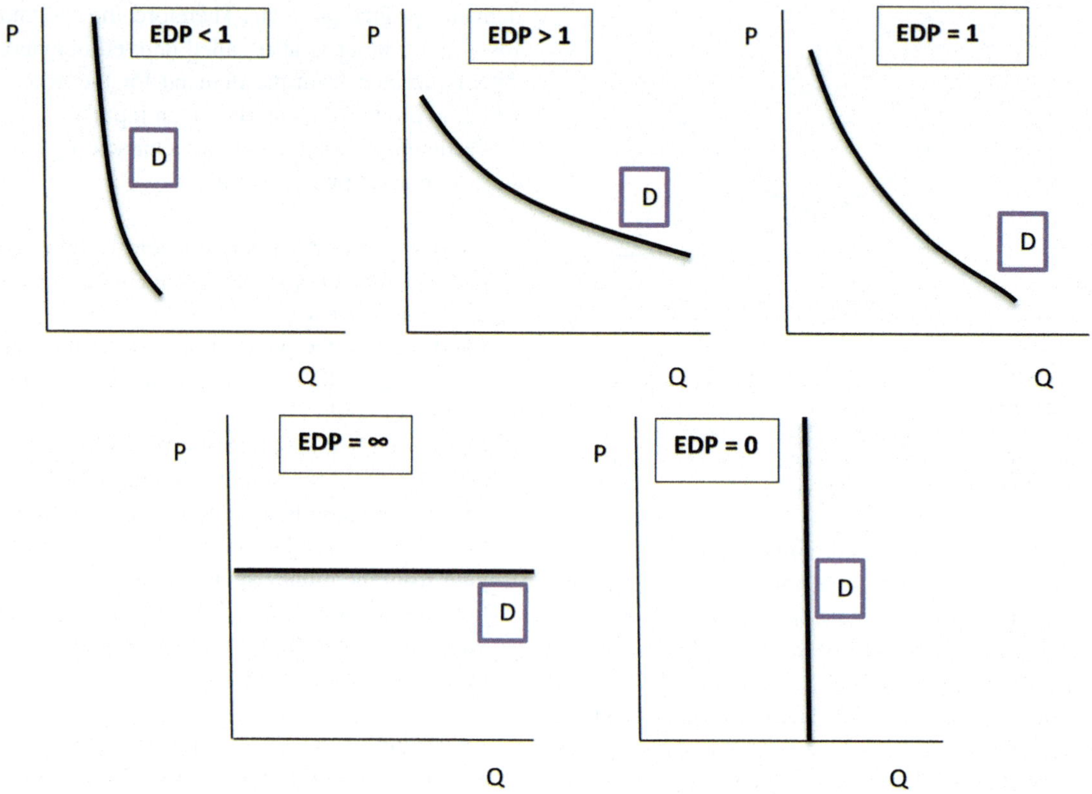

Fig. 5.2 **Examples of price elacticity of demand**. (Source: Own processing based on Samuelson and Nordhaus 2010; *P* price, *Q* quantity, *D* demand, *EDP* coefficient of price elasticity of demand)

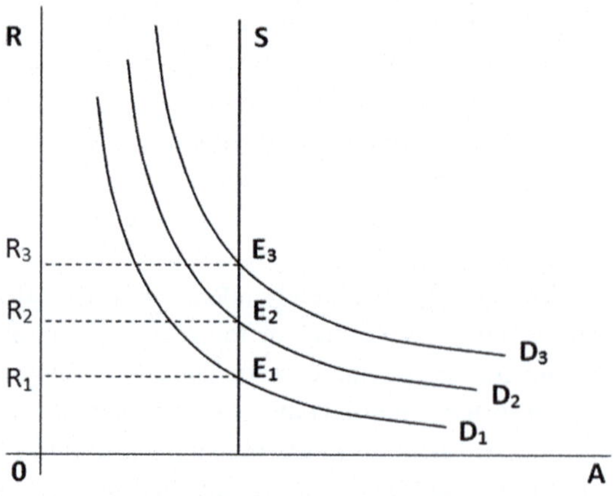

Fig. 5.3 **The land market equilibrium**. (Source: Own processing based on Samuelson and Nordhaus 2010; R_1, R_2, R_3 represent the price of land – the level of rent. D_1, D_2, D_3 represent the level of demand)

demand is higher. In poor countries, the demand for the same sort of land will be probably lower.

Labour Market

The demand for labour is determined by its marginal productivity in producing final output. The marginal productivity of labour can rise in the following cases:

- workers have more or better capital goods to work with,
- the marginal productivity of better-trained or better-educated workers is higher than that of workers with less "human capital."

Real wages differ among countries, Table 5.2 shows the example of wages in manufacturing in the year 2006.

Labour costs differ among countries, Picture 5.2 shows the overview of estimated hourly labour costs in EUR in EU countries in the year 2017. The labour costs cover both wages and salaries costs and other costs.

The spatial view can also include migration. The critical aspect is the character of immigrants if they are legal immigrants or political refugees, their skills and education. From the point of view of labour supply, the overall effect of immigration can be an increase in the supply of low-skilled workers relative to high-skilled workers. Studies have estimated that this change in supply has contributed to the decline in the wages of less-educated groups relative to the college-educated.

Table 5.2 Comparison of wages in manufacturing in the year 2006

Region	Wages and fringe benefits in manufacturing, 2006 ($ per hour)
Germany	34.21
Italy	25.07
The United States	23.82
Japan	20.20
South Korea	14.72
Mexico	2.75
China	1.37
Philippines	1.07

Source: Own processing based on Samuelson and Nordhaus (2010)

5.1.1.6 Competition

We can distinguish two key categories of competition within the economy.

The first one is perfect competition. It represents the case, where no firm or consumer can affect prices. Imperfect competition exists whenever a market, hypothetical or real, violates the abstract tenets of neoclassical pure or perfect competition. Since all real markets exist outside of the plane of the perfect competition model, each can be classified as imperfect. The contemporary theory of imperfect versus perfect competition stems from the Cambridge tradition of post-classical economic thought.

The second one is imperfect competition. It represents the case where a buyer or seller can affect a good's price. Regarding companies, we can distinguish the following cases:

- Monopoly: a single seller with complete control over an industry.
- Oligopoly: a few sellers, each firm can affect the market price.

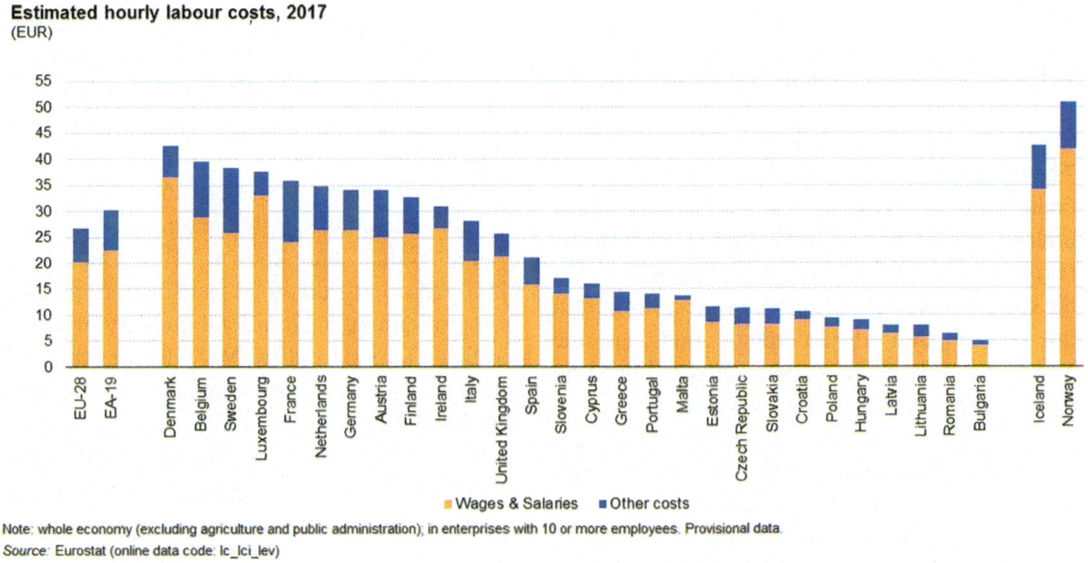

Picture 5.2 Comparison of hourly labour costs in EU countries in 2017. (Source: EUROSTAT 2018)

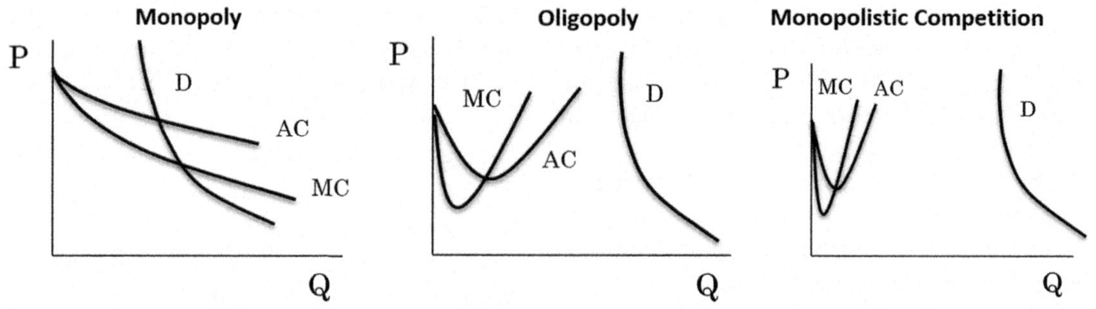

Fig. 5.4 Examples of imperfect competition. (Source: Own processing based on Samuelson and Nordhaus 2010; *P* Price, *Q* Quantity, *D* Demand, *MC* Marginal Costs, *AC* Average Costs

– Monopolistic competition: a large number of sellers produce differentiated products; there are many sellers, none of whom has a large share of the market.

Selected examples of demand curves of different kinds of imperfect competition are shown in Fig. 5.4.

Generally, curves of marginal costs (MC) and average costs (AC) represent individual company's costs in the market; on the contrary, demand curve (D) represents the whole market demand. Consequently, the company – monopolist is a price maker in the market, the company in the monopolistic competition is rather a price taker in the market.

5.1.2 Decision-Making Issues

5.1.2.1 Decision-Making of Consumer

The behaviour of both individuals and economic entities can be explained by comparison of the effects of economic activity and a "detriment" (expenses, costs) associated with this activity. In the case of individuals, the utilities resulting from the consumption of individual goods are the effect; the "detriment" is connected to spending the incomes to purchase these goods.

The concept of utility was developed by philosophers/economists – Jeremy Bentham and John Stuart Mill. In microeconomic theory, it was believed a consumer would buy goods depending on the marginal utility (satisfaction) they get from the good. This theory assumes consumers are rational and seeking to maximise the satisfaction they get. Rationally acting consumers maximise the utility. Making decisions, however, is limited by their income. At the same time, utilities are influenced by consumer preferences.

The starting point for the consumer theory is a consideration that an individual chooses from different consumer goods baskets. The result of consumer decision-making is then the choice of such a consumer basket that brings the maximum utility. Consumers compare individual consumer situations from the perspective of their preferences.

Since the development of the utility theory, economic theory faces the problem of how to measure the utility and whether it is measurable. Based on an approach to the utility measurability, we distinguish cardinal and ordinal theory. In both cases, the utility of one good depends not only on its quantity but also on the quantity of other goods.

Cardinal theory considers the utility to be directly measurable. In this case, specific values of the utility are known. The total utility (TU) represents the total satisfaction received from consuming a given total quantity of a commodity. Marginal utility (MU) denotes the additional utility you get from the consumption of an additional unit of a commodity.

Since the total utility is dependent on the quantity of all goods, in unaltered conditions, the utility is the function of the number of goods consumed:

$$\mathbf{U} \doteq \mathbf{f}\ (\mathbf{X1}, \mathbf{X2}, \ldots, \mathbf{Xn}) \qquad (5.2)$$

where $X_1, X_2, \ldots, X_n$ are quantities of individual goods.

According to the ordinal theory, the utility is not directly measurable. Consumers can say what their preferences are, but not to assess the utility. Consumers are able to arrange combinations of goods according to their utility, but not to determine the amount of the utility of such combinations. The curves showing combinations with the same utility are called indifference curves.

When deciding on the purchase of goods, however, consumers are limited by their income and the prices of the products they buy.

Budget line is indicating the combination of commodities X and Y that a consumer can buy with a given income at a given set of prices.

Consumer Optimum

Consumers choose optimal combinations of goods depending on their preferences and market options. These options are affected by both their income and market prices of goods. The way of determining the consumer optimum depends on the possibility to measure the utility.

Rationally acting consumers maximise the utility within their budgetary constraints. Consumer surplus is the difference between the total utility of the consumed quantity of a given good and the total amount spent on it.

5.1.2.2 Decision Making of Producer

Corporate behaviour is limited mainly by the technological possibilities of production and financial capacity of a firm.

To be able to analyse the decisions of a firm, whose main activity is the transformation of inputs into an output, i.e., production, it is useful to create an abstract model of production depicting the relations between input and output as simply as possible. ***The production function*** serves as this model. The production function is a relation between the number of inputs used in production in a given period, and the maximum volume of output the inputs created by their functioning in the given period.

The inputs used in production are labour, land, capital and the entrepreneurial spirit. We can simplify the real situation and assume that goods X are being produced (with the output labelled Q) from two inputs - capital (K) and labour (L), sufficient for the realisation of X. Same as with the production of goods which is considered to be the flow of output, inputs are also considered to flow in the production process.

- K/t = machine hours per time,
- L/t = hours worked per time.

After these simplifications, we can write the production function in the following form.

$$\mathbf{Q} = \mathbf{f}\ (\mathbf{K}, \mathbf{L}) \qquad (5.3)$$

where Q = output, K = input of capital per time, L = input of labour per time. A production function defined this way has the following properties:

(a) expresses the fact that an output can be produced through different combinations of inputs,
(b) shows the technological constraints of production, because it depends on the level of technology,
(c) does not work with unnecessary or inefficient production processes, which stems from the emphasis on maximum output in the definition, i.e., firms use the most efficient combination of inputs to create output.

If a firm uses the most efficient technology available, its output will depend mainly on the number of inputs used and the efficiency of their use.

Time horizon in which a firm operates is also essential for the further analysis of corporate behaviour. Short run (SR) is characterised as a period in which the services of at least one factor of production a firm uses are fixed as a result of previous choices. In the case of two factors of production, capital is considered to be the fixed input, because it physically exists, for instance in the form of machinery, which is fixed at a specific location. A firm can own it or lease it but cannot change its volume in order to change the output. On the other hand, the volume of labour involved in the production process can be easily reduced or increased if necessary, usually through short-term employment contracts. We, therefore, consider labour as a variable input in the short run.

Since there is at least one fixed input, in our case capital, in the short run, the relation between input and variable output at a given level of capital is characterised by a short-run production function. In other words, it shows how the output changes as a result of changes in one input - labour. That means that the returns from only one variable factor of production are the property of the production function in the short run.

Long run (LR) is a period sufficient for the change in the amount of all inputs used, i.e., it is characterised by the fact that all inputs are variable. In the long run, the firm can mutually substitute the two inputs we use. A long-run production function depicts the relation between the change in volume of both inputs used and the subsequent change in output. If we focus only on the currently proportionally equal increase in the amount of all inputs and the change in output in the long-run production function, we deal with returns to scale. The essential characteristics of a production function in the long run, therefore, are the substitution of inputs and return to the scale of input.

5.1.2.3 Costs Overview

Total costs represent total expenses needed to produce output. In most cases, the total costs increase together with the increase in output. Total costs (TC) in the short run are the sum of fixed costs (FC) and variable costs (VC):

$$\mathbf{TC} = \mathbf{FC} + \mathbf{VC} \tag{5.4}$$

Fixed costs represent expenses that are paid out even when no output is produced; unaffected by any variation in the quantity of output. Variable costs represent expenses that vary with the level of output - such as raw materials, wages, and fuel - and includes all costs that are not fixed. If the output is zero, variable costs are zero as well. The development of variable costs is a significant element for the development of total short-run costs.

Opportunity costs (alternative costs) - assuming the best choice is made, it is the "cost" incurred by not enjoying the benefit that would have been had by taking the second best available choice.

Marginal costs (MC) shows additional costs connected with the production of one extra unit of output.

Average costs AC (unit costs) represent the sum of total costs (TC) divided quantity (Q):

$$\mathbf{AC} = \mathbf{TC}/\mathbf{Q} \tag{5.5}$$

The Costs of a Firm

Before we analyse the costs, recall the difference between the economic and accounting concept of costs. In the narrower sense - accounting - costs are all actual costs incurred the movement of which is recorded in the accounting books. These are the explicit costs. The economic concept of costs is broader: economists take into account not only explicit costs but also implicit costs. Implicit costs are costs that the firm does not actually pay. The existence of these is based on the principle of alternative costs, i.e., opportunity costs. Implicit costs represent the costs the firm loses by using limited resources in a certain way and not any other. For better understanding, let's have a look at the specific differences between accounting and economic concept of costs of labour and capital (inputs used in our analysis).

The costs of labour do not differ much as an accounting and economic concept: both approaches consider these to be explicit costs. From an accounting perspective, these form a part of the actual costs incurred. From an economic perspective, the costs of labour are derived from the wage rate, which forms a part of the employment contract. It is assumed that this wage rate is the same as the best alternative return of this input for its owner.

Costs of capital are perceived completely differently by accountants and economists. From the accounting perspective, the costs of capital are determined by the price of the capital goods which is used for determining the concrete share of capital costs on the costs of a given output. These are thus the actual, explicit costs incurred. Economists, however, consider the costs of capital to be implicit; their amount per hour is determined by the price anyone would be willing to pay for using the given capital goods if he/she rented it for an hour. The firm thus loses the alternative return from renting the capital goods to someone else, i.e., the rent, because it is the only subject using the capital goods. The costs of capital are thus determined by the amount of rent based on the nest alternative usage of the given capital goods.

When analysing the costs, we use the simplified situation in which the firm produces only goods X and uses only two inputs for the production, the prices of which do not change with the quantity purchased (in other words, we assume that there is perfect competition on the market of labour and capital). Substantial simplification is the assumption of a completely homogeneous labour and completely homogeneous capital. Speaking of the prices of inputs, remember that these are considered to flow in the production process. Based on what has already been said about the costs of labour and capital:

- the price of labour is the wage rate (w), the amount of money for one hour of work,
- the price of capital is the rent (r) corresponding to the amount of money for one machine hour.

The firm can compare the price of capital to the interest it could gain from the money spent on the purchase of the capital goods if it had put it in a bank. The interest thus represents the alternative cost of the ownership of the capital goods.

The starting point for the analysis of costs is the functional relationship between the costs and the output per unit of time. Since we know that the amount of output is a function of the inputs used and if we know the prices of the inputs used by the firm in the production process, we can calculate the costs of production of a specific output. The level and development of the costs due to changes in the output of the firm thus depend on two important factors:

1. The character of the production function in question (which determines the shape of the cost functions of the firm),
2. The prices of inputs (these determine the costs).

A cost function can be expressed as:

$$\mathbf{TC} = \mathbf{f}\ (\mathbf{Q}, \mathbf{w}, \mathbf{r}) \tag{5.6}$$

If we assume that the firm behaves rationally, this cost function expresses the minimum costs of a firm for the production of various amounts of output, using various combinations of labour and capital.

The character of costs in the short run is different in many ways from the character of costs in the long run. Since the firm cannot increase the output in the short run by changing the production premises or the technologies used, it can increase it only by changing the variable inputs used to be able to change the output by increasing the amount of any input.

5.1.3 Market Failures

Neo-classical economics has become associated with a belief in the efficiency of markets. The microeconomic theory has also incorporated the criticisms and limitations of free-markets. The obstacles that prevent the price mechanism from the efficient allocation of resources are described as "market failures". A perfectly functioning market mechanism can be described by both the demand side and the supply side provide objective information on the market situation. The carrier of this information is the price that is created at the market. In the real economic world, there are plenty of obstacles of perfect competition that can cause market failures. The most important of these may be classified as the following:

- imperfect competition,
- externalities,
- public goods and
- imperfect information.

Imperfect competition and its features is described above in the Sect. 5.1.1.6. Competition.

Externalities occur when a firm or people impose costs (negative externalities) or benefits (positive externalities) on others outside the marketplace. Negative externalities occur when production and/or consumption impose external costs on third parties outside of the market for which no appropriate compensation is paid, for example:

- Harmful impact of toxic 'passive smoking' on non-smokers
- Air pollution from road use and traffic congestion and the impact of road fumes on lungs
- The external cost of food waste
- The external costs of cleaning up from litter and the dropping of chewing gum
- The external costs of the miles that food travels from producer to the final consumer

Public goods are commodities, which can be enjoyed by everyone and from which no one can be excluded. Public goods provide an example of market failure resulting from missing markets. Pure public goods are non-excludable and non-rival in consumption. Public goods are also known as collective consumption goods, for example, police, defence, crime control, sanitation infrastructure.

Regarding spatial view, it can be interesting to observe the spatial distribution of market failures in the economy. Good examples can be the distribution of monopolist or oligopolist power in EU countries or the whole World, the distribution of negative externalities and negative environmental impacts on a local, regional, national and global level or different placement of public goods within the society and particular countries/regions.

5.2 Macroeconomics and Relationships Between Variables

5.2.1 Macroeconomic Issues

5.2.1.1 General Macroeconomic Model

Macroeconomics studies the behaviour of the economy as a whole. The following Picture 5.3 shows relationships between particular markets and economic subjects – resource market, product market, households, companies and government.

A macroeconomic model is an analytical tool designed to describe the operation of the econ omy of a country or a region. These models are usually developed to examine the dynamics of aggregate quantities such as the total amount of goods and services produced, total income earned, the level of employment of productive resources, and the level of prices.

Macroeconomic models may be logical, mathematical or computational; the different types of macroeconomic models serve different purposes and have various advantages and disadvantages. Macroeconomic models may be used to clarify and illustrate basic theoretical principles; they may be used to test, compare, and quantify different macroeconomic theories; they may be used to produce "what if" scenarios (usually to predict the effects of changes in monetary, fiscal, or other macroeconomic policies); and they may be used to generate economic forecasts. Thus, macroeconomic models are widely used in academia, teaching and research, and are also widely used by international organisations, national governments and larger corporations, as well as by economics consultants and think tanks.

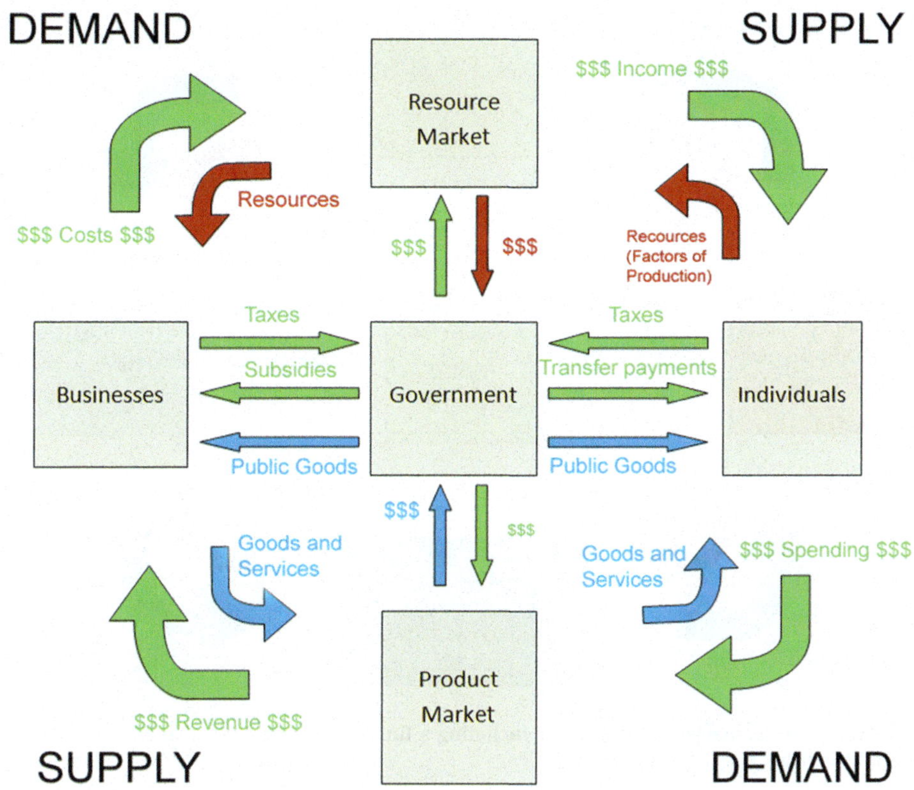

Picture 5.3 A general macroeconomic model. (Source: Own processing, based on Samuelson and Nordhaus 2010)

Households (individuals) behave rationally, and therefore they do not spend their entire income to purchase goods and services. The remaining part of their income they save as they assume that in the future their savings (S) will bring additional higher income.

Firms (businesses) set their business objectives and to meet them, they need additional funds to renew or expand their production, which means that they need to get a loan for purchasing the means of production. This leads to the need for creating a financial market (see Picture 5.4), where household savings are transformed into the investment resources of firms. The transformation of household savings into the investment resources of firms has two forms:

1. Households save their savings in financial institutions (primarily banks), which provide loans to businesses. The money market mediates the relationship between households and firms.
2. Households use their savings to purchase securities issued by firms and therefore directly, without the intermediary role of banks, provide firms with the necessary investment funds. In this case, the relationship between households and firms is mediated by the securities market.

Household spending on goods and services, i.e. consumption expenditure (C) is supplemented by investment spending (expenditure) of firms (I). The equality of the total income and the total product is then kept, and therefore the total (national) income and the total product are usually identified by the same symbol (letter) - Y (yield). The quantity of the total product is equal to the total expenditure (E), which was spent in the economy ($Y \equiv E$).

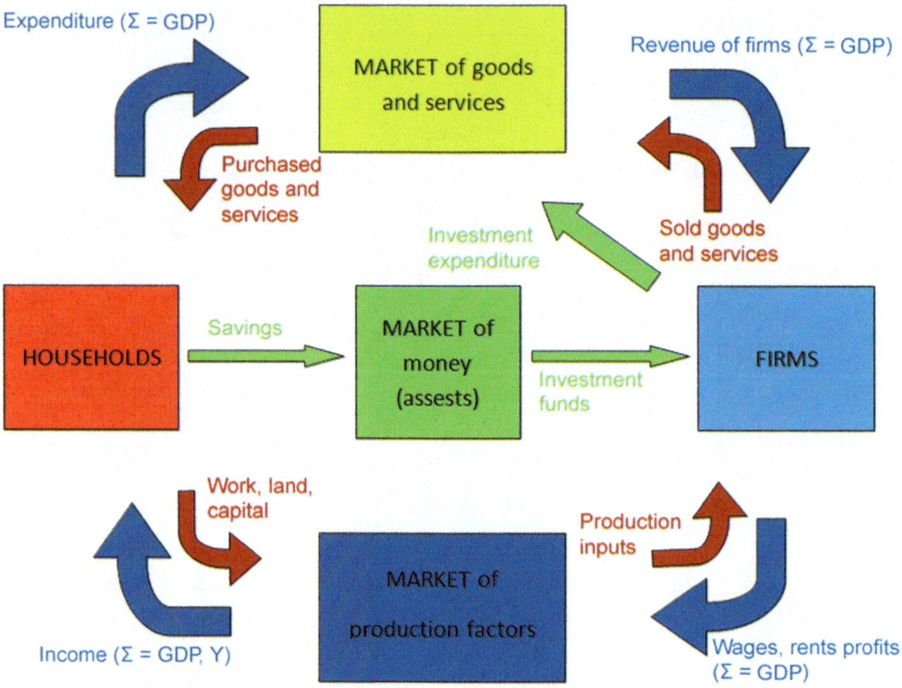

Picture 5.4 Model of the macroeconomic cycle including a financial market. (Source: Own processing, based on Samuelson and Nordhaus 2010)

5.2.1.2 Macroeconomic Indicators

Basic Macroeconomic Indicators and Their Characteristics

Gross domestic product (GDP) is the measure of the market value of all final goods and services produced in a country during a year. There are two ways to measure GDP - nominal GDP is measured in actual market prices, real GDP is calculated in constant or invariant prices.

Gross domestic product (GDP) is the most frequently used measure for the overall size of an economy, while derived indicators such as GDP per capita — for example, in euro or adjusted for differences in price levels — are widely used for a comparison of living standards, or to monitor the process of convergence across the European Union (EU). Moreover, the development of specific GDP components and related indicators, such as those for economic output, imports and exports, domestic (private and public) consumption or investments, as well as data on the distribution of income and savings, can give valuable insights into the main drivers of economic activity and thus be the basis for the design, monitoring and evaluation of specific policies.

Inflation occurs when the general price level of prices is rising. We calculate inflation by using price indexes - weighted averages of the prices of thousands of individual products.

Inflation is the increase in the general level of prices of goods and services in an economy; the reverse situation is deflation when the general level of prices falls. Inflation and deflation are usually measured by consumer price indices or retail price indices. Within the European Union (EU), a specific consumer price index has been developed — the harmonised index of consumer prices (HICP). Other factors (such as wages) being equal, inflation in an economy means that the purchasing power of consumers falls as they are no longer able to purchase the same amount of goods and services with the same amount of money. Purchasing power parities estimate price level differences between countries and can be used to calculate price level indices, which may, in turn, be used as a starting point for analysing price convergence between countries or regions.

The unemployment rate is defined as the number of people who are unemployed expressed in relation to the total labour force (persons who are employed or unemployed). Focusing on the example published by EUROSTAT (2018), we can compare development in particular countries and also in the EU as a whole. At the start of the financial and economic crisis in 2008, there were 16.8 million unemployed persons in the EU-28, which gave an unemployment rate of 7.0%. Five years later — in 2013 — this figure had risen to 26.3 million unemployed persons, an overall increase of 9.5 million. The number of unemployed persons in the EU-28 fell in both 2014 and 2015, to 22.9 million (ora rate of 9.4%). As such, the total number of people who were out of work in 2015 was more than one third (36.5%) higher than at the onset of the crisis, while the unemployment rate was 2.4 percentage points higher.

The foreign exchange rate is the price of one currency in terms of another currency. The foreign exchange rate is determined in the foreign exchange market, which is the market where different currencies are traded.

Current Data vs. Constant Data

Data reported in the current (or "nominal") prices for each year are expressed in the value of the currency for that particular year. For example, current price data shown for 1990 are based on 1990 prices, for 2000 are based on 2000 prices, and so on.

Other series in statistics can show data in "constant" or "real" terms. Constant series show the data for each year expressed in the value of a particular base year. Thus, for example, data reported in constant 2010 prices show data for 1990, 2000, and all other years in 2010 prices.

Current series is influenced by the effect of price inflation. Data reported in "constant" or "real" terms (constant series) are used to measure the true growth of a series, i.e. adjusting for the effects of price inflation. For example (using year one as the base year), suppose nominal Gross Domestic Product (GDP) rises from 100 billion to 110 billion, and inflation is about 4%. In real prices, the second year GDP would be approximately 106 billion, reflecting its true growth of 6%. Except for rare instances of deflation (i.e. negative inflation), a country's current price series on a local currency basis will be higher than its constant price series in the years succeeding the constant price base year. (World Bank 2018).

Table 5.3 shows the example of official source of macroeconomic data published by the Ministry of Finance of the Czech Republic in 2018.

Economic Development – Different Possibilities and Expressions

A gross domestic product, GDP, is a basic measure of the overall size of a country's economy.

As an aggregate measure of production, GDP is equal to

1. the sum of the gross value added of all resident institutional units engaged in production, plus any taxes on products and minus any subsidies on products. Gross value added is the difference between output and intermediate consumption,
2. the sum of the final uses of goods and services (all uses except intermediate consumption) measured in purchasers' prices, minus the value of imports of goods and services,
3. the sum of primary incomes distributed by resident producer units.

We can use different possibilities of GDP development presentation. It is possible to show GDP growth in a selected period as a % change compared with previous years or as GDP development in current or constant prices. The examples, published by EUROSTAT in the year 2018, show different GDP development presentation in China, USA, EU-28, Euro area and Japan in the period 2007–2017 (Picture 5.5 and Picture 5.6).

The Eurostat (2018) comments the GDP development in the following way: "The global financial and economic crisis resulted in a severe recession

Table 5.3 Main macroeconomic indicators in the Czech Republic

		2013	2014	2015	2016	2017	2018	2019	2018	2019
							Current forecast		*Previous forecast*	
Gross domestic product	*bill. CZK*	**4098**	**4314**	**4596**	**4768**	**5045**	**5296**	**5568**	5300	5589
Gross domestic product	*real growth in %*	**−0.5**	**2.7**	**5.3**	**2.5**	**4.3**	**3.0**	**2.9**	3.2	3.1
Consumption of households	*real growth in %*	**0.5**	**1.8**	**3.7**	**3.6**	**4.3**	**3.6**	**3.6**	4.3	3.9
Consumption of government	*real growth in %*	**2.5**	**1.1**	**1.9**	**2.7**	**1.3**	**2.5**	**2.1**	2.1	2.0
Gross fixed capital formation	*real growth in %*	**−2.5**	**3.9**	**10.2**	**−3.1**	**3.3**	**7.8**	**3.2**	7.5	3.2
Net exports	*contr. to GDP growth, pp*	**0.1**	**−0.5**	**−0.2**	**1.4**	**1.1**	**−0.4**	**0.0**	−0.7	0.1
Change in inventories	*contr. to GDP growth, pp*	**−0.7**	**1.1**	**0.8**	**−0.4**	**0.1**	**−0.8**	**0.0**	−0.4	0.0
GDP deflator	*growth in %*	**1.4**	**2.5**	**1.2**	**1.3**	**1.5**	**1.9**	**2.2**	1.8	2.3
Average inflation rate	*%*	**1.4**	**0.4**	**0.3**	**0.7**	**2.5**	**2.2**	**2.3**	2.2	2.3
Employment	*growth in %*	**1.0**	**0.8**	**1.4**	**1.9**	**1.6**	**1.3**	**0.2**	1.3	0.2
Unemployment rate	*average in %*	**7.0**	**6.1**	**5.1**	**4.0**	**2.9**	**2.3**	**2.3**	2.3	2.3
Wage bill (domestic concept)	*growth in %*	**0.5**	**3.6**	**4.8**	**5.7**	**8.2**	**9.5**	**8.4**	9.3	8.3
Current account balance	*% of GDP*	**−0.5**	**0.2**	**0.2**	**1.6**	**1.1**	**0.4**	**0.4**	−0.3	−0.2
General government balance	*% of GDP*	**−1.2**	**−2.1**	**−0.6**	**0.7**	**1.5**	**1.6**	**1.0**	1.6	.
Exchange rate CZK/EUR		**26.0**	**27.5**	**27.3**	**27.0**	**26.3**	**25.5**	**24.9**	25.6	25.2
Long-term interest rates	*% p.a.*	**2.2**	**1.4**	**0.6**	**0.4**	**1.0**	**2.1**	**2.6**	2.1	2.5
Crude oil Brent	*USD/barrel*	**109**	**99**	**52**	**44**	**54**	**74**	**77**	73	72
GDP in Eurozone	*real growth in %*	**−0.2**	**1.4**	**2.1**	**1.9**	**2.4**	**2.0**	**1.7**	2.1	1.8

Source: Ministry of Finance of the CR (2018)

in the EU in 2009 (see Picture 5.5), followed by a recovery in 2010. The crisis started earlier in Japan and the United States, with negative annual rates of change for GDP (in real terms) already recorded in 2008, deepening in 2009, before rebounding in 2010. By contrast, economic output in China continued to grow at a relatively rapid pace during the crisis (close to 10% each year), slowing somewhat in subsequent years, but remaining considerably higher than in any of the other economies shown in Picture 5.5."

Cross-country comparisons are often made using purchasing power standards (PPS) which adjust values to account for differences in price levels between countries. The data shown in Picture 5.6 are in current prices and should not be used for comparisons over time because of inflation and exchange rate fluctuations.

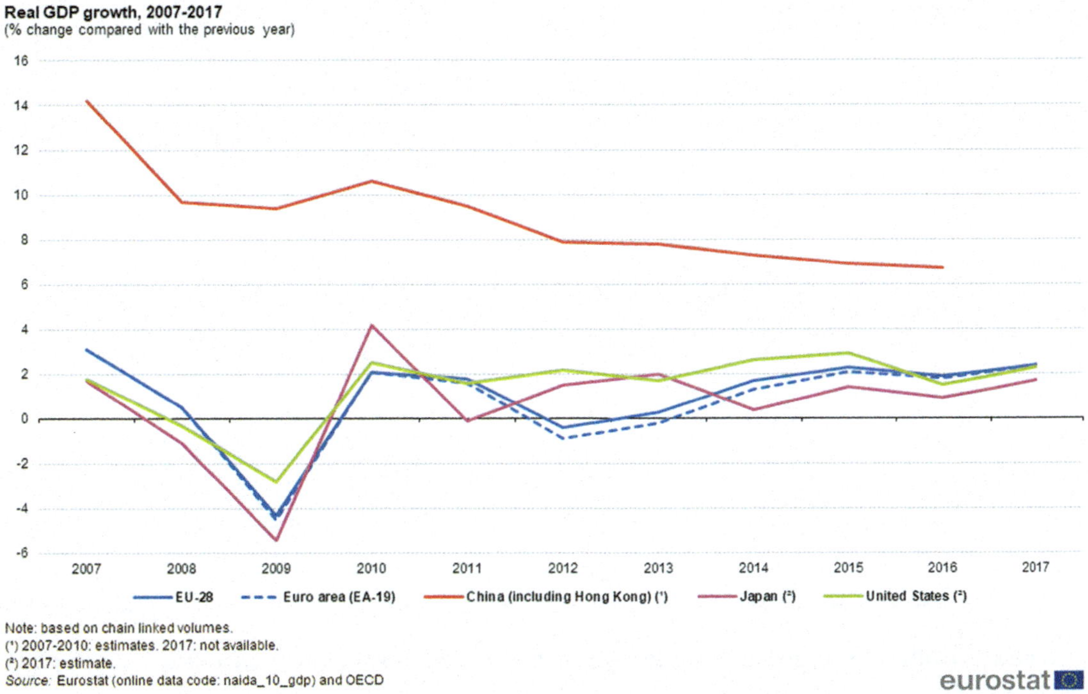

Picture 5.5 **Real GDP growth, 2007–2017 (% change compared with previous year)**. (Source: Eurostat 2018)

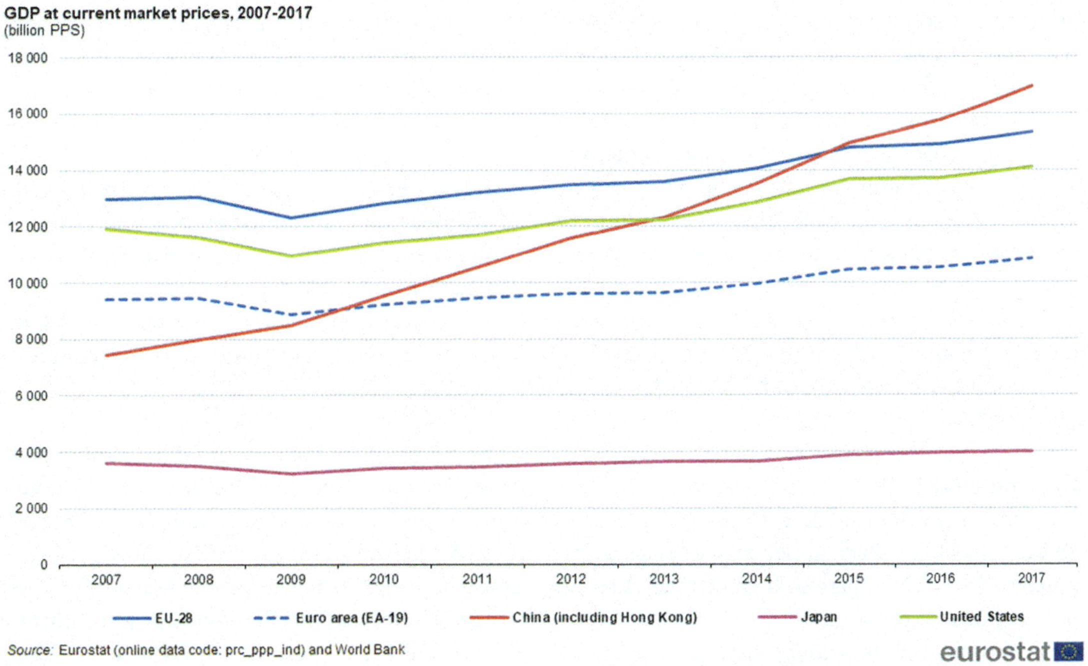

Picture 5.6 **GDP development in current market prices, 2007–2017 (billion PPS – purchasing power standard)**. (Source: Eurostat 2018)

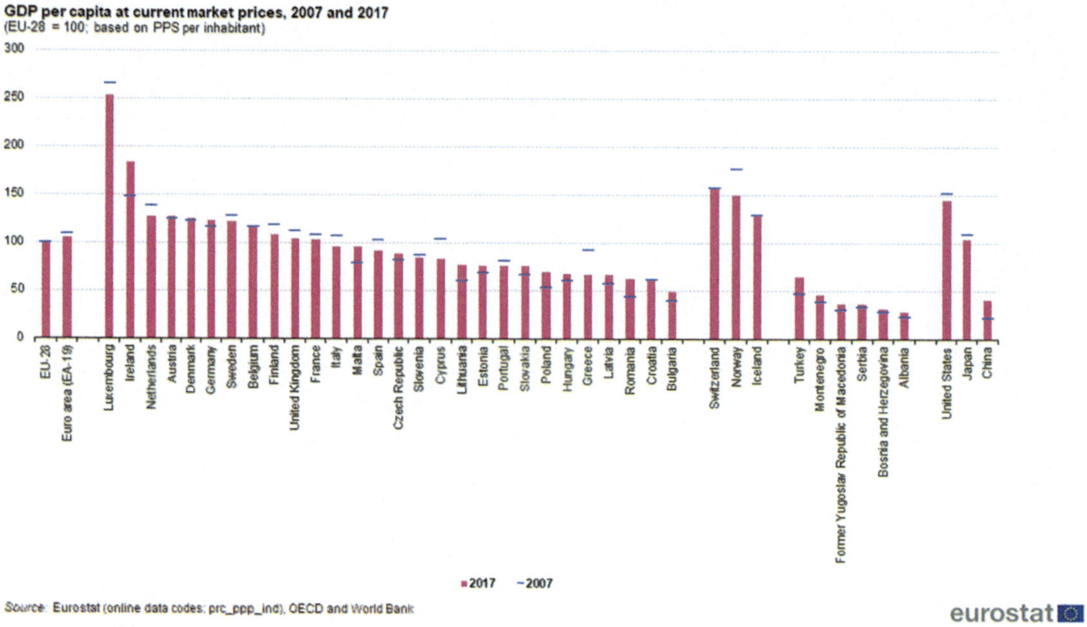

Picture 5.7 **GDP per capita, current prices 2007 a 2017 (EU-28 = 100; base PPS per capita)**. (Source: Eurostat 2018)

GDP per Capita

There are also other possibilities for GDP presentation. It is suitable to show GDP per capita, for regional comparison. Inequalities that exist between different regions can be attributed to a wide range of factors, including changes brought about by globalisation (such as the relocation and outsourcing of manufacturing and service activities), the legacy of former economic systems, socioeconomic developments, geographic remoteness, and the availability of resources, including human resources.

Picture 5.7 presents the differences between GDP per capita in the years 2007 and 2017 in EU countries, published by EUROSTAT in the year 2018.

Economic Growth

Economic growth is an increase in the capacity of an economy to produce goods and services, compared in the selected period. It can be measured in nominal or real prices.

Economic growth sources are represented by:

- Increase in the number of production factors,
- Increase of total productivity of production factors.

Extensive economic growth can be observed in the case that inputs are increasing faster than outputs. Intensive economic growth occurs when outputs are increasing faster than inputs.

Economic growth indicators are the following:

- Economic power indicator, described by the absolute amount of final products and services in the economy. It is presented as GDP in mil. EUR or other selected currency.
- Economic level indicator, represented by the efficiency of production. It is presented as GDP per capita.

Similarly, as in the case of GDP development, there are different possibilities of economic growth presentation. The examples, published by EUROSTAT in the year 2018, show economic power indicator (Picture 5.8) and economic level indicator (Picture 5.8). Economic power indicator (Picture 5.8) is expressed as GDP per inhabitant in PPS in relation to the EU-28 average in the year 2015, focusing on NUTS 2 regions. Economic level indicator (Picture 5.9) is expressed as a

Gross domestic product (GDP) per inhabitant, by NUTS 2 regions, 2016
(based on data in purchasing power standards (PPS) in relation to the EU-28 average, EU-28 = 100)

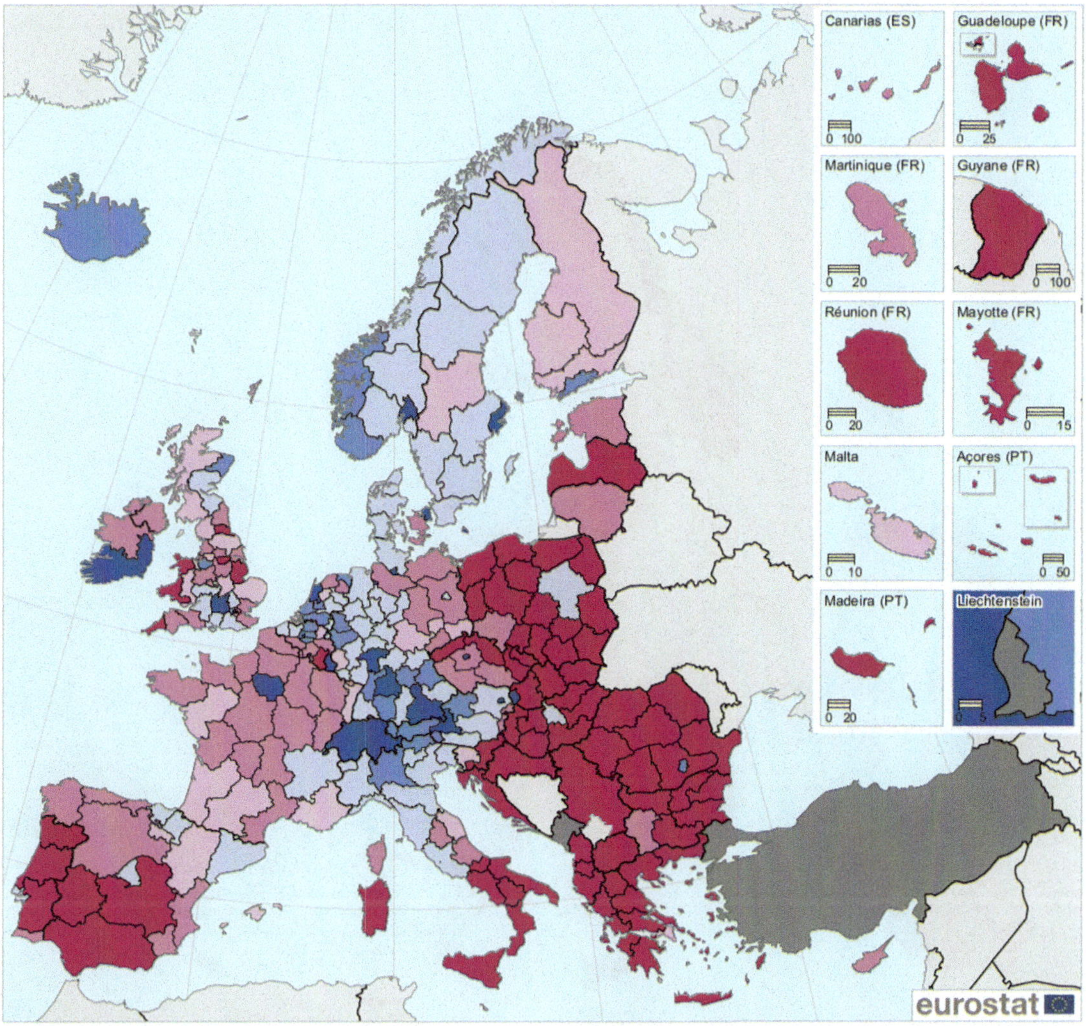

(based on data in purchasing power standards (PPS) in relation to the EU-28 average, EU-28 = 100)

EU-28 = 100

- **< 75**
- **75 - < 90**
- **90 - < 100**
- **100 - < 125**
- **125 - < 150**
- **≥ 150**
- **Data not available**

Administrative boundaries: © EuroGeographics © UN-FAO © INSTAT © Turkstat
Cartography: Eurostat — GISCO, 04/2018

0 200 400 600 800 km

Note: Norway, 2015. Former Yugoslav Republic of Macedonia and Albania: 2014. Switzerland and Serbia: national data. Switzerland: provisional.
Source: Eurostat (online data codes: nama_10r_2gdp, nama_10_gdp, nama_10r_3popgdp and nama_10_pe)

Picture 5.8 Economic power indicator. (Source: Eurostat 2018)

Change of gross domestic product (GDP) per inhabitant, by NUTS 2 regions, 2007-2016
(percentage points difference for 2016 minus 2007; based on data in purchasing power standards (PPS) in relation to the EU-28 average, EU-28 = 100)

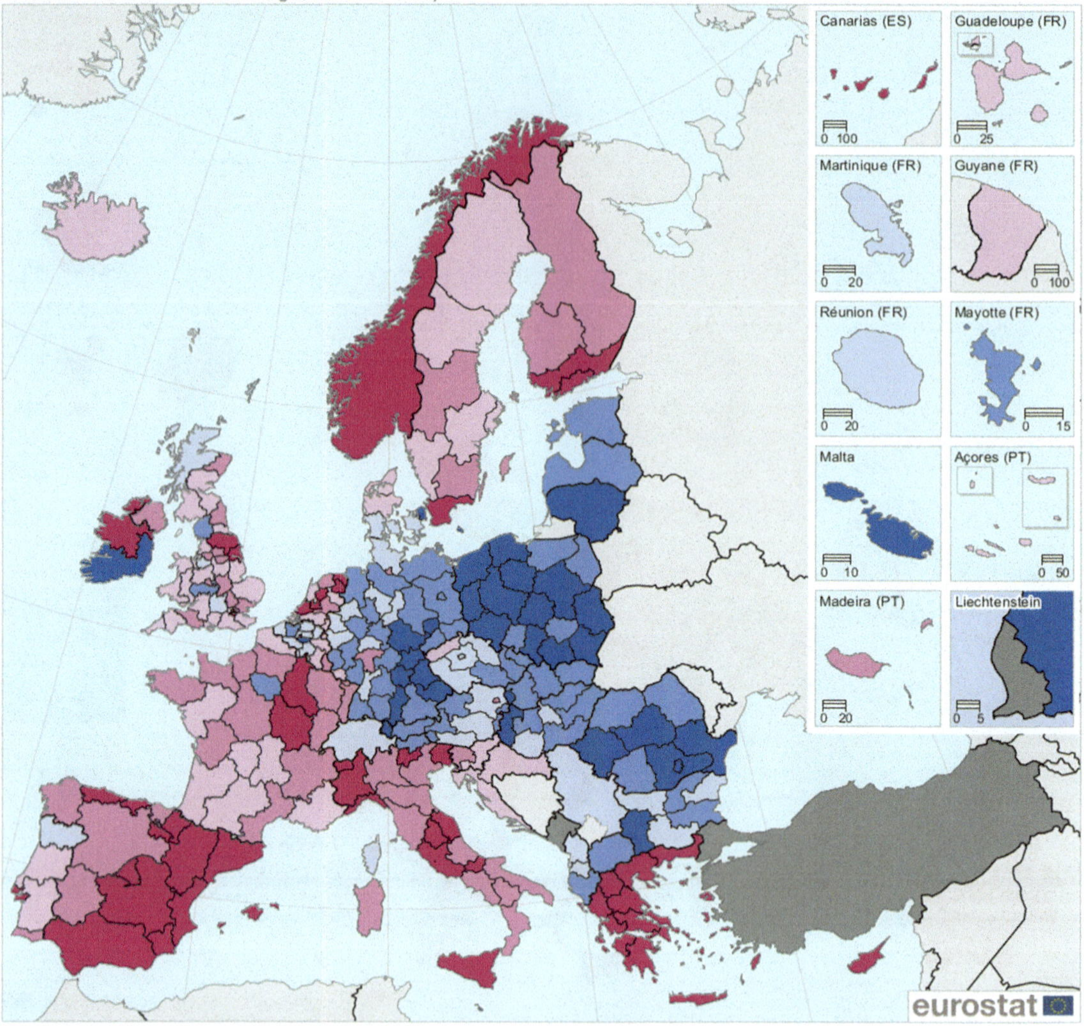

(percentage points difference for 2016 minus 2007; based on data in purchasing power standards (PPS) in relation to the EU-28 average, EU-28 = 100)

EU-28 = 0

- < -10
- -10 - < -5
- -5 - < 0
- 0 - < 5
- 5 - < 10
- ≥ 10
- Data not available

Administrative boundaries: © EuroGeographics © UN-FAO © INSTAT © Turkstat
Cartography: Eurostat — GISCO, 04/2018

0 200 400 600 800 km

Note: Albania, 2008-2014. Norway, Switzerland and Serbia: national data. Switzerland: provisional. Bulgaria, Germany and Romania: break in series.

Source: Eurostat (online data codes: nama_10r_2gdp, nama_10_gdp, nama_10r_3popgdp and nama_10_pe)

Picture 5.9 Economic level indicator. (Source: Eurostat 2018) change of GDP per inhabitant in PPS in relation to the EU-28 average in the period 2007–2015, also dealing with NUTS 2 regions.

The "poorest" regions in the EU, with GDP per capita less than 75% of the EU-28 average are shown in the darkest shade of purple in Picture 5.7. On the other hand, the darkest shade of blue colour represents the "richest" regions in the EU.

Picture 5.9 shows changes in regional GDP per inhabitant relative to the EU-28 average for 2007–2016; the comparison covers the period associated with the global financial and economic crisis which has had a lasting impact on several regions. Among the multi-regional EU Member States, GDP per capita grew at a faster pace than the EU-28 average in every region of Bulgaria, Hungary, Poland, Romania, Slovakia and all three of the Baltic Member States (each of which is a single region at this level of detail), as well as every region except one in Austria and the Czech Republic. The majority of regions in Germany also recorded an increase in their relative living standards. By contrast, average GDP per capita grew at a slower pace than the EU-28 average in every region of Greece, Spain, Croatia, Italy, the Netherlands, Slovenia, Finland and Sweden, while a similar pattern was repeated in all but one region of mainland France and Portugal (Eurostat 2018).

5.2.1.3 State Budget Indicators and Public Finance

We can identify three essential instruments or tools that the government uses to influence economic activity:

1. Taxes on incomes, goods and services. These reduce private income, and consequently decrease private expenditures (on automobiles or restaurant food) and provide resources for public expenditures (on education and healthcare). The tax system also serves to discourage certain activities by taxing them more heavily (such as smoking cigarettes) while encouraging other activities by taxing them lightly or even subsidising them (such as environmental protection).
2. Expenditures on specific goods or services (such as roads, education, or police protection), along with transfer payments (like social security and contributions) that provide resources to individuals.
3. Regulations or controls that direct people to perform or refrain from certain economic activities. Examples include rules that limit the amount firms can pollute, or that divide up the radio spectrum, or that mandate testing the safety of new drugs.

Fiscal policy regulates the use of taxes and government expenditures. Government expenditures come in two distinct forms. First, there are government purchases. These comprise spending on goods and services—purchases of tanks, construction of roads, salaries for judges, and so forth. Also, there are government transfer payments, which increase the incomes of targeted groups such as the elderly or the unemployed. Government spending determines the relative size of the public and private sectors, that is, how much of GDP is consumed collectively rather than privately. From a macroeconomic perspective, government expenditures also affect the overall level of spending in the economy and thereby influence the level of GDP.

The other part of fiscal policy, taxation, affects the overall economy in two ways. Firstly, taxes affect people's incomes. By leaving households with more or less disposable or spendable income, taxes affect the amount people spend on goods and services as well as the amount of private saving. Private consumption and saving have essential effects on investment and output in the short and long run. Also, taxes affect the prices of goods and factors of production and thereby affect incentives and behaviour. Many provisions of the tax code have an important impact on economic activity through their effect on the incentives to work and to save.

5.2.1.4 Monetary Indicators and Monetary Policy Instruments

Regarding monetary indicators and monetary policy instruments, national banks have more possibilities. The following overview of monetary policy instruments is based on the instruments currently used by the Czech National Bank (find more in CNB 2018).

Open Market Operations
Open market operations are used for steering interest rates in the economy. Open market operations are mostly executed in the form of repo operations (based on a general agreement on trading on the financial market). Concerning their aim and regularity, the central banks open market operations can be divided into the following categories:

The main monetary policy instrument takes the form of repo tenders. The central bank accepts surplus liquidity from banks and in return transfers eligible securities to them as collateral. The two parties agree to reverse the transaction at a future point in time when the central bank as borrower repays the principal of the loan plus interest and the creditor bank returns the collateral to the CNB. The first duration of these operations is 14 days; the two-week repo rate (2 W repo rate) is therefore considered to be crucial importance regarding monetary policy.

The additional monetary instrument is the three-month repo tender. Here, the central bank accepts liquidity for a three-month period. Fine-tuning tools (foreign exchange operations and securities operations) are used ad hoc, mainly to smooth the effects on interest rates caused by unexpected liquidity fluctuations in the market. These instruments are rarely used.

Automatic Facilities
Automatic facilities are used for providing and depositing liquidity overnight. As, from the banks' point of view, these represent standing facilities for collecting or borrowing money, the interest rates applied to them form the corridor for short-term money market rates (as well as for the two-week repo rate).

Minimum Reserves
In general, the minimum reserves are generally one of the main monetary policy instruments through which the central bank can influence the amount of liquidity (free funds) in the banking system.

The application of the reserve requirement in practice involves several areas (obliged entities, the reserve requirement rate, maintenance periods, the reserve base, fulfilment of the reserve requirement, remuneration, the reserve requirement where statements are not submitted, etc.) whose individual parameters can change flexibly, reflecting the need to react to changes in trend in the banking system.

FX Interventions
FX interventions are purchases or sales of foreign currencies against selected basic currency on the foreign exchange market. They are aimed at dampening foreign exchange market volatility and easing/tightening monetary policy. FX interventions are not a regularly used instrument in the inflation targeting regime. The standard instrument is interest rates.

Nevertheless, FX interventions may be used under certain circumstances. An example of such a situation is a reduction in monetary policy interest rates to "technical zero", where further monetary policy easing can be achieved by weakening the selected basic currency exchange rate. The CNB faced this situation between autumn 2013 and spring 2017 when it used an exchange rate commitment to intervene on the foreign exchange market if necessary to weaken the Czech koruna to maintain the exchange rate close to CZK 27 to the euro.

5.2.1.5 Aggregate Demand and Supply

Aggregate demand refers to the total amount that different sectors in the economy willingly spend in a given period.

An aggregate demand curve is the sum of individual demand curves for different sectors of the economy. The aggregate demand is usually described as a linear sum of four separable demand sources:

$$\mathbf{AD} = \mathbf{C} + \mathbf{I} + \mathbf{G} + (\mathbf{X} - \mathbf{M}) \tag{5.7}$$

where C is consumption, I is investments, G is government spending, NX = X (export) - M (import) is net export.

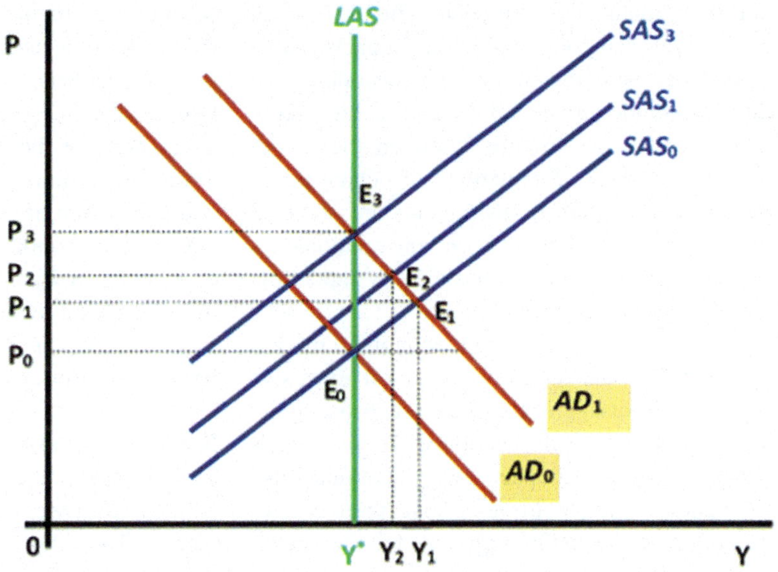

Fig. 5.5 **Aggregate supply and demand model**. (Source: Own processing, based on Samuelson and Nordhaus 2010; *AD* (AD_0, AD_1) aggregate demand, *SAS* (SAS_0, SAS_1, SAS_3) short-time aggregate supply, *LAS* long-term aggregate supply, *Y* product, $Y*$ potential product, *P* price level, *E* (E_0, E_1, E_2, E_3) equilibrium)

Aggregate supply refers to the total quantity of goods and services that the nation's businesses willingly produce and sell in a given period.

In the long run, the ability of an economy to produce goods and services to meet demand is based on the state of production technology and the availability and quality of factor inputs.

The AD-AS or aggregate demand-aggregate supply model is a macroeconomic model that explains price level and output through the relationship of aggregate demand and aggregate supply. It is based on the theory of John Maynard Keynes presented in his work The General Theory of Employment, Interest, and Money. It is one of the first simplified representations in the modern field of macroeconomics and is used by a broad array of economists, from libertarian, Monetarist supporters of laissez-faire, such as Milton Friedman, to Post-Keynesian supporters of economic interventionism, such as Joan Robinson.

The conventional "aggregate supply and demand" model is a Keynesian visualisation that has come to be a widely accepted image of the theory (see Fig. 5.5). The classical supply and demand model, which is mostly based on Say's law (supply creates its demand) depicts the aggregate supply curve as being vertical at all times (not just in the long-run).

5.3 Economic Modelling

5.3.1 Overview of Applied Economic Models

5.3.1.1 Possibilities of Economic Models

At present, many studies are dealing with predictions and modelling of the impacts of economic policy instruments. In the given area, ex-ante and ex-post analysis of the government policies impacts on the economy are discussed. We can find analysis of the government policies impacts on the economy as a whole, as well as on households, firms, sectors of the economy, the environment, macroeconomic variables, individual markets or mutual links between sectors and markets. For analysing the economic policy impacts, a wide range of both quantitative and qualitative methods has been used so far. Since appropriate tools are rarely introduced separately, the final selection of the method used depends on the required outputs.

Abstract models are the basis of understanding real systems and are used to study and predict their behaviour in a wide range of tasks of both the system analysis and system synthesis. The most widespread types of abstract models are numerical mathematical models. The analyses of particular policies and regulations impacts can be carried out with the use of macroeconomic or microeconomic models (Bork 2006). Macroeconomic models are usually based on the use of aggregated data. Simulation of changes in policies and regulatory instruments are obtained in particular by modelling economic relationships between different sectors and also by modelling their behaviour changes

On the other hand, microeconomic models (Lund 2007) are typically based on extensive data files of disaggregated data, such as family accounts or corporate accounts. These models simulate the tax impact on individual units (households or companies), and their microeconomic results can be aggregated at the macroeconomic level to estimate the effects in the context of the entire economy. The creation of a microeconomic model depends on the availability of the necessary disaggregated data (Bach et al. 2002; Bork 2006).

When modelling the macroeconomic impacts, it is necessary to distinguish the time horizon of the model. When modelling the short-term consequences, it cannot be expected that the adjustment mechanisms in the economy began to take effect (Sahlin et al. 2007). If the aim is to evaluate the short-term impacts, it is possible to ignore the medium-term adjustment mechanisms of the primary state variables (stock capital, the external position of the economy, government debt). For modelling the medium-term impacts (in 2–5 years), it is necessary, by the economic theory, to make assumptions about the effects of the dynamic adjustment mechanisms. As an example, these mechanisms can be used to specify changes in relative prices of domestic and foreign goods in the context of a cost-push shock.

Furthermore, the parameters of these mechanisms (for example, the price elasticity of exports, the possibilities of substitution in the production function, etc.) should be calibrated (i.e. numerically quantified). Economists are not consistent in this area; individual designs of macroeconomic models differ from one another. Currently, the standard macroeconomic models are for example the dynamic models of general equilibrium. In the general equilibrium model, the economy is interpreted as a system of interdependent markets. A change, which at first sight affects only one market, may, in practice, affect all markets in the economic system. The long-term modelling focuses mainly on the endogenisation of mechanisms which are in standard macroeconomic modelling considered as exogenous. It is, in particular, modelling of the induced technological progress. Today, the standard tool for such modelling is the so-called integrated assessment models (IAM).

5.3.1.2 Agent-Based Modelling

The optimal behaviour of particular economic entities is multicriterial and dependent on many factors. The rules of their behaviour are different depending on their individuality or on their integration into a superior economic structure.

The behaviour of economic entities is a favourite issue of particular economic, statistical and econometric models and simulations. From the time period view, there are short, medium or long-term models, both static and dynamic. Conventional econometric and statistical models fail to capture cooperation and decision-making of economic entities under conditions of a large number of businesses that try to optimise their economic situation. Conventional models cannot distinguish the differences between individual decision-making and cooperative decision-making within a superior economic structure. The real behaviour and decision making of particular economic entities can be different in the situation with or without interactions with other entities – in other words, the rules within a group of economic entities can be different than individual entity rules.

The approach, which also includes interaction rules, is called ABM - agent-based modelling. Applying agent-based modelling, the researcher explicitly describes the decision making processes of particular actors at a micro level. The structure emerges at the macro level as a result of the actions of the agents and their interactions with each other (Janssen and Ostrom 2006).

The modelling based on the agent-based modelling or complex multiagent modelling has been historically used mainly in the field of engineering and information sciences; however, the importance of this kind of models has been rapidly increasing in the economic sciences and management. It is visible mainly in the area of financial markets management, corporate management, water management, waste management, land management, transportation and energy sources management.

A simple multiagent economic system can be based on the basic economic entity – the broker – with the only goal – the profit. Therefore we should mention scientific studies based on financial markets - for example agent-based model with multi-level herding for complex financial systems (Chen et al. 2015), consentaneous agent-based and stochastic model of financial markets (Gontis and Kononovicius 2014), agent-based double auction markets (Cai et al. 2014) and synthesis of agent-based financial markets and New Keynesian macroeconomics (Lengnick and Wohltmann 2013).

Regarding interesting studies in management, there should be mentioned mainly multi-agent systems for the simulation of land-use and land-cover change (Parker et al. 2003), ecosystem management (Bousquet and Le Page 2004), urban traffic management and planning (Fiosins et al. 2011) or energy management (Lagorse et al. 2010).

There are also studies focused directly on multiagent models connected with climate change or carbon emissions reduction, for example, the study focused on estimating the impacts of climate change policy on land use (Morgan and Daigneault 2015) and exploration of the carbon emissions trading scheme in China (Tang et al. 2015).

Companies can mutually cooperate and coordinate their behaviour and their targets in the area of carbon emission reduction and costs optimisation. After selection of representation and formalisation of the agent's behaviour and relationships between particular agents, it is necessary to design new methods for investigating their behaviour. It is possible to study not only the behaviour of the individual agents, but also the behaviour of the groups of agents, or even of the whole economic system. It will allow us to identify cooperation, coordination and joint actions. It also offers the use of cognitive analysis within the multi-agent system. A multi-agent system (MAS) can be defined as a group of connected autonomous systems (agents), cooperating to achieve a common goal – in our case the maximal decrease of carbon emissions with the minimal costs. Activities of particular agents are based on the principles of cooperation, coordination and optimisation of their behaviour to achieve both individual and global goals. Particular agents can influence their behaviour to each other; their activities may be carried out by the others or merely avoid interfering with the work of the others. However, they may also act against others. For MAS creating, there are available the following types of agents: cooperative agents - they have common goals; competitive agents – they have conflicting goals, and agents displacing each other.

For multiagent model creation, particular agents must be placed into the structure of the multiagent system, which is also associated with a precise definition of mutual relations and hierarchies. Defined links will also serve as the communication channels for sending the messages - the type of ACL.

Both adaptation and learning can be designed for the following three levels: the level of the agent, the level of groups of agents and the global level.

A multi-agent system (MAS) can be defined as a group of connected autonomous systems (agents), cooperating to achieve a common goal. Activities of particular agents are based on the principles of cooperation, coordination and optimisation of their behaviour to achieve both individual and global objectives. Particular agents can influence their behaviour to each other; their activities may be carried out by the others or simply avoid interfering with the work of the others. However, they may also act against others. For MAS creating, there are available the following types of agents: cooperative agents - they have common goals, competitive agents – they have conflicting goals and collaborative agents - displacing each other.

MAS will be performed in the specific conditions of economic systems. From this perspective, the important characteristics of particular agents can be examined, precisely their autonomy (the ability to achieve their objectives without outside interference, ie. their interaction with the environment), reactivity (the ability to continuously respond to changes in the environment), intentionality (the ability to think about their long-term objectives) and social intelligence (the ability to communicate with other agents). Particular attention will be focused on deliberative agents (intentional and intelligent), whose behaviour is approaching the autonomous behaviour of humans. This kind of agent has its symbolic representation of the external environment, which is implemented in the form of knowledge base assertions about the world and enables him to realize rational and purposeful behaviour.

The multi-agent system is based on the following procedures:

1. Communication - the process, which is a crucial base for cooperation and coordination. The communication is a process during which the two or more agents exchange information in the form of simple communication messages in the appropriate language.
2. Cooperation – the process, when particular agents negotiate to solve their problems or conflicts. Precisely, it is a process of achieving goals. The management of negotiating agents can be centralised or decentralised. The role of each agent within the MAS has to be precisely defined.
3. Coordination – the process, when the limited resources and tasks are divided to particular agents in such way, that the whole behaviour of the system is both effective and economical and the system optimises the way for the optimal output achieving. This communication process results in a rational way of the behaviour of the whole system.
4. Emergence - the process of spontaneous formation of macroscopic characteristics and structures of complex systems, which cannot be derived from the characteristics of their components. Associations of particular agents and their mutual interactions can lead to emergent behaviour, which can positively affect the behaviour during the evolution of the system.

The solution of cooperation and coordination procedures will use specific characteristics of particular agents, such as the possibility of their complex, nonlinear or discontinuous behaviour and interactions, in which the role is represented by the spatial and social structure. It is obvious that it is not possible to predict precisely the future on the base of historical development, due to external disturbances and different levels of information available to particular agents. Therefore each agent uses learning ability and both behaviour adaptations and modifications. An essential aspect of the solution is represented by the purposeful optimisation of individual agents' decision-making in conditions of uncertainty, unexpected changes and disturbances. Particular agents will be based on the expert models, which will consist of the set of rules of behaviour.

5.3.1.3 Input - Output Models

The adjustment mechanisms in the economy cannot be expected during modelling of short-term impacts. Focusing on the main objective of the analysis (short-term impacts), it is possible to ignore the medium-term adjustment mechanisms of the following variables: capital stock, external position of the economy, government debt. In this case, a so-called structural analysis also referred to as the input-output table analysis, is a useful tool. This analysis makes it possible to predict short-term measures of economic policy or exogenous shocks to individual sectors and individual types of households in the economy. Fundamentals of structural analysis were formulated in 1936 by Nobel Prize economist Vasili Leontief. He was inspired by two theoretical bases: the neoclassical theory of national economic equilibrium and the principles of compiling national economic balances. Theoretical conclusions have been demonstrated by a detailed analysis of the US economy in the years 1919, 1929 and 1939 (Leontief 1966).

The structural analysis represents a set of models and methods that serve to find an equilibrium solution for a particular economic system. An equilibrium solution means a solution that presupposes the balance of resources and the needs of this system. Under the economic system, we can understand the national economy and particular interconnected sectors of the national economy.

The input-output method is an adaptation of the neoclassical theory of general equilibrium to an empirical study of quantitative interdependence between different economic sectors. Initially, this method was developed to analyse the relationships between production and consumer sectors within the national economy but was also applied to the study of smaller economic systems such as a region or even for a private enterprise. The analysis is also used to identify international economic relationships (Leontief 1966).

Regardless the purpose of input-output analysis, the concept is essentially the same - a series of linear equations are describing the dependence between the different sectors of the national economy and then their specific structural characteristics are reflected in the numerical expression of the coefficients to these equations. Coefficients must be determined empirically. In the analysis of the structural characteristics of the national economy, they are usually determined from statistical input-output tables.

This kind of analysis enables to predict the short-term impacts of economic policy measures, or exogenous shocks, on individual sectors of the economy. Input-output tables are suitable for the description of the relationships between particular sectors of the national economy. It makes it possible to analyse not only the direct impacts on these sectors (for example, the impact of new taxation) but also the "second order" effects between the different sectors.

5.3.1.4 General Equilibrium Models - CGE

To model medium-term impacts (2 to 5 years), it is necessary to make assumptions about the effects of dynamic adaptation mechanisms, based on economic theory. Dynamic models of general equilibrium represent standard tools, which assume that economic entities can maximise their profit in the form of decisions about investment activities (companies) or consumption (households). This assumption sets the dynamics of the model. Within the CGE model, the economic system is interpreted as a system of interdependent markets. A change that primarily affects only one market can influence in longer time period all markets in the given economic system (Kriström, OECD 2006).

Computable General Equilibrium models (CGE models) are often used for empirical work. These models are used for several reasons. For example, these models can be used to evaluate who is the "winner" and who is the "loser" in the case of tax reform. The model makes it possible to clarify the link between different tax bases. If fossil fuels are the subject of energy tax and carbon tax, raising carbon tax will reduce energy tax revenues. Changing carbon tax may also affect VAT and other tax revenues. All these tax interactions can be easily influenced by the CGE model (Kriström, OECD 2006).

CGE models simulate the behaviour and mutual interactions of individual economic entities operating in different markets. These models are based on neoclassical microeconomic assumptions, primarily on the premise of rational (optimal) behaviour of economic actors. The data source of these models is the Social Accounting Matrix (SAM). In a very simplified way, SAM can be described as a table aggregating flows of goods, services and money in the economy in a given period of time, usually one year.

The main areas of application of CGE models are the analyses of the impacts of significant changes in environmental, tax and foreign trade policies. Because the economy is composed of interconnected markets, a general equilibrium perspective is needed to determine the distributional impacts of economic policy (Kristrom, OECD 2006).

Some uncertainties are also associated with the CGE model. For example, models are based on demand and supply curves, whose gradients cannot be determined with 100% confidence. Regarding ex-ante models, it is not possible to determine how consumers and companies will finally react to the changes in economic policy. If the supply and demand in the model are identified incorrectly, this error may expand within the model and will further complicate predictions about the levels of particular variables. Also, CGE models rarely include a complete description of the tax system, which is a disadvantage in many cases, for example, in interpreting the regression or progressivity of tax reforms (Kristrom, OECD 2006).

5.3.1.5 Long-Term Models

Long-term modelling focuses mainly on the endogenisation of mechanisms that are taken as exogenous in standard macroeconomic modelling. It can be in particular the modelling of induced technological development. The standard tool for such modelling is Integrated Assesment Models (IAM).

W. Nordhaus created one of the best-known IAM models (Nordhaus 2017). It is a model of climate change DICE (Dynamic Integrated Model of Climate and Economy) and a slightly modified DICE-99 model. It is a highly aggregated global model combining global production, energy consumption and the climate sector (producing carbon dioxide emissions). The DICE model builds on a more detailed RICE model (Regional Dynamic Integrated Model of Climate and the Economy) model, also created by W. Nordhaus, but this model is disaggregated and covers eight regions. Technological change in the DICE model is expressed as the development of technologies involving climate change (reducing carbon dioxide emissions as greenhouse gas per unit of production). Therefore it is involved as an exogenous factor. The economic damage caused by greenhouse gas emissions is modelled as a variable that is dependent on the global average temperature increase due to CO_2 emissions produced by industry.

Standard macroeconomic models focus mainly on the description of the main medium-term dynamic links within the economy (payment balance equilibrium, capital investments development, public debt dynamics of public finances); however, long-term growth mechanisms (change in factor productivity, interactions between demographic and macroeconomic variables etc.) are not included, or included merely, for example, using an exogenous model. On the contrary, IAM models explicitly seek to focus on these long-term mechanisms consistently.

References

Bach, S., Kohlhaas, M., Meyer, B., Praetorius, B., & Welsch, H. (2002). The effects of environmental fiscal reform in Germany: A simulation study. *Energy Policy, 30*, 803–811.

Bork, C. H. (2006). Distributional effects of the ecological tax reform in Germany: An evaluation with a microsimulation model. In Y. Serret & N. Johnstone (Eds.), *The distributional effects of environmental policy* (1st ed., pp. 139–170). Cheltenham, UK: Edward Elgar. OECD. ISBN 978-1-84542-3.

Bousquet, F., & Le Page, C. (2004). Multi-agent simulations and ecosystem management: a review. *Ecological Modelling, 176*(3–4), 313–332. https://doi.org/10.1016/j.ecolmodel.2004.01.011.

Cai, K., Niu, J. Z., & Parsons, S. (2014). On the effects of competition between agent-based double auction markets. *Electronic Commerce Research and Applications, 13*(4), 229–242. https://doi.org/10.1016/j.elerap.2014.04.002.

Chen, J. J., Tan, L., & Zheng, B. (2015). Agent-based model with multi-level herding for complex financial systems. *Scientific Reports, 5*, 8399. https://doi.org/10.1038/srep08399.

CNB (Czech National Bank). (2018). *Monetary policy instruments.* Available at https://www.cnb.cz/en/monetary_policy/instruments/

EUROSTAT. (2018). *Statistics explained.* Available at https://ec.europa.eu/eurostat/statistics-explained/.

Fiosins, M., Fiosina, J., Muller, J. P., & Gormer, J. (2011). Agent-based integrated decision making for autonomous vehicles in urban traffic. Advances on practical applications of agents and multi-agent systems. In Y. Demazeau, M. Pechoucek, J. M. Corchado, & J. Bajo (Eds.), *Advances in Intelligent and Soft Computing* (Vol. 88, pp. 173–178). https://doi.org/10.1007/978-3-642-19875-5_22.

Gontis, V., & Kononovicius, A. (2014). Consentaneous agent-based and stochastic model of the financial markets. *Plos One, 9*(7), e102201. https://doi.org/10.1371/journal.pone.0102201.

Janssen, M. A., & Ostrom, E. (2006). Empirically based, agent-based models. *Ecology and Society, 11*(2), art.37.

Kriström, B. (2006). Framework for assessing the distributional of financial effects of environmental policy. In Y. Serret & N. Johnstone (Eds.), *The distributional effects of environmental policy* (pp. 139–170). Cheltenham: Edward Elgar.

Lagorse, J., Paire, D., & Miraoui, A. (2010). A multi-agent system for energy management of distributed power sources. *Renewable Energy, 35*(1), 174–182. https://doi.org/10.1016/j.renene.2009.02.029.

Lengnick, M., & Wohltmann, H. W. (2013). Agent-based financial markets and New Keynesian macroeconomics: a synthesis. *Journal of Economic Interaction and Coordination, 8*(1), 1–32. https://doi.org/10.1007/s11403-012-0100-y.

Leontief, W. (1966). *Input - output economics*. New York: Oxford University Press.

Lund, P. (2007). Impacts of EU carbon emission trade directive on energy-intensive industries — Indicative micro-economic analyses. *Ecological Economics, 63*, 799–806.

Ministry of Finance of the Czech Republic. (2018). *Macroeconomic forecast – November 2018*. Available at https://www.mfcr.cz/en/statistics/macroeconomic-forecast/2018/macroeconomic-forecast-november-2018-33462

Morgan, F. J., & Daigneault, A. J. (2015). Estimating impacts of climate change policy on land use: An agent-based modelling approach. *PLoS One, 10*(5), e0127317.

Nordhaus, W. (2017). *DICE/RICE models*. Yale Economics. Available here https://sites.google.com/site/williamdnordhaus/dice-rice

Parker, D. C., Manson, S. M., Janssen, M. A., Hoffmann, M. J., & Deadman, P. (2003). Multi-agent systems for the simulation of land-use and land-cover change: A review. *Annals of the Association of American Geographers, 93*(2), 314–337. https://doi.org/10.1111/1467-8306.9302004.

Sahlin, J., Ekvall, T., Bisaillon, M., & Sundberg, J. (2007). Introduction of a waste incineration tax: Effects on the Swedish waste flow. *Resources, Conservation and Recycling, 51*(4), 827–846.

Samuelson, P. A., & Nordhaus, W. D. (2010). *Economics*. 19th ed. The McGraw-Hill series economics ISBN-13: 978–0–07-351129-0 (alk. paper), ISBN-10: 0-07-351129-3 (alk. paper).

Tang, L., Wu, J., Yu, L., & Bao, Q. (2015). Carbon emissions trading scheme exploration in China: A multi-agent-based model. *Energy Policy, 81*, 152–169.

World Bank. (2018). *What is the difference between current and constant data?* Available at https://datahelpdesk.worldbank.org/knowledgebase/articles/114942-what-is-the-difference-between-current-and-constan

Business and Finance

6

Michal Menšík

Abstract
What is the goal of a business? Is it even a good question to ask? Even the smallest companies have mission and vision. Whether it is in the head of the sole owner of small entrepreneur or is it stated in the strategic documents of the large global company, there always is the answer to the questions "Why are we here?" and "What we want to achieve in future?". These questions are the same as the famous "Where Do We Come From? What Are We? Where Are We Going?" None of these questions is related to making money. However, if the mission and vision are to be sustainable, the enterprise definitely cannot lose money while working on the mission and vision. The goal is to achieve the mission and vision while earning a decent profit. And as profit itself is too narrow measure of success and usually related to shareholders (Friedman 1970), also the perspective of other stakeholders (Freeman 2010) is taken into account while answering the question "Is this company a good one?" or "Is this business sustainable?" Within this perspective and perspective of Spationomy, issues such as the following will be asked and answered:

- Can spatial data and their analysis be a sustainable business itself?
- Can spatial data and their analysis provide the company with relevant information?
- Is it interesting only for specific companies or broad area of business?

Keywords
Business · Finance · Profit · Cash flow · Case studies

6.1 What Is the Goal of a Business?

Frequent answer to that question is that business is focused on earning money or profit. However, it is not valid. While studying the stories of some legendary (and successful) companies, the people who started Harley Davidson, Facebook, Hewlett Packard, Daimler Benz etc., they were not interested in earning money that much, at least at the beginning. They were interested in bringing or developing something new. A lot of companies started as a hobby project, which evolved into something more.

Example 6.1: TILAK
Tilak is a Czech company, producing sporting goods, especially using Gore-Tex®. This company has been set up by

(continued)

M. Menšík (✉)
Department of Business Economics, Moravian Business College Olomouc, Olomouc, Czech Republic
e-mail: michal.mensik@mvso.cz

V. Pászto et al. (eds.), *Spationomy*, https://doi.org/10.1007/978-3-030-26626-4_6

Example 6.1 (continued)

Mr Roman Kamler. Roman Kamler is Czech rock climber who was missing the "right sleeping bag for rock climber" on the market in the 1980s.In that era, there was no free market in Czech, so it was quite challenging to get high-quality equipment, especially those produced in so-called western countries.

So he took his mother's sewing machine and sewed a sleeping bag for himself based on his design, because who knows better, how a sleeping bag for a hardcore rock climber should look like. He then went on a trek with his friend and the friend's reaction was obvious – what kind of sleeping bag it is, where did he get it etc. remember, Czech rock climbers cannot simply buy high tech equipment. So everyone from the community was curious about each other material and ways how to improve it. At the end of the trek, Roman Kamler was asked to sew one more sleeping bag. He did, and his friend was taking the bag to another trek with other rock climbers. When he returned, he had the orders for eight more sleeping bags. At that moment, Roman Kamler started to think about the possibility of shifting his hobby into the business. Nowadays, the company TILAK belongs to the category of small and medium enterprises (SME). It has more than 60 employees with a turnover of around two mil EUR, exporting their products worldwide (Fig. 6.1).

The pattern of this story has been repeated again and again through history. It started as a hobby, then it got bigger, and now it is business. In the beginning, it is usual to spent own money on the hobby, after a while, when the "hobby" scales up, it has to be financially self-sufficient and even create some financial benefit for the founders – profit.

Fig. 6.1 Roman Kamler is opening a new store of his TILAK company. (Source: www.tilak.cz)

Profit is an essential side effect, crucial and necessary, but it is not the primary one. This opinion is not entirely new, as it goes back to 1999 (Denning Steve 2013a, b; Murray Sarah 2013; Montier 2014).

The real goal of the business is to fulfil a specific mission, create added value and to deliver the value to the customers. In microeconomics, we are talking about "satisfying needs". And of course, this has to be done in the fashion of long term sustainability, including financial sustainability.

The conclusion is simple – business focuses on customers and satisfying their needs while being long-term sustainable (including financially sustainable). Which is not entirely new idea, since it has been declared in 1973 – "The only valid purpose of a firm is to create a customer." – by Peter Drucker (Drucker 2006).

6.1.1 What Is Profit?

If profit is important, what is the profit and how it is measured? From the accounting perspective, profit is the increase in the equity[1] of a firm other than those relating to contributions from equity participants. Simply written, profit is, when the company has higher equity. And the increase of the equity is not because investors have invested more in the company, this increase has to be achieved by earning it. The most common way how to calculate profit is

$$\text{PROFIT} = \text{REVENUES} - \text{COSTS}$$

Total revenues (what has been sold)

– Costs (what we need to spent) such as:
  - Labour costs such as wages
  - Inventory required for production
  - Raw materials
  - Transportation costs
  - Sales and marketing costs
  - Production costs and overhead

For a hot dog stand, revenues are the total amount of sales (sold hot dogs), usually for some period (year, month, etc.). Costs are the value of hot dogs when bought from a supplier, rent of the stand, salary, electricity bills, phone bills etc. again for the same period. Hot dog stand is generating profit only in case the total revenues are higher than total costs. If they are lower, hot dog stand business is suffering the loss.

There are various levels of profit such as:

EBIT – Earnings before interests and taxes
EAT – Earnings after taxes
NOPAT – Net operating profit after taxes
Etc.

For more information about the profit and its various categories, please study more about accounting and financial management such as the classic Principles of Corporate Finance (Brealey et al. 2016). However, any textbook for a basic university course of finance will do.

Example 6.2: Basic Accounting Categories
What are the basic accounting categories and how to understand the profit? This topic is already covered by the theory of accounting, yet it is useful to mention it here.

An **asset (A)** is a resource controlled by the enterprise as a result of past events and from which future economic benefits are expected to flow to the enterprise.
A **liability (L)** is a present obligation of the enterprise arising from past events, the settlement of which is expected to result in an outflow from the enterprise of resources embodying economic benefits.
Equity (E) is the residual interest in the assets of the enterprise after deducting all its liabilities. Hence the equity interlinks the Assets and Liabilities so that we can define the basic accounting equation $E = A - L$.

The other two basic accounting categories are focused on the changes in equity.

(continued)

[1] To be clear in basic accounting categories please read the IFRS Conceptual Framework or the GAAP Conceptual Framework.

Example 6.2 (continued)

Income is an increase in economic benefits during the accounting period in the form of inflows or enhancements of assets or decreases of liabilities that result in increases in equity, other than those relating to contributions from equity participants.

Expense is decreased in economic benefits during the accounting period in the form of outflows or depletions of assets or incurrences of liabilities that result in decreases in equity, other than those relating to distributions to equity participants.

It is interesting that in the definitions of income and expense, there is not mentioned the cash – even though both categories are usually strongly related to the movement of money. It is not a necessary condition, and both events (income, expense) can occur without the immediate impact on cash. As a result, even the company with high profit (incomes are higher than expenses) can be in financial problems due to the not having enough cash to pay the liabilities and vice versa – a company with loss (expenses are higher than income) can have enough of cash.

Now which company is better, the one with a profit of 1 mil EUR or the one with a profit of 100 mil EUR? If hot dog stand makes a profit of 1 mil EUR per year, it is an excellent performance. If the whole Volkswagen Group makes a profit of 100 mil EUR per year, it is a disaster, as in 2017 Volkswagen Group generated a profit of 11 638 mil EUR (more than hundred times higher than 100 mil EUR), see the appendix 1 – Accounting statements of Volkswagen Group 2017. So merely the profit itself is not a perfect measure for performance measurement if we want to compare various companies or enterprises. That is why the need to compare the profit (the monetary measure of how successful we are) with the sacrifice (the monetary measure of what we have to give up). Therefore, the use of measures such as ROA (Return On Assets), which compares EBIT and total assets:

$$ROA = \frac{EBIT}{A}$$

or ROE (Return On Equity), which compares EAT and shareholders' equity:

$$ROE = \frac{EAT}{E}$$

Another traditional metrics for measuring the performance is ROS (Return On Sales) which shows, how much of the sales (S) is retained in the company as a profit:

$$ROS = \frac{EBIT}{S}$$

All these metrics can be expressed as the indices or in per cent whereas per cent is more common and easy to understand. If we go back to the Volkswagen Group and their annual report for 2017, we see the values as follows[2] (in mil EUR):

A (Total Assets)	422,193
E (Equity)	109,077
S (Sales)	230,682
Operating result (EBIT)	13,818
Earning after tax (EAT)	11,638

Based upon these values the profitability metrics are:

$$ROA = \frac{13\ 818}{422\ 193} = 0,0327$$

$$ROE = \frac{11\ 638}{109\ 077} = 0,1067$$

$$ROS = \frac{13\ 818}{230\ 682} = 0,0599$$

The usual interpretation of these traditional performance measures is:

[2] Even though EBIT is not the same as Operating Result, there is usually no significant difference and Operating Result (or Operating Profit) can be used as a very good approximation to EBIT.

Each 1 EUR of assets generates 3.27 cents to EBIT. Hence the ROA is 3.27%.
Each 1 EUR of equity invested by shareholders generates 10.67 cents of EAT. Hence the ROE is 10.67%.
Each 1 EUR of sales generates 5.99 cents to EBIT. Hence the ROS is 5.99%.

These ratios are not only measuring the performance but since these ratios put into relations the size of the company (measured by the total asset, equity or sales) and the profit, they allow to measure the adequacy of the profit and therefore to compare different companies. These metrics are offering the answer to the question:

"If we invest our money in this company – and therefore become the shareholders of the company – how much do we get back for every 1 EUR invested?"

This perspective – those, who are investing money are expecting the adequate profit – is based upon the shareholder approach, shareholder theory (Friedman 1970). However, the ideas about the profit adequacy are quite old, even though the Pacioli's (Fischer 2000) approach was more focused on profits "too high to be reasonable" and Friedman's approach is more "... increase its profits so long as it stays within the rules of the game, which is to say, engages in open and free competition without deception or fraud."

6.1.2 Profit or Money?

As the accounting perspective – based upon the definitions of accounting categories –income and expense are not necessarily connected with money, and yet if there is the need to invest and gain some return, cash is needed as well as the taking into the account the aspect of time.

Example 6.3: Money and Time
Money can be invested and bring back the return or interest. Let's suppose that there is the opportunity to invest money into the government bonds (investment with almost no risk) and within one year gain the interest of 8% p.a. So if the investment of 100 EUR today is done in bonds, within one year, the value of the investment is 108 EUR. That is why it is better to have 100 EUR today rather than in 1 year, as having the 100 EUR today is not equal to having 100 EUR within one year. 100 EUR today (while having the opportunity to invest in government bonds at 8% p.a.) is equal to 108 EUR in 1 year. In 2 years the investment is $100 * 1.08 * 1.08 = 100 * 1.08^2 = 116.64$ EU. From this perspective, there is (while considering the interest rate of 8% p.a.) no difference between having 100 EUR today or having 116.64 EUR in 2 years.

Principles and relations, which are briefly suggested in Example 6.3, are connected with the time value of money. Time value is focused on future value and present value.

Future value is:

$$FV = PV * (1 + i)^n$$

where

FV is the future value of the present investment,
PV is present value of the investment (what we invest today),
i is the interest rate and,
n is the number of years (duration of the investment).

If the investment is 528 625 EUR at the interest rate of 12% and the investment will be due in 5 years, the future value is calculated as:

$$\begin{aligned} FV &= 528\ 625 * (1 + 0,12)^5 \\ &= 931\ 617,87\ EUR \end{aligned}$$

Present value is the opposite perspective – it calculates the value of future payments from today's perspective.

$$PV = \frac{FV}{(1+i)^n}$$

What is the present value of the investment, if the investment design is two cash flows worth 1 000 EUR in the second and fourth year, while the interest rate is 8% p.a.? Present value is calculated as

$$PV = \frac{1000}{(1+0,08)^2} + \frac{1000}{(1+0,08)^4} = 1\,592,37\ EUR$$

The present value of the investment is 1 592,37 EUR. Present value can be compared with the required investment. For example, if the investment of 1 800 EUR will generate two cash flows worth 1 000 EUR at the end of the second and fourth year at 8% p.a., it is clear that it is not a good investment, as the present value is only 1 592,37 EUR and it is less than 1 800 EUR. Thus this investment brings negative effect.

Summarising the present values of all cash flows (in and out) while taking the time value of money into account is called Net Present Value (NPV).

$$NPV = \sum_{t=0}^{n} \frac{CF_t}{(1+i)^t}$$

where

NPV is Net Present Value,
n is the number of periods,
t is a certain period,
$\mathbf{CF_t}$ is Cash Flow in the period t and,
i is the interest rate.

Investment is accepted as long as it generates positive NPV.

What is the NPV of investment into the hot dog stand, if the price of purchase the stand is 10,000 EUR and the stand will generate yearly cash flow of 2000 EUR for 7 years while interest is 4% p.a.?

$$NPV = -\frac{10\,000}{(1+0,04)^0} + \frac{2\,000}{(1+0,04)^1} + \frac{2\,000}{(1+0,04)^2} + \frac{2\,000}{(1+0,04)^3} + \frac{2\,000}{(1+0,04)^4} + \frac{2\,000}{(1+0,04)^5} + \frac{2\,000}{(1+0,04)^6} + \frac{2\,000}{(1+0,04)^7} = 2\,004,11\ EUR$$

A project of the hot dog stand generates positive NPV worth 2004.11 EUR and therefore is acceptable.

From the mathematical perspective, it is clear that the higher the interest (i), the lower the NPV, hence the relation is negative. In economic terms – future cash flows are deteriorated by the interest rate (the more the distant future it is). Figure 6.2 shows the relation between the i and NPV for the hot dog stand investment project.

One of the implications of Fig. 6.2 is – there is interest rate when NPV is zero. This interest rate is called Internal Rate of Return (IRR) and is also used as a criterion for the investment evaluation. For further reading on IRR and NPV see (Brealey et al. 2016).

The profit approach is used mostly in the short term decision making, and the cash flow approach is used for long term decision making. Despite what has been written at the beginning of this chapter – the main goal of business is not making money – profit or positive cash flow is an important requirement for a business to be long term sustainable.

Based upon this interim conclusion two questions arise:

Are Spatial knowledge and related techniques, tools and applications business itself? Can the Geoinformatics, Remote sensing, Urban planning and all the area related to the spatial sciences generate profit and positive cash flow?

And the second question:

Can SPATIONOMY benefit (other) business in general? Is there need and opportunity forthe SPATIONOMY to be part of other (non-geographic area) businesses? Can it help to generate profit and cash flow?

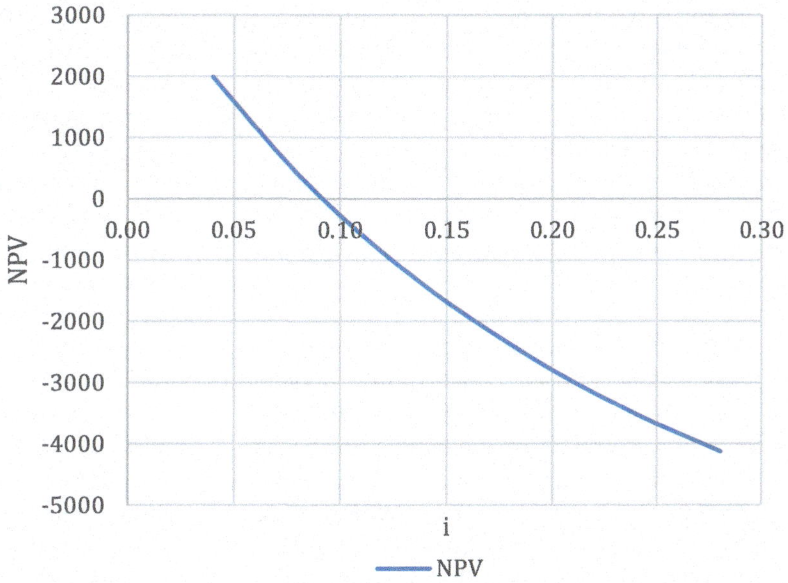

Fig. 6.2 Relation between NPV and I. (Source: authors)

6.2 Spatial Business

Let's start with an example:

Example 6.4: TomTom

TomTom is a Dutch origin company founded in 1991. Today, it is broadly known for its navigation software and applications for car navigation (started in 2001), yet their product and service range is broader (Table 6.1).

What are the numbers for Tom Tom, according to their 2016 annual report?

Based on the numbers it is clear that TomTom is generating profit in both observed years as well as positive cash flow. Relative performance measures, such as ROA and ROE are (Table 6.2)

Is this enough or not? There is no general rule on how high the ROA or ROE should be. Comparison with other companies may reveal at least the position towards them. Volkswagen AG for 2016 achieved ROA 1.73% and ROE 5.79%, Microsoft in 2016 achieved ROA 10.42% and ROE 23.33%. Volkswagen and Microsoft performed better; on the other, both companies are the world top leaders in their fields.

Overall conclusion for this example is – yes, there can be a business built on the spatial information.

Table 6.1 TomTom annual report data

	Thousands of EUR	
Year	2016	2015
Revenues/sales	987,329	1,006,607
Operating result/EBIT	8894	600
Net result/EAT	11,958	18,293
Total assets	1,628,828	1,653,587
Equity	996,737	970,533
Cash flow from operating activities	144,317	118,775

Source: (TomTom 2017)

Table 6.2 TomTom return on assets and return on equity

Year	2016	2015
ROA	0.55%	0.04%
ROE	1.20%	1.88%

Source (TomTom 2017)

Example 6.4 shows that there is a possibility/opportunity for a viable and sustainable business based upon the geo field. There is no final list of all business activities possible; the opportunity to create one will be based upon the innovative thinking, creativity, business vision, scientific and technological development etc. What examples or types of businesses are present on the market so far?

1. Mapmaking – probably the oldest activity, map making was often related to the military and exploring activities. Nowadays commercially produced maps are both printed and digital, sold as an atlas, maps or applications. Examples of such companies are Electronic Chart Centre, Tele Atlas, Ausway or Carta.

   What is the value-added for customer/user? Nowadays (traditional) maps are still used for navigation. Despite the possibility for more advanced digital navigation (see next), traditional paper maps are still in use for navigation because they do not need batteries, they are highly functional in extreme hot or cold conditions, and of course, they are also used as a backup for digital applications. Precise and correct navigation saves time, fuel, improves security (e.g. avoiding collision), decreases the depreciation etc. All these effects will reflect in lower costs hence improving the financial result (profit) and also reducing expenditures, thus increasing the total Cash Flow. This value-added is the primary purpose why customers/end users are willing to pay money for traditional maps.

   Traditional maps (atlases) are also used for educational purposes at any level of the education process. However, this is usually not connected with the business. Still, it is part of the market. There is also a market for traditional maps because of aesthetic reasons – art map can be used as a decoration.
2. Navigation applications – a combination of digital maps, electronic devices and GPS allows developing navigational applications for personal and commercial use. These applications are focused on car navigation, touristic navigation, boat charting and navigational applications and more. Examples of such businesses are TomTom, Navigator, Waze, MX Mariner or BackCountry Navigator.

   The value added is pretty much the same as in the previous case. However, the combination of modern advanced technologies brings many new features to utilise abovementioned benefits. Whether it is the use of mobile digital devices, Artificial Intelligence, Swarm Intelligence etc. all these (especially in combination) can create more value-added to users/customers. The navigation is based upon the current situation (road obstructions, car accidents, road constructions, traffic, etc.) thus overcome the problems of traditional maps – static character and (after a while) obsolete information. Digital navigation can easily update map sources (new roads, new bridges etc.) and in combination with an online connection, the navigation also is improved by the most up to date information.

Example 6.5: Waze

Waze started as the FreeMap Israel in 2006, aiming at – based upon community project and crowdsourcing – creating a free digital map of Israel with up to date information. In 2009 it was renamed to Waze Mobile Ltd. And became commercial. In January 2012 application has been downloaded by 12 million people, on July 20 million. The nature of community project is still present in the application, as it collects data from users to create, update and modify maps as well as to collect actual data related to traffic jams, accidents etc. Well, known is the example of Tour de France, where Waze updated traffic info, especially for this event.

From a business perspective, the year of 2013 is interesting, as Google bought Waze for one billion of USD and employees have been paid on average 1.2 M USD (Teig Amir 2013) which shows how lucrative the business is as Google paid that sum only in faith to earn even more.

3. Location devices or applications – these devices and applications are used when there is the need for fast localisation of person or object, especially when the time is a critical factor. These devices and applications can be suitable in a vast area of situations, from really hazardous situations (such as avalanche beacon) to a simple time- and cost-saving (localising your keys or cell phone). Examples of such devices and applications are PIEPS, Arva, Your Devices at Google Account, Angel Sense, PocketFinder etc.

   Value-added for this application is time saved. The time saved can be just because of the comfort of the users (finding your keys or cell phone) or – in case of life-threatening events – the time is a critical factor. For example in the avalanche, if the victim is dug out in the first 5 min, the survival rate is 90%. After 45 min survival rate is somewhere between 20% and 30%. The probability of surviving the avalanche drops to nearly 0% after 2 h. A similar situation is in the case of a heart attack, where the probability of successful recovery is strongly correlated to the time. If the professional aid is not available within the first 3 or 4 h, almost half of patients do not survive. Therefore users (whether they are rock climbers, skiers, athletes, hospitals, police or law enforcement) are paying for the better probability of survival or saving the time.

As to make an interim conclusion, spatial business is strongly connected with time. To get somewhere faster, to find/localise something or someone faster – and the time is the most significant value-added. Other relevant factors (in business expenses and expenditures) are usually correlated with the time, e.g. the longer the journey takes, the more fuel is consumed, the higher the costs.

In case the business is not entirely based upon the spatial information, there are still potential benefits from using the spatial data, however before exploring them, it is needed to discuss the profit and cash flow more.

Generally accepted axiom for financial management is that higher profit is better than lower profit, ceteris paribus. What means ceteris paribus? It is from Latin, and it means "all other things being equal". What are the other things? In financial management, it usually is risk and liquidity. Sometimes it is called investment triangle (Valach 1997), magic triangle or investing trinity – the risk, return and liquidity. Why triangle? It is often depicted as the following figure.

The triangle depicts three axioms of financial management:

1. It is better to have a higher return (profit, cash flow) than a lower return.
2. It is better to have a lower risk than higher risk.
3. It is better to have higher liquidity than lower liquidity.[3]

Also Fig. 6.3 shows the inherent laws of "2 out of 3" – while investing, it is possible to achieve only two of the three vertices. Thus it is possible to have an investment with:

1. Low risk, high liquidity BUT low return – e.g. government bonds.
2. High liquidity, high return BUT high risk – e.g. shares of new tech companies, such as Waze in the past.
3. High return, low risk BUT low liquidity – e.g. investment in the real property, such as buying a flat and renting it.

Since the axioms mentioned above are taken as valid, the financial management usually aims to increase return, increase liquidity and decrease

[3] Liquidity is the feature how quickly can be something purchased or sold on the market. Money are the asset with the best liquidity, land is the asset with very low liquidity.

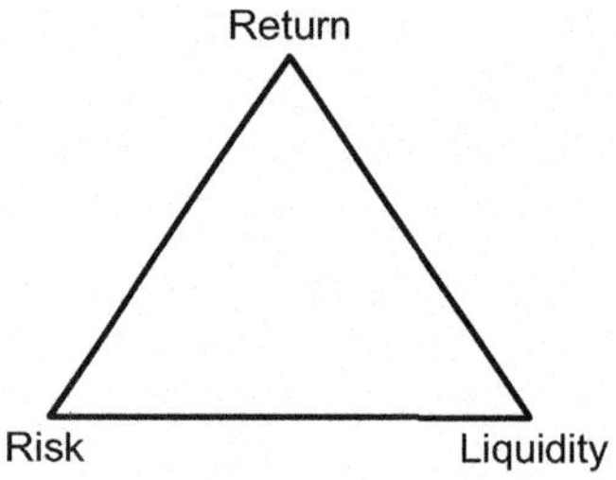

Fig. 6.3 Investment triangle. (Source: authors)

or reduce risk. How to increase the return? As has been demonstrated at the beginning of this chapter, profit is connected with revenues and expenses. Revenues are generated through sales – the more we sell, the higher the revenues. Costs may or may not be influenced by the volume and profit is the difference between total revenues and total costs.

Example 6.6
It is possible to apply a more analytical approach to the Hot dog stand example from the beginning of the chapter. This information related to the hot dog stand is available:

For a hot dog stand, revenues are the total amount of sales (sold hot dogs), usually for some period (year, month, etc.). Costs are the value of hot dogs when bought from a supplier, rent of the stand, salary, electricity bills, phone bills etc. again for the same period. Hot dog stand is generating profit only in case the total revenues are higher than total costs. If they are lower, hot dog stand business is generating the loss.

Revenues are a function of the volume, total revenues (sales) are the equal price for unit multiplied by the number of units (hot dogs) sold. Costs are slightly more complicated. Some costs, such as hot dogs bought from suppliers are a function of volume – the more hot dogs are sold, the more costs of buying hot dogs are expensed. These type costs are labelled as **variable** costs. On the other hand, rent is not influenced by the volume of hot dogs sold; these type of costs are labelled as **fixed** costs.

As this kind of decision making is based upon the **c**osts, **v**olume and **p**rofit, it is often called CVP analysis. CVP analysis can be based upon the non-linear model. However, this model is more difficult to use, and the benefits of better accuracy do not overweight the problems. That is why the linear CVP model is used. Linear CVP model is based on the relation:

$$\text{Profit} = \text{Revenues} - \text{Costs or } P = R - C$$

This model is called linear because it assumes all the relations within are linear. Hence the model can be transformed into:

$$P = P_u * Q - VC_u * Q - FC$$

where,

P_u is the price per unit,
Q is the volume of units sold,
VC_u is the variable costs per unit,
and FC are fixed costs.

As rational business is selling for more than variable unit cost ($P_u \geq VC_u$), the positive difference per unit is called margin per unit or contribution per unit. The total sum of these unit contributions generates total margin or contribution margin. This is how the business generates profit. In the beginning, the total margin is lower than fixed costs and business creates a loss; after covering the fixed costs, the business generates profit.

Example 6.7
In the extension of the previous example with a hot dog stand, the following assumptions will be introduced:

(continued)

Example 6.7 (continued)

P_u for one hot dog is 3 EUR, VC_u for one hot dog is 2 EUR, FC is 1200 EUR per month. What is the profit for various volumes of sold hot dogs?

Conclusions made based upon the Table 6.3 are apparent – until a specific volume the business is losing money, it is not worth to start the business unless the demand for the product or service is big enough. After a certain volume, business is making money. At a precise moment, business is not losing nor making money – it is called the breakeven point.

Table 6.3 Linear CVP model of hot dog stand

Q	R	VC	FC	P
0	0	0	1200	−1200
300	900	600	1200	−900
600	1800	1200	1200	−600
900	2700	1800	1200	−300
1200	3600	2400	1200	0
1500	4500	3000	1200	300
1800	5400	3600	1200	600
2100	6300	4200	1200	900

Source: Authors

The breakeven point is a moment when previous conclusions (losing money) are changing, yet new conclusions (earning money) are still not valid. From the perspective of the CVP analysis, breakeven point (often abbreviated to BEP) is the volume of units sold, where total revenues are equal to total costs. Hence profit is zero. In Example 6.7, the breakeven point is 1200 units (hot dogs).

How to calculate the breakeven precisely? BEP is defined as the volume (Q) when total revenues are equal to total costs:

$$\mathrm{TR} = \mathrm{TC}$$

which is the same as

$$\mathrm{P_u} * \mathrm{Q_{BEP}} = \mathrm{VC_u} * \mathrm{Q_{BEP}} - \mathrm{FC}$$

So basically it is equation with one variable (Q_{BEP}). The solution of the equation is, therefore (based on the standard mathematical operation):

$$Q_{BEP} = \frac{FC}{P_u - VC_u}$$

The volume for the breakeven point (Q_{BEP}) is essential information for business, as it is the answer to the question "How many customers we need to have not to lose money?" On the other hand, the business usually expects some profit for further development, research, corporate social responsible action etc. Hence the question is modified "How many customers we need to have to earn a certain profit?". In Example 6.7, hot dog stand business needs to have 1800 sold hot dogs per month, to earn a profit of 600 EUR. Precise calculation of the volume for specific profit is:

$$Q_{PROFIT} = \frac{FC + P}{P_u - VC_u}$$

CVP analysis is not strictly limited only on business decision making; it can be as well applied in other areas.

Example 6.8

Health insurance system pays to the hospital 1500 EUR for the non-complicated upper limb fracture treatment.

The treatment requires variable costs (plaster or fibreglass cast, etc.) of 60 EUR. Fixed costs of the Trauma department (the simplifying assumption is that it is strictly devoted to non-complicated fractures, providing no other treatment) are 2,500,000 EUR per year. The statistical probability of upper limb fracture is $p = 0.0439$. Now, how populated has the hospital gravity field to be, so the total average costs per treatment are within the limit of the 1500 EUR?

Formally, the total average costs (AC) – total costs divided by the quantity - have to be less or equal to 1500 EUR, therefore inequality:

(continued)

Example 6.8 (continued)

$$AC = \left(VC_u + \frac{FC}{Q * p}\right) \leq 1500$$

As the only unknown variable is Q, the inequality can be transformed (logical assumptions are VC_u, Q and p are positive values) into inequality:

$$\frac{FC}{(1500 - VC_u) * p} \leq Q$$

After filling in the numbers, the inequality gets this form:

$$\frac{2500000}{(1500 - 60) * 0,0439} \leq Q$$

Hence the result is:

$$Q \geq 39546,95$$

The gravity field has to have at least 39,547 people, so the average costs per treatment are within the limit of the health insurance system. Even though the assignment is very simplified (based upon simplifying assumptions), the underlying philosophy is correct – certain public services, such as hospitals, city public transport etc. are connected with a critical mass of the population – there have to be enough people demanding the service to provide the service.

CVP analysis also shows why there are the urge and effort for growing – the bigger the volume, the higher the margin and profit. On the other hand, it also shows the other way how to improve profit – to reduce unit costs or fixed costs. Both these ways – increasing the volume and reducing the costs – can be supported and applied. CVP analysis can be also used for another type of decision making.

Example 6.9

The present situation with the hot dog stand is 1800 sold hot dogs per month for the unit price of 3 EUR, while hot dogs are purchased from suppliers for 2 EUR. Fixed costs are 1200 EUR per month.

The hot dog business is considering the opportunity to improve marketing by creating posters and running the banners on social networks. Expected costs of these activities are 300 EUR. Expected result of these activities is increasing sales by 25%.

Is the money in the marketing well spent?

Incremental costs (or marginal costs) of decision are 300 EUR increase in fixed costs and also the extra costs spent on extra hot dogs sold. The increase is 25% which corresponds to 450 extra hot dogs. Hence incremental variable costs are 900 EUR (450 * 2). The total increase in costs is 900 + 300 = 1200 EUR, that is how the costs will change in case the decision is made. An alternative decision is to do nothing, and costs remain unchanged.

Revenues will be changed (increased) due to the decision by 1350 EUR (450 * 3). As the increase in the revenues is higher than the increase in the costs, the decision – spent money on marketing – will benefit the business.

As each sold hot dog causes extra 3 EUR in revenues and additional 2 EUR in variable costs, the benefit of each sold hot dog is 1 EUR in the margin. The decision can be made based upon the unit margin and the incremental margin. As marketing should bring an extra 450 units, it should also bring additional 450 EUR in the margin. As the increase in margin is 450 EUR, it is higher than the increase in fixed costs of 300 EUR due to the marketing expenses. This is enough information to decide that the extra 300 EUR is money well spent, the impact on the profit (and cash flow as well) will be extra 150 EUR.

In Example 6.9, the decision problem is defined as the trade-off between increased fixed costs and increased volume, where volume will result in increasing the margin. This type of decision making can be described as "What – if". In this case, the question is "What will happen if the company spend more on marketing? Is it a good decision?" Of course, it is challenging to estimate the impact of the marketing expenses, and it is almost sure that the effect will not be precisely 25%. However it is possible to reverse the logic of the example and redefine the decision making into the question: "In case we spent 300 EUR on marketing, how big the impact on the volume has to be so we do not lose the money?" As already explained in Example 6.9, each hot dog sold brings extra 1 EUR of margin. Hence the minimum impact of the marketing must be extra 300 hot dogs sold. Each hot dog above the 300 units is generating profit.

Example 6.10
The hot dog stand is growing and being successful; there is the offer to join – based on the franchise – a worldwide network of the fast-food chain. The franchise has to be paid every month 400 EUR. However, the franchise chain will supply its hotdogs for the 15% discount. The current volume is 1800 sold units (hot dogs). However, there is the expectation of increasing our customers by 5% based upon the fact of becoming a brand.

Is it – from a financial perspective – right decision?

The present situation is based upon the equation:

$$P = P_u * Q - VC_u * Q - FC$$

Hence the profit is:

$$P = 3 * 1800 - 2 * 1800 - 1200 = 600 \text{ EUR}$$

The alternative situation is based upon a modified equation:

$$P = P_u * Q * (1 + 0,05) - VC_u * (1 - 0,15) * Q * (1 + 0,05) - (FC + 400)$$

The profit of the alternative option is:

$$P = 3 * 1800 * 1,05 - 2 * (1 - 0,15) * 1800 * 1,05 - (1200 + 400) = 857 \text{ EUR}$$

As the profit for the alternative option is higher than the present situation, from a financial perspective, it is a good decision to join the franchise.

All these examples (Examples from 6.6 to 6.10) show how even the simple linear CVP model can be used for various decision making, analysing different situations and comparing them. Also based upon the linear CVP analysis – from the perspective of the mathematic equation – profit (or cash flow) can be improved by increasing revenues or decreasing expenses. This conclusion is not surprising. However, it suggests how the Geo part can play a useful and significant role for companies – either it will increase revenues or decrease expenses.

6.3 Business and Spatial issues

How the spatial methods can be used in a business is shown based upon the two case studies. A first case study is based upon the Cyclone Kyrill; second case study is based upon the application Urban Planner.

6.3.1 Case 1 – Cyclone Kyrill

Cyclone Kyrill (or Storm Kyrill) was extratropical cyclone with hurricane-strength winds which formed on the 15th January 2007 and dissipated the 24th January 2007. This winter storm was very destructive and caused significant damages and disruptions across ten countries in Europe. Kyrill claimed 47 lives (13 of them in Germany) and economic losses are estimated to USD 12 billion (Tatge, 2017).

Kirill hit Ireland and Great Britain in the evening the 17th of January 2007, crossed the North Sea and in the afternoon of the 18th of January hit Germany and Dutch coasts. Kirill than moved to Poland and Russia moving over the Baltic Sea.

During the storm, the wind speed was high above the 100 km/h with a record of 202 km/h on the Wendelstein in Germany.

More than two millions of households were without electricity across the affected parts in Europe, trains completely stopped, hundreds of flights have been cancelled, people were sleeping in stranded trains etc.

Among other damages and problems, Kirill caused devastation on forests in the affected area. According to the Beaufort scale – scale for measuring the wind speed – the storm of the strength 10 is strong enough to uproot trees. Kirill was up to the 12, which classifies it like a hurricane. Thus the damage is described as devastation.

From the perspective of the company which operates in forestry, this type of event will cause damage to the assets. Based upon the strength/wind speed, the damage could range from broken/fallen twigs and branches to broken trees up to complete devastation. The issue is to evaluate the loss in economic terms – how significant the damage is in the monetary unit. The need is to get precise information about the damage. The more we spent our resources (time, money, labour) on the information, the better (precise, relevant, etc.) the information is. The traditional approach has been based upon the personal reconnaissance; someone must walk through the area or observe it from higher ground and estimate the damage. The problems were of several kinds. First, the damaged forest is unstable, some trees clinched, locked, weakened etc. and therefore even light breeze can cause another collapse, hence endanger the safety of workers, who were providing the estimations.

The second problem is the timeliness – precise information is valued if it is provided on time. The more accurate information is required, the more time it takes to get it. So there is the trade-off between the precision and timeliness and costs. It is not a wise decision to spend 5000 EUR to get the information "Damage is worth 500 EUR" nor is it wise to get precise information after a couple of months or years.

Modern approaches offer more sophisticated tools to solve the problem. The tool or the field which can be applied is called remote sensing. For more information on remote sensing, see subchapter 1.5. The company, instead of sending workers in a dangerous area, can get aerial images or satellite images of the area and based upon the images, the estimation can be done.

The first picture (Fig. 6.4) is taken before the Kirill occurred, and it is a random picture from Germany.

The areas with forest can be easily identified by the plain eye, more precise yet very simple estimation can be done by putting a rectangular grid pattern on the image and estimate the % of the forested area. The grid is shown in Fig. 6.5.

The smaller the dimension of a rectangle, the more precise the measurement, however also the more challenging to calculate. Of course, if more advanced tools are applied, such as image analysis, than the tedious work can be done by the computers (especially if the image is a digital one).

The image of the post – Kirill situation is shown in Fig. 6.6.

The deforestation because of the Kirill Storm is visible. Whether the image is gathered via drone or downloaded from free or paid source, the benefits of this attitude are apparent – it is fast, cheaper, precise and much safer for the workers. Thus the remote sensing and image analysis saves costs and thus improving the profit and cash flow.

The damage estimation itself is quite simple: Value of 1 hectare of the forest (may depend on the age of the forest. However this information is available for the company) x number of hectares damaged or destroyed. Thus the company, thanks to the remote sensing, has better information in less time and for less money (lower costs).

Fig. 6.4 Image of the landscape before the Kirill Storm. (Source: Land NRW (2018/2009/2006)) Datenlizenz Deutschland-Namensnennung-Version 2.0 (source www.govdata.de/dl-de/by-2-0)

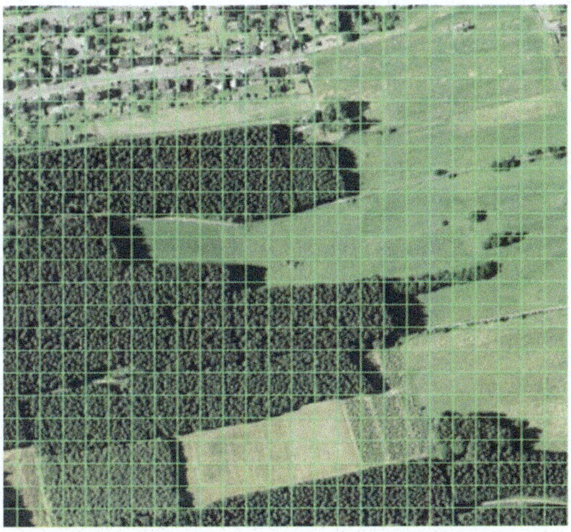

Fig. 6.5 The rectangular grid on the landscape image. (Source: Land NRW (2018/2009/2006)) Datenlizenz Deutschland-Namensnennung-Version 2.0 (source www.govdata.de/dl-de/by-2-0)

The company can approach this problem also form different attitude and demand this information from a specialised company as a service. This is called **outsourcing,** and the basic idea (applied on the forestry business) is: "We are experts in forestry, we are not experts in other fields, we will find experts on remote sensing and image analysis and hire them to do the job." The outsourcing approach suggests that viable business based upon the remote sensing may be possible and of course, it is.

6.3.2 Case 2 – Urban Planner

When companies are considering the location of its headquarter, production site, shop etc. they are considering different variables, as this decision is rarely one-dimensional. The issue of choosing the best location is therefore based upon the multidimensional decision making.

Fig. 6.6 Image of the landscape after the Kirill Storm. (Source: Land NRW (2018/2009/2006)) Datenlizenz Deutschland-Namensnennung-Version 2.0 (source www.govdata.de/dl-de/by-2-0)

Example 6.11: Multidimensional Decision in Sport

Every year there is a sporting event when shooters are competing in three different disciplines. The first discipline is rimfire pistol, distance is 25 meters, and the target is standard 50/20 according to the ISSF where the best score is 10 point, the circle with 10 is 50 mm wide. Competition is ten rounds theoretical maximum is 100 points. The second discipline is rimfire rifle, and the distance is 50 meters, the target is the ISSF air rifle target with the ten as maximum, and the ten circles is 11,5 mm. Competition is again ten rounds, and the theoretical maximum is 100 points. Finally, the third discipline is centerfire pistol steel targets, and the main aim is to knock down five steel targets as fast as possible, best shooters finish this discipline between 5 and 6 s.

Any shooter can be immediately disqualified for safety violations – as for this sport, this is extremely important – and for unsportsmanlike conduct.

To determine who is the best shooter in a separate discipline is quite easy. In first two, it is the shooter with the highest points, in the last discipline, it is the shooter with the shortest time. Table 6.4 presents the model results for 20 shooters in each discipline and to find the best one in each is quickly done by the naked eye.

However, to determine, who is the best shooter of the day, that is a multidimensional problem. Also other issues are how to measure and how to compare various metrics and measures as some metrics are positive (it is better to have high values such as points), some are negative (it is better to have lower values, such as time), some have limited values (it is impossible to get 160 points based upon 10 shots), some have unlimited values (theoretically there is no time limit for steel targets), etc.

(continued)

Example 6.11 (continued)

Table 6.4 Results of the competition

		Rimfire pistol	Rimfire rifle	Steel targets
	Name	Points	Points	Time
1	Alice	82	74	7.7
2	Brenda	60	65	11.54
3	Cecil	67	78	6.73
4	Donna	80	80	12.07
5	Eve	41	42	14.03
6	Frank	93	78	7.86
7	Gina	93	92	8.81
8	Hana	86	87	9.05
9	Ivan	86	80	11.2
10	John	49	54	9.05
11	Karl	47	83	7.53
12	Lena	64	64	11.34
13	Mike	86	91	6.2
14	Nora	75	64	6.81
15	Oscar	56	76	8.45
16	Peter	85	89	4.63
17	Quinn	61	61	8.5
18	Robert	76	86	11.28
19	Stefania	88	92	9.95
20	Tim	89	94	6.62

Source: Authors

The first possible approach towards the problem is ranking. The ranking is based upon ranks to determine who is first, second, third etc. in each discipline separately. This is usually easy and is based upon simple measurement and determining ranks. After that, the sum of ranks for each competitor is calculated. Theoretically, if there is a competitor, who is best in all three disciplines, the best (minimum) value is 3. The best competitor is the one with the lowest rank-sum. Ranks are shown in Table 6.5.

Based on Table 6.5, it is possible to determine the best shooter of the whole contest, and it is Tim, who has the lowest sum of ranks. There are two third places (Gina and Peter), etc.

In case there are exactly 20 competitors, the relative difference between each rank is exactly 5%, which is not necessarily the truth. The difference between Tim (rank 3 in the first discipline) and Stefania (rank 4 in the first discipline) is one point. Because the maximum is 100 points, the difference is 1%. The two competitors have practically the same result. However, the ranking approach conclusion is that the gap between 3rd and 4th place is 5%. Theoretically, if there are two competitors (A and B), two disciplines (I and II) and results are as shown in Table 6.6, there is the wrong conclusion.

The ranking approach suggests both competitors equal, which is not valid, as they are (almost) equal in the discipline II.

(continued)

Example 6.11 (continued)

Table 6.5 Ranking

	Name	Rimfire pistol Rank	Rimfire rifle Rank	Steel targets Rank	Rank Sum	Final rank
1	Alice	9	14	7	30	10
2	Brenda	16	15	18	49	18
3	Cecil	13	11	4	28	8
4	Donna	10	9	19	38	14
5	Eve	20	20	20	60	20
6	Frank	1	11	8	20	5
7	Gina	1	2	11	14	3
8	Hana	5	6	12	23	7
9	Ivan	5	9	15	29	9
10	John	18	19	12	49	18
11	Karl	19	8	6	33	11
12	Lena	14	16	17	47	17
13	Mike	5	4	2	11	2
14	Nora	12	16	5	33	11
15	Oscar	17	13	9	39	15
16	Peter	8	5	1	14	3
17	Quinn	15	18	10	43	16
18	Robert	11	7	16	34	13
19	Stefania	4	2	14	20	5
20	Tim	3	1	3	7	1

Source: Authors

Table 6.6 Ranking issue demonstration

	Disciplines					
Competitor	I	II	Rank I	Rank II	Rank Sum	Final rank
A	100	99	1	2	3	1
B	35	100	2	1	3	1

Source: Authors

However, different in discipline I. However ranking system is assuming an equal difference between ranks, thus smoothing the differences. One possible solution for this issue is the relative approach to ranking.

The relative approach is based upon comparing the competitors to the best result. The best result can be the theoretical best (100 points for the first discipline in this example) or the real best (the best result achieved by a competitor). The best result represents 100%, and the other results are also calculated in % to show the relative difference between competitors. The relative rank is calculated based upon simple mathematical formulas, different for positive and negative metrics.

(continued)

Example 6.11 (continued)

Table 6.7 Relative ranks

	Name	Rimfire pistol (%)	Rimfire rifle (%)	Steel targets (%)	Sum	Final rank
1	Alice	88.17%	78.72%	60.13%	227.03%	8
2	Brenda	64.52%	69.15%	40.12%	173.79%	18
3	Cecil	72.04%	82.98%	68.80%	223.82%	9
4	Donna	86.02%	85.11%	38.36%	209.49%	13
5	Eve	44.09%	44.68%	33.00%	121.77%	20
6	Frank	100.00%	82.98%	58.91%	241.88%	5
7	Gina	100.00%	97.87%	52.55%	250.43%	4
8	Hana	92.47%	92.55%	51.16%	236.19%	7
9	Ivan	92.47%	85.11%	41.34%	218.92%	10
10	John	52.69%	57.45%	51.16%	161.30%	19
11	Karl	50.54%	88.30%	61.49%	200.32%	14
12	Lena	68.82%	68.09%	40.83%	177.73%	17
13	Mike	92.47%	96.81%	74.68%	263.96%	3
14	Nora	80.65%	68.09%	67.99%	216.72%	11
15	Oscar	60.22%	80.85%	54.79%	195.86%	15
16	Peter	91.40%	94.68%	100.00%	286.08%	1
17	Quinn	65.59%	64.89%	54.47%	184.96%	16
18	Robert	81.72%	91.49%	41.05%	214.26%	12
19	Stefania	94.62%	97.87%	46.53%	239.03%	6
20	Tim	95.70%	100.00%	69.94%	265.64%	2

Source: Authors

- Positive metric

$$\frac{\textit{evaluated entity value}}{\textit{best (highest) value}} * 100\%$$

- Negative metric

$$\frac{\textit{best (smallest) value}}{\textit{evaluated entity value}} * 100\%$$

Relative rankings calculated for the model data are presented in Table 6.7.

In this approach, there are no issues having two competitors with the same score, and in this approach, the differences are calculated more precisely. That is why there is very little difference between Tim and Stefania in the first discipline. This approach also changed the winner, this time it is Peter with an overall score of 286.08%. Former winner Tim is second, third place is Mike.

There are several other options, how to calculate the final score or rank from various variables and choose the best option. The important lesson learned based upon this simple example is

- Measurement is not neutral, as various measures can cause varying results.
- Methods of measurement should reflect our needs, preferences, values, etc.
- To represent sets of parameters or variables by one measure (number) is tricky.

What can be the variables, considered by companies (or institutions) while choosing the location?

(a) Transport infrastructure – based upon the nature and field of business, requirements to be within a certain distance from the highway, international airport, train station etc. can be made. Being closer to the transport infrastructure will reduce costs of transport thus improving cash flow and profit, on the other hand, transport is usually caused of air pollution, and for some businesses, this can be harmful (health care, wellness, ecological agriculture etc.). That is why this variable can be positive or negative, depending on the needs.

    The economic impact of this variable is usually projected in expenses. The bigger the distance between the producer/provider and customer/consumer, the higher the transportation costs. Producers generally prefer a location closer to the transport infrastructure; residential areas usually prefer a more distant location.

(b) Power supply – certain companies have specific requirements for power supply. In Prague (Czechia) the company Prusa Research, who is the producer of 3D printers, needed the high power supply for the biggest print farm in the world and this requirement (among others) was the primary motivation to change location in 2017.

    The economic impact of this variable is almost binary (and thus similar to the Example 6.10 safety violation rule), the location either offer sufficient power supply or not. In case there is not enough power supply, the site is either unsuitable, or it requires investment (expenditures and costs) to improving it.

(c) Employees – companies require employees with a specified set of skills and opening the branch in a specific location can be motivated by this factor. It is not a coincidence that important innovative hubs (such as Silicon Valley) are close to excellent research and educational institutions (in the case of Silicon Valley it is Stanford University).

    The economic impact of this variable is also focused on expenses. If the employees (with a needed set of skills, education, experience etc.) are scarce in a certain location, they require higher salaries (for example to leave their current jobs and move to the new one), employee benefits etc. and all these will reflect in higher costs. On the other hand, if the unemployment rate is higher (there is a lot of skilled worker in the labour market, willing to work) than of course salaries can be lower.

(d) Customers – business, which delivers products and services to the consumers or end-users and business which is required to supply in Just In Time manufacturing also need to consider the distance from customers. At the beginning of this chapter, the hot dog stand is a typical example of a business, which needs to be close to the customers.

    The economic impact of this variable can be on revenues; this is valid for the businesses focused on personal services or selling goods to the consumers (restaurant, hot dog stand, supermarket, etc.). The area with more customers is positively correlated with higher revenues.

    In the case of Just In Time manufacturing the distance between the producer (supplier) and the customer will affect transportation costs.

(e) Ecological and safety restrictions – certain business cannot be located nearby water sources, protected areas, populated areas etc. One example is EXPLOSIA, a.s., which is a company producing explosives. Historically it was located outside the city of Pardubice (Czechia) in 1920, however as the city of Pardubice grows (today the population of Pardubice is around 90 000 inhabitants), the current situation is that the risky business is operating in the suburb of a highly-populated area. On the other hand, in Pardubice, there is also a university, which initially started as Chemical Institute in

1950, thus confirming the paragraph ad c) Employees.

The economic impact of these regulations and restriction can be total – certain plants cannot be located in certain areas. In other cases the effect is relative – companies must spend extra money (thus increasing costs) on environment protection, such as sewage plant, noise barrier etc.

This is not the complete list of all considered variables. However, it shows that the variables are miscellaneous, sometimes contradictory, sometimes positively correlated, sometimes negatively correlated. The final decision is always the result of considering all relevant and important variables, evaluating them and choosing the best option, whether based upon ranking, relative ranking or other multi-criteria decision methods.

Each variable will, in the end, somehow project into the cash flow and profit; the company will either be harmed or benefited.

A lot of the variables mentioned above are connected with the location of a certain phenomenon. That is why the tool, which helps companies answer the question "Where to locate new plant/shop/ service centre . . .?" can save a lot of money for companies or earn a lot of money for the companies. Such tools exist, and they are based upon combining different map layers. Layers cover the information for the location-based decision making and the first step in the analysis is usually the selection of variables. One of the possible set of variables (or factors) is displayed in Fig. 6.7.

In this approach, it is possible to fine-tune the importance of each variable to adjust it to the needs of the decision-makers. Each factor can also be set more precisely by adjusting the weights to the actual needs of the decision-makers as seen in Fig. 6.8.

After the setting and adjusting the variables (factors), the analysis is done by the software automatically. Output can be presented to the users in numerical/digital format (as a set of tables with determined values) or in graphical form, as shown in Fig. 6.9.

White areas are unacceptable; they do not meet the required parameters. Blue areas are the areas which fit the required parameters, the darker the

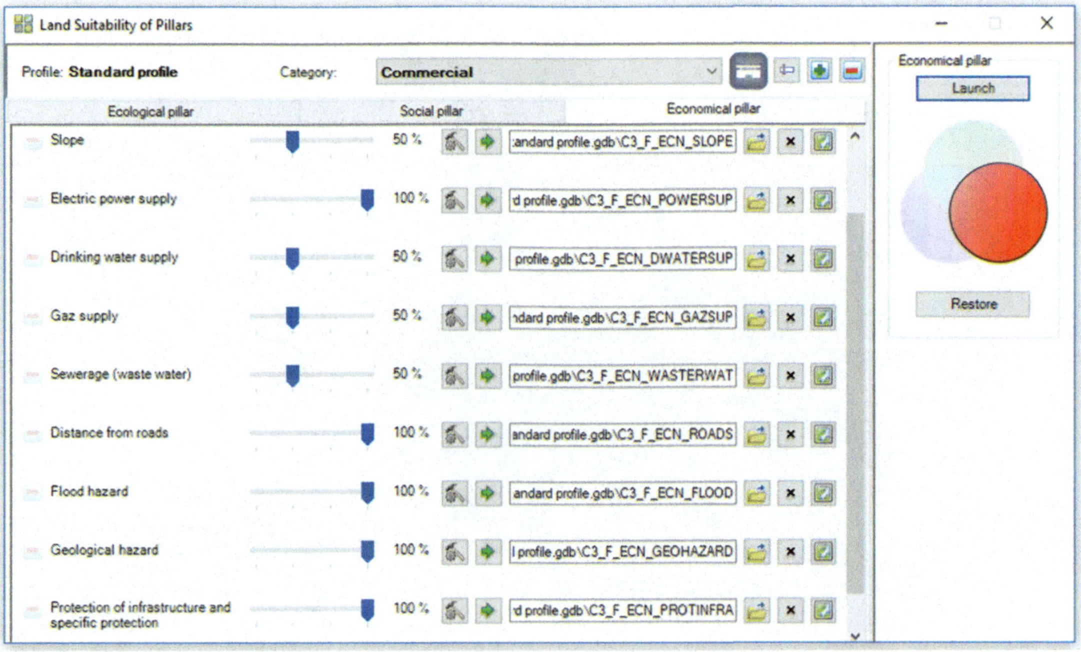

Fig. 6.7 Factors and variables are setting. (Source: authors)

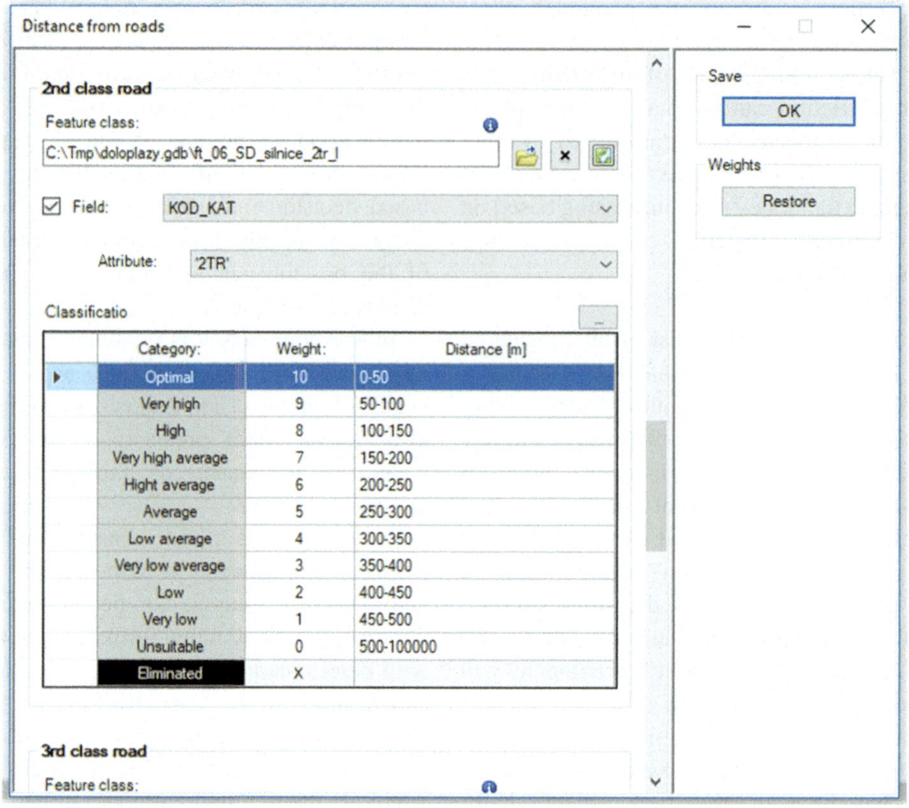

Fig. 6.8 Adjusting the weights. (Source: authors)

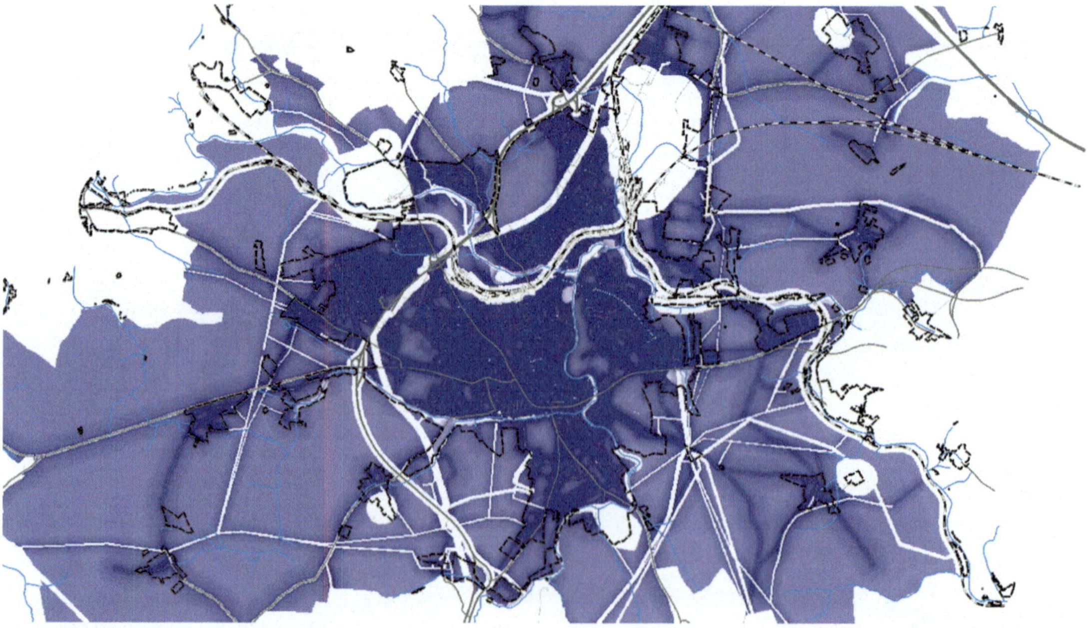

Fig. 6.9 Output presented in the visual image. (Source: authors)

blue, the better the fit. Based upon the outputs, the decision regarding the localisation can be made, in this case, the dark blue areas (lines) are the best options.

6.4 Summary

In this chapter, it has been shown how the business is connected with generating added value for customers. Generating profit and cash flow, even though it is not the true goal of business, is an important and crucial condition of the long term sustainability. Various measures of the profit and cash flow have been used and demonstrated.

In the second part of the chapter, the connection between business and spatial methods has been demonstrated. The first attitude was based upon creating or running the business on spatial expertise. The second attitude has shown how companies from other areas can benefit from the usage and involving of spatial methods.

References

Brealey, R., Myers, S., & Allen, F. (2016). *Principles of corporate finance*. New York: McGraw-Hill College.

Denning Steve. (2013a). *Don Tapscott: We need fundamental change in all our institutions*. Available at: https://www.forbes.com/sites/stevedenning/2013/07/07/don-tapscott-we-need-fundamental-change-in-all-our-institutions/#19588de93577. Accessed 27 Nov 2018.

Denning Steve. (2013b). *FT urges business schools to stop teaching 'The world's dumbest idea'*, FORBES. Available at: https://www.forbes.com/sites/stevedenning/2013/07/09/ft-urges-business-schools-to-stop-teaching-the-worlds-dumbest-idea/#5f34d0225548. Accessed 27 Nov 2018.

Drucker, P. F. (2006). *The practice of management*. New York: Collins.

Fischer, M. J. (2000). Luca Pacioli on business profits. *Journal of Business Ethics, 25*(4), 299–312. https://doi.org/10.1023/A:1006281415873. Kluwer Academic Publishers.

Freeman, R. E. (2010). *Strategic management : A stakeholder approach*. Cambridge: Cambridge University Press.

Friedman, M. (1970). The social responsibility of business is to increase its profits the New York Times magazine. *The New York Times Magazine* (pp. 32–33 122–124). Available at: http://umich.edu/~thecore/doc/Friedman.pdf. Accessed 27 Nov 2018.

Montier, J. (2014). *The world's dumbest idea*. Available at: https://www.lawschool.cornell.edu/academics/clarke_business_law_institute/corporations-and-society/upload/JM_TheWorldsDumbestIdea_1214.pdf. Accessed 27 Nov 2018.

Murray Sarah. (2013). MBA teaching urged to move away from a focus on shareholder primacy model. *Financial Times*. Available at: https://www.ft.com/content/e392f12c-adac-11e2-82b8-00144feabdc0. Accessed 27 Nov 2018.

Tatge, Y. (2017). *Kyrill, the winter storm that walloped most of Europe*. Available at: https://www.air-worldwide.com/Blog/Kyrill,-the-Winter-Storm-That-Walloped-Most-of-Europe/. Accessed 28 Jan 2019.

Teig Amir (2013). *Waze employees clinch the most lucrative exit in Israeli history – Business – Haaretz.com, Haaretz*. Available at: https://www.haaretz.com/israel-news/business/.premium-waze-workers-sharing-in-google-buyout-1.5278721. Accessed 5 Jan 2019.

TomTom. (2017). *TomTom annual report 2016*. Available at: https://corporate.tomtom.com/static-files/765e69fb-27ef-4ae2-8ec0-a415c3a12d20. Accessed 14 Dec 2018.

Valach, J. (1997). *Finanční řízení podniku*. Ekopress.

www.tilak.cz. https://www.tilak.cz/files/old_clanky_gal_d752502688d4a51e427a36ab2854a722.jpg. 10.1.2019.

Economic Geography

7

Vít Pászto

Abstract

Geography as an independent scientific branch comprises of a great variety of subjects with lots of methods, tools and approaches. For instance, when studying human migration flows, we need to be able to use pieces of knowledge from (1) geopolitics to understand the initial motives of migration, (2) (geo)demography to analytically describe, e.g. age structure of migrants, (3) behavioural geography to find out how migrants inclusion would work, (4) urban (spatial) planning to cope with unexpected migrant inflows, or (5) economic geography to model, e.g. how the labour market will react to a sudden increase of potential workforce. On this example, we demonstrate how interdisciplinary the geography is. Moreover, if we add specialised, accompanied disciplines, such as GIScience and cartography, an amazing box of geographical analyses opens.

Economic geography – geography with a focus on various economic aspects – is usually understood as a sub-branch of human geography. Definitions of economic geography are given differently by different scholars, experts and practitioners. Sometimes, we talk about the spatial economy or spatial econometrics; sometimes, it is merely geography applied to some economic theme. In this chapter, we try to shed light on the terminology of "economic geography" in the first part. Following sub-chapter will give an overview of the basic concepts of location theories, which has been broadly used in both economic and geographical theories and frameworks.

Keywords

History · Terminology · Definitions · Location theories · Spationomy

7.1 Definitions and History in Brief

In this subchapter, we will provide a reader with the most used and common definitions of economic geography by various scholars, experts and also scientific-popular sources. We then attempt to summarise the key points from these definitions, helping the reader to understand the main subject of economic geography. We also guide the reader through the (modern) history of economic geography to imagine all the consequences leading to the contemporary state-of-the-art in this field.

7.1.1 Definitions

Taking the definitions by Barnes (2009, 2013a) the economic geography represents a subfield of

V. Pászto (✉)
Department of Informatics and Applied Mathematics, Moravian Business College Olomouc, Olomouc, Czech Republic

Department of Geoinformatics, Palacký University Olomouc, Olomouc, Czech Republic
e-mail: vit.paszto@gmail.com

V. Pászto et al. (eds.), *Spationomy*, https://doi.org/10.1007/978-3-030-26626-4_7

human geography concerned with describing and explaining the varied places and spaces in which economic activities are carried out and circulate. According to the Barnes (2009, 2013a), it was a subject that was more empirically grounded, concerned with context, less abstract, formally theoretical than economics, and it has been subject to so much change due to its empirical basis. Barnes (2009, 2013a) continue with his conceptual thoughts about the discipline and mentions that "since around 1990, after the economic geography absorbed many outer approaches (e.g. from spatial science, post-structuralism and post-Fordism as well as regional science), the topics as labour and work, financial and business services, consumption, retailing, and the firm became prominent subjects of (economic) geographical research." In the line with contemporary issues in the World's' globalised economy, we think that together with the above-mentioned topics, the scope of economic geography can also cover transformational economies (e.g. post-soviet, post-Arab spring, Chinese economy) and their geographical context, big (economic) data analysis, multi-polarity of economic world, increasing mobility of business activities and financial flows, technological businesses, progress and online environments, which create entirely new (virtual) domain to be studied, and many more.

Back to the formal definitions. As mentioned in the introductory chapter to The SAGE Handbook of Economic Geography (Leyshon et al. 2011), practices in economic geography are not independent of the times and places, thus sensitive to the circumstances in which they happen and very sensitive to geographical and historical context. Leyshon et al. (2011) also note that there is a lack of economic geography canon (no core texts or seminal bodies of work) due to the interdisciplinary nature of geography. Therefore, it is rather a diversity characterising today's economic geography. However, there can be identified two main approaches to the field of economic geography:

1. economy as a separate spatial object
2. economy as a delimiter for sub-fields in geography

For the first approach, the most general definitions are valid, i.e. those focusing on geographical aspects of economic topics with the use of a mix of (geographical) methods. One such definition is given by Aoyama et al. (2011): economic geography studies geographically specific factors that shape economic processes and identify key agents (such as firms, labour and the state) and drivers (such as innovation, institutions, entrepreneurship and accessibility) that prompt uneven territorial development and change (such as industrial clusters, regional disparities and core-periphery). Another definition by Maryáš and Vystoupil (2004) states that the primary goal of economic geography is to shed light on a spatial organisation and differentiation of social-economic system and to understand particular economic phenomena in a geographical context. However, Maryáš and Vystoupil (2004) also use the approach of individual sub-fields of geography (e.g. geography of industry, services, tourism geography) to be part of "superior" discipline of economic geography. Castree et al. (2013) define economic geography as a subdiscipline of geography that seeks to describe and explain the absolute and relative location of economic activities, and the flows of information, raw materials, goods, and people that connect otherwise separate local, regional, and national economies. Taking less scholarly approached definitions found on the internet, one of the definitions says that economic geography looks at where economic activities occur, and how they vary by location and interact between places; studies the location, distribution, and spatial organisation of economic activities across the world (Kimutai 2017). According to the Merriam-Webster dictionary, the economic geography is a branch of geography that deals with the relations of physical and economic conditions to the production and distribution of commodities (Merriam-Webster Dictionary 2019). Market Business News website (Market Business News 2019) mentions description of economic geography from the University of Washington's Department of Geography; economic geography is as a field studying the (locational, organizational and behavioural) principles and processes associated with the spatial allocation of scarce (human, man-made and natural) resources (which are also distributed spatially) and the spatial patterns and (direct and indirect, social, environmental and economic) consequences resulting from such allocations.

The second point of view of economic geography distinguishes several sub-fields, where various geographical approaches are applied to specific themes with an emphasis on economic aspects, such as tourism geography, transport geography, labour geography, the geography of resources, rural geography etc. These come mostly from regions with former or current centralised economies (former Soviet bloc, or China) focusing on surveys of natural resources, the selection of sites for industrial plants and railways, land use planning in agriculture, the integrated planning of industrial sectors, and spatial distribution of industry (Coe et al. 2013). As Coe et al. (2013) also noted, scholars and researches from these "geographical schools" were limited by the doctrine of respective political regimes, thus constrained by political conditions not allowing critical thinking about the economic geography. To be fair, this does not mean that the second approach is not acceptable; but quite the opposite – it confirms the statement of Leyshon et al. (2011) mentioned previously – economic geography is sensitive to the circumstances in which happens, especially to the geographical and historical context. The most common categories within the second "sectoral" approach are (to name a few):

- Geography of Agriculture
- Geography of Industry
- Transport and Communication Geography
- Geography of Leisure and Tourism
- Geography of Services, Trade and Retail
- Geography of Resources
- Labour Geography
- Health Geography
- Geography Entrepreneurship
- Environmental and Development Geography
- Rural and Urban Geography (economic aspects)
- Globalisation
- and many more

These categories are not strictly bounded and are interconnected and interrelated, which is typical of geography as such. However, it is important to emphasise that broadness of these categories does not implicitly evoke that economies are the main subject. Generally speaking, any sub-branch of geography potentially carries the "economic aspect" inside. That is why, it is always important to define, what (economic) aspect will be studied and what geographical methods and knowledge will be deployed.

Also, it is worth to show how individual courses and modules of economic geography at higher education institutions (universities) are composed. For instance, London School of Economics in its 2018–2019 course information sheet (see http://www.lse.ac.uk/study-at-lse/uolip/Assets/documents/course-information-sheets/gy2164-cis1.pdf) on economic geography offers pure economical topics (e.g. neo-classical, Marxist and evolutionary/institutionalist views), key concepts and theories (e.g. central place theory, urban hierarchy, core-periphery theories of economic change, agglomeration economies; divisions of labour; cycle theories, and more), economic geographies of the contemporary world (e.g. geographies of economic globalisation in agriculture, manufacturing and services, geographies of ICT and knowledge economies), and economic geography and policy challenges (e.g. uneven development and inequality in the global age, alternative economic approaches). Department of Geography at University College London on economic geography brings for academic year 2018/2019 topics to understand the spaces and spatiality of economies across the Global North, South and emerging economies, e.g. production, exchange, consumption, work, finance and emergent economic activities in connection to urban geography, political ecology and development studies (see https://www.geog.ucl.ac.uk/study/undergraduate/current-students/modules/geog0023). To be more specific, the course will go through lectures about Corporations and Global Production Networks, Resource Geographies, Global (and Gendered) Financial Centres, The Sharing Economies, or Digital Capitalism to name a few. By looking at the syllabus of the

English version of economic geography course at Masaryk University (Czechia), we can find topics such as Population Geography, Geography of Settlements, Geography of Trade and Services, Transportation, Tourism, Recreation, Agriculture and similar. This illustrates the less critical (in terms of thinking) and more structured sectoral approach, as mentioned earlier. On the other hand, the economic geography module offered by Department of Social Geography and Regional Development at Charles University in Prague (Czechia) does not follow this approach and contains cross-sectoral topics such as Labour Force, Natural Environment and Economics, Commodity Chains and Globalisation, Economic Geography of Consumption, Agglomeration Cluster etc.

To illustrate how colourful and diverse the economic geography is, it is the best to list all the topical entries from the International Encyclopedia of Human Geography (Kitchin and Thrift 2009) that are contained in the "economic geography" theme. This variety of 96 entries in Table 7.1 speaks for itself.

All in all, common denominators for most of the economic geography definitions are:

- It is a sub-field of geography with the focus on location and distribution of human economic activities,
- it studies not only immediate geographical manifestations of such activities but also their interplay with other social-political phenomena,
- it is an open discipline, very flexible to ongoing changes and development, with no distinct paradigm,
- there exist space and time dependence of theories and approaches.

Krugman (1991b) states that economic geography is the study of the location of factors of production in space. Although it is probably the most straightforward definition of economic geography, the author Paul Krugman (awardee of 2008 Nobel Prize in Economic Sciences) can be treated as a guarantee of its validity. To conclude this part, it is the best to quote Barnes's notes (Barnes 2009, 2013a) – that still, economic geography has not its own shaped and bounded orthodoxy or paradigm to be typical for. It is rather the discipline that is intellectually open, eclectic, pluralist and very flexible in terms of "breath in" various temporary trends and notions.

7.1.2 Historical Overview

This historical overview of modern economic geography is fuelled mainly from the extensive books authored Coe et al. (2013) and Aoyama et al. (2011). The authors mention that the modern economic geography started after World War II, when the colonial tradition of major countries (especially the United Kingdom), established for several centuries, started to shatter. This is an important moment since the colonialism geographies and history strongly influenced leading geographical schools. Aoyama et al. (2011) describe that the early history of economic geography was formed and defined by various approaches. Firstly, it is in line with Coe et al. (2013) that economic geography was closely related to British colonialism with a focus on commodities, transportation modes and trade routes (Barnes 2000). Secondly, a very influential stream was linked with German location theories formalised by notable persons such as Heinrich Von Thünen, Alfred Weber, Walter Christaller, August Lösch, and Walter Isard from North American geography (see more about the theories in Sect. 7.2). Later on, these theories were the basis for a new geographical approach called regional science. Another understanding of economic geography can be represented by Alfred Marshall, who formed a concept of industrial agglomerations and emphasises so-called economies of scale. His work continues to influence some of the current economic research on agglomerations and clusters. The last lineage of economic geography, as Aoyama et al. (2011) mentions, had its root in North American geography. The main concerns of American researches at the time were connected with their human-environment approach in geography – with human adaptations to natural resources in the process of industrialisation. On

Table 7.1 Topics in the economic geography theme listed in the contents of international Encyclopedia of Human Geography (2009)

Topic			
Agglomeration	Environmental regulation	Information technology	Professional services
Business services	Ethnic economies	Innovation	Public policy
Capital and space	Export processing zones	International financial centres	Radical political economy
Capitalism	Feminism and work	International organizations	Regional integration
Commodity chains	Feminist political economy	International trade	Regional production networks
Competitiveness	Financial exclusion	Internationalization of education	Relational economic geography
Consumption	Financial knowledge	Investment promotion	Resource and environmental economics
Corporate responsibilities	Financial risks and management	Knowledge communities	Resource industries
Creativity	Firms	Knowledge economy	Retail geographies
Cultural economy	Food networks	Knowledge intensive business services	Spatial division of labour
Cultural turn	Food regimes	Labour control regime	State theory
e-business and e-commerce	Fordism	Labour flexibility	Technological change
Economic crises	Fordism, post-Fordism and flexible specialization	Labour geography	Technology and regional development
Economic geography	Foreign direct investment	Labour market	Technology industries
Economic geography and the internet	Global commodity chains	Labour unionism	Telecommunications
Economic geography, quantitative	Global production networks	Local economic development (politics of)	Trade blocs
Economies, alternative	Globalization and transnational corporations	Locality debates	Transitional economies
Economies, borderland	Governance, corporate	Location theory	Transnational corporations in developing countries
Economies, branch plant	High-tech industry	Migrant workers	Transnational elites
Economies, imagined	Industrial location	Natural resources	Transnational ethnic networks
Economy, informal	Industrial organization	Neoliberal economic strategies	Transnationalism and labour geography
Embeddedness	Industrial parks	Networks	Transnationalism and technological transfer
Enterprise discourse	Industrial restructuring	Offshore finance	Uneven development
Entrepreneurship	Industrialization	Privatization	Venture capital

top of it, as a universal discourse, there has been and will be contradictory methodologies – deductive scientific approach leading to abstraction and universal applications (we can call it "nomothetic"), and descriptive approach gathering evidence and concrete information "outside a laboratory" with strong emphasis on humanity (we can call this approach as "idiographic").

Coe et al. (2013) divide the modern era of economic geography with regards to the most influential post-war philosophical and geographical trends of the twentieth century – positivism, structuralism, and post-structuralism. All three trajectories are too complex to be described in detail here; however, the most important remarks will be mentioned.

7.1.2.1 Positivism

In the era of positivism, a scientific approach in which the emphasis is put on the universal (nature) laws and quantitative methods, the systematic and deductive methodologies were developed and also applied in economic geography. Typically, as mentioned in Coe et al. (2013), economic geographers were looking for universal principles lying under spatial patterns of economic activity and using quantitative data and methods (statistics in particular) to find and proof such spatial patterns. According to Scott (2000), these quantitative methodologies were used in two ways – (1) spatial analysis using mathematical models (with the advances in computer science), and (2) integrating space and location into neoclassical models of economic theory. During this era, scholars were following the famous German location theories, applying them and building new concepts on these classical works. Typical applications included searching and evaluating optimal sites for various facilities, urban system models (e.g. with the use of physical concepts such as entropy, chaos theory and fractal geometry), accessibility of urban functions, innovation diffusion models, or optimisation of land use patterns (Coe et al. 2013; Aoyama et al. 2011). Although this positivistic approach was not universally used throughout the whole economic geography community, it certainly helped to encapsulate and specify by then not really bounded research agenda of economic geography. Widespread enthusiasm about new opportunities in spatial and regional science and economic geography led to attracting new young scientists who are nowadays called as a "young cadets" (Aoyama et al. 2011), such as Brian Barry, William Bunge, Waldo Tobler, or Arthur Getis. Their "father", William Garrison, is treated as the central quantitative (economic) geography figure after World War II. Considering that the generation of quantitative geographer was significantly influencing the field (they become a new classics from today's point of view), their heritage is still present in contemporary approaches in economic geography, in its quantitative branch respectively. According to Coe et al. (2013), quantitative methodologies are evident and needed in research where large datasets are analysed to find and describe hidden patterns of studied phenomena represented by the data.

7.1.2.2 Structuralism

The second major trend in economic geography identified by Coe et al. (2013) is called structuralism. Generally, structuralism represents a theoretical concept based on a presumption that various phenomena and processes visible "on a surface" have their causations hidden deep in their (invisible) structure (Daněk 2013). Therefore, for structuralists, the key research task is to unveil and understand these structures. It is also worth to mentions two (other) important subsets of structuralism that penetrated economic geography, such as dependency theory (Frank 1966) and world system theory (Wallerstein 1974), which were then mostly applied in social sciences (Aoyama et al. 2011). One of the most important structures linked with people's actions were a social class, race, or gender (Coe et al. 2013), which raised general and scientific awareness of social issues. Since the social aspects of humankind and their spatial (geographical) manifestations and consequences were barely possible to be grasped by the quantitative approach, a new trajectory had to be found. As Coe et al. (2013) noted, quantitative geography and location theories reacted to the economic expansion of late 19th, and up to the mid-twentieth century, in the late 1960s and early 1970, the economies (at least in the West) were starting to decelerate. This resulted in slowly emerging social problems and new issues (e.g. urban segregation, gender inequalities, international deindustrialisation, changing labour markets), which required the understanding deeper structure of ongoing economic processes. It is one of the most important (economic) geographers – David Harvey – who shifted geographical research interests from its quantitative approach to Marxism and political activism (Harvey, 1974). As noted by Barnes (2011), this Harvey's twist took only 3 years. While in 1969 he celebrated quantitative geography in his book Explanation in Geography (Harvey 1969), in 1972 Harvey (1972) began by attacking the usefulness of the theory and statistical techniques (Barnes 2011). Under the umbrella of Marxist theories, Harvey and his students were convinced that economies based on capitalism and class-inequalities would lead to the crisis. The Marxian concept was not "comfortable" for many geographers, however, identification of class-based power relations in economic processes, and to a conceptualisation of uneven development became acceptable in geographical mainstream (Coe et al. 2013). Another approach, rather positive in terms of preventing economic crisis, is known as regulation theory focusing on the role of institutions or state itself in order to avert a crisis. A wide range of topics within the structuralism is summarised by Coe et al. (2013), i.e. Marxism, institutionalism, feminism and anti-racism, by identifying a common denominator – the existence of underlying structures of power.

7.1.2.3 Post-structuralism

The last major sub-field of modern economic geography described by Coe et al. (2013) is influenced by philosophical concept post-structuralism, which generally allows the existence of multiple "truths" depending on the researcher's circumstances (the knowledge is time and place-specific). Taking the explanations of post-structuralism from Daněk (2013), we can say that it is an approach of several "small" theories considered as toolboxes which can geographers select to achieve their goals. Post-structuralism denies its predecessor - structuralism – i.e. complete rejection of "hidden essence", fully scientific representation of the world and one universal truth (Daněk 2013). Coe et al. (2013) identified several ways of how post-structuralism approach penetrated economic geography:

- Economic geographers started to think about how they understand and represent economic processes. This lead to a proposal to imagine alternative and diverse forms of economic life (Gibson-Graham 2006),
- questions about the power of representation and discourse in economic life were raised,
- studies about the individual identities in economic processes were conducted in order to "pull out" individuals from existing structures (e.g. worker, immigrant, woman),
- and realisation of the importance who has the power to produce knowledge. Questioning the Western understanding of economic development and poverty has been one of the main issues for post-structuralist studies in economic geography.

Coe et al. (2013) summarise that post-structural approaches to Economic Geography ask how we are constructing our knowledge about the economic world and what are the consequences of understanding things in that way.

7.1.2.4 Future Directions

Even though the economic geography is a subdiscipline with no exact border and with a great ability to absorb knowledge and methods from other disciplines, several future trajectories can be found. These future pathways presented in this part is a fusion from the books by Coe et al. (2013), Aoyama et al. (2011) and chapter in International Encyclopedia of Human Geography by Barnes (2013a). Starting with Coe et al. (2013), authors identified following topics which will resonate in the next few years: (1) a shift in global economic power (although there is uneven distribution of global wealth, its reorganisation is underway – e.g. with significantly rising economies of China, India and other developing countries in Asia); (2) new forms of global integration (in terms of migration and adaptation to new conditions and labour force, and also in terms of increasing mobility of financial capital. This will lead to new forms of regulations and collective control on this issues); (3) continued dynamism in the development of new technology (information and communication technologies significantly changed how we consume, what we consume, and how we work. It will raise the needs of regulation and control of cyberspace, e.g. when it comes to personal data protection); and (4) the need to address environmental challenges (there is global interdependence on the natural environment. Thus it will be inevitable to cope with climate change. Economic costs of a changing climate will be significant, while new green technologies, carbon emissions, or smart land use will be more and more a subject to study).

Aoyama et al. (2011) predict that the knowledge economy steadily gaining significance will be the basis for newly emerging sectors of economic activities of knowledge-based resources. Financialisation of economies will gain more importance with the financial crisis of 2008, while consumption is a relatively new focus for economic geographers (formerly preoccupied with production). Finally, Aoyama et al. (2011) pinpoint a concept of sustainable development (emerged in the 1980s) that is on the rise due to the intersection of climate change and economic change in cities, suburbs and in developing countries.

Barnes (2009, 2013a, b) looks at the economic geography as a subject and notes that the current discipline's inconsistency is the only constant. Influences from other disciplines seem to have a greater effect on the economic geography, which makes its future not assured and challenging. However, Barnes (2013a) identified four major areas in contemporary debates and research in economic geography: (1) Discussion of methods (from quantitative, through qualitative, to in-depth case study and action research) and their consequent liberal and open applications; (2) along methodological openness, also theories are somewhat permissive – Thrift and Olds (1996) speak of "polycentric" economic geography in this sense; (3) various grasping approaches to globalisation (international division of labour, multinational corporations, international financial capital, communications, and global commodity chains are just a few examples of the globalisation issue coverage); and (4) primary resources and nature become the focus of disciplinary discussions (mainly due to their socio-economic-cultural impact). As Barnes (2013a, b) concludes, the unusual challenge for the discipline will be less breaking new ground, than holding its existing one.

In order to demonstrate how all these future outlooks, with the newest from 2013 by Coe et al. (2013), fits the keywords of the current agenda, we included a word cloud from the last Global Conference on Economic Geography (GCEG) held in Cologne (Germany) in summer 2018 (Fig. 7.1). This word cloud is composed out of 100 most cited words in the conference paper

Fig. 7.1 Word cloud of the top 100 used words in the conference paper titles at Global Conference on Economic Geography. (Source: detailed programme PDF file at https://www.gceg2018.com)

titles which speaks for itself. At first glance, the most visible words are "global" and "development", which is interesting since it is in line with some of the future directions mentioned in the previous text. The umbrella topic of the conference was "Dynamics in an Unequal World", so the issues such as Poverty, Inequality, and the Global South were intensively discussed. To provide a bit more complete picture of the flagship conference on economic geography, the topic of the fourth GCEG held in Oxford (UK) in 2015 was "Mapping Economies in Transformation".

7.2 Location Theories

It is hard to select a topic following the part with definitions and historical development of economic geography. There are so many exciting issues from economic geography (historical or contemporary), which could be discussed and analysed in this part. So why location theories? The inclusion of this part was mainly driven by the author's liking for their beautiful artistic representation of a landscape-economic pattern, however, based on geometrical assumptions, therefore unrealistic when applied to real-life situations. This is by the purely subjective notion of the author. However, there is also an objective reason why to present location theories – they and their authors are meant to be one of the foundation stones of economic geography. Without understanding the fundamental basics of economic geography, it would be like building a house from the top. As Aoyama et al. (2011) note, every student who claims to know something about economic geography must know its disciplinary roots. Some of the principles of location theories remain true today, several decades after they were proposed. Moreover, location theories can also be inspiring for the creation of new concepts, or a modification of existing ones – those theories that come from other disciplines (e.g. physics or biology) to be infiltrated into economic geography.

As a synthesis of a various sources (Aoyama et al. 2011; Barnes 2013b; Leyshon et al. 2011; Murray 2009) a term "location theory" in economic geography stands for a concept that (in its simplest definition) works with two spatial-economic features – distance and area (e.g. transportation costs, the cost of overcoming distance, affect the price of products, the location of production facilities, the geographic extent of markets) – that are put into formal and abstract models in order to set ideal patterns of the space economy, and to develop a generalisable framework explain industrial localities. All with the assumption that the actors are rational and maximising their economic gain, and decisions are made regardless of social, cultural and environmental factors. Location theories try to answer the questions of 'why' and 'how' spatial patterns of (economic) activity have evolved. Approaches that strive to build idealised patterns of whatever socio-economic issue are called general equilibrium analysis (Ponsard 1983).

In this part, we briefly introduce basics of the four major location theorists – Heinrich von Thünen (1783–1850), Alfred Weber (1869–1958), August Lösch (1906–1945) and Walter Christaller (1893–1969); interestingly, all of them were Germans. We also name a few other theories and their authors, mostly build upon the classical ones, but this will serve more as a reference for self-study of those who are interested in the topic.

7.2.1 Von Thünen Location Theory

Heinrich von Thünen is treated as the author of the first spatial economic theory Der isolierte Staat in 1826. He was dealing mainly with land-use functions of a city defining their particular usage based on distance from the market. The market itself was located in the centre of the modelled area (city). He worked with three assumptions (Aoyama et al. 2011): (1) an isolated state with a single central city (market) surrounded by agricultural land on a uniform plain, (2) farmers are rational profit maximizers who all face the same production costs and market prices, and (3) transport cost is proportional to distance. He developed the main notion for such spatial organisation, which is an economic/

location rent (net profit). This rent should be maximised, and from the farmer's point of view (and based on assumptions as mentioned earlier) it generated a zonal land-use model (Fig. 7.2a). It differentiates the land use into four main categories based on the consumers' needs, and distance – dairying and basic food gardening was placed closest to the market (city centre) since its products spoil quickly; forests as a source of wood for construction and heating is another layer; extensive farming producing "long-growing and long-lasting" products (e.g. wheat) is the third general land-use; and extensive area used mainly for livestock and grazing was the last. This idealistic model was to be then used in comparison with reality. Generally, the model assumes land-use functions as it was seen to ideal in the nineteenth century; thus, the model is not applicable in today's reality. Throughout the description of Von Thünen model including geographical and historical context is given in Leyshon et al. (2011). However, after more than a century, Alonso (1964) developed a bid-rent model applied to urban areas (Fig. 7.2b). This was somehow a more suitable model for cities at the time and is with some limitations applicable to specific urban systems today as well. It separates city zones into four main categories based on their accessibility/bid rent - retailing (shopping), commerce/industry (manufacturing/offices), apartments (high-density residential), single houses (low-density residential). Again, this zoning is based on a distance from a city centre, or it is better to say, in this case, from Central Business District (CBD).

7.2.2 Weber Location Theory

After more than 70 years after the first location theory, a new concept was developed by another German scholar – Alfred Weber. He elaborated the theory about the optimal placement (most profitable) of a firm/factory based on so-called "factors of production" or "Standortfaktoren" (Weber 1909). These factors are composed mainly of the three production items - land (as a resource factor), labour, and capital (Smith 2013). His theory on industrial production also takes the accessibility or transportation costs into account in the sense of their minimising. As a result, Weber came out with a location triangle, nowadays known as Weber's triangle, calculating transportation costs as a product of Euclidean distance (i.e. straight line) and the amount of material transported (Murray 2009). According to Leyshon et al. (2011), Weber

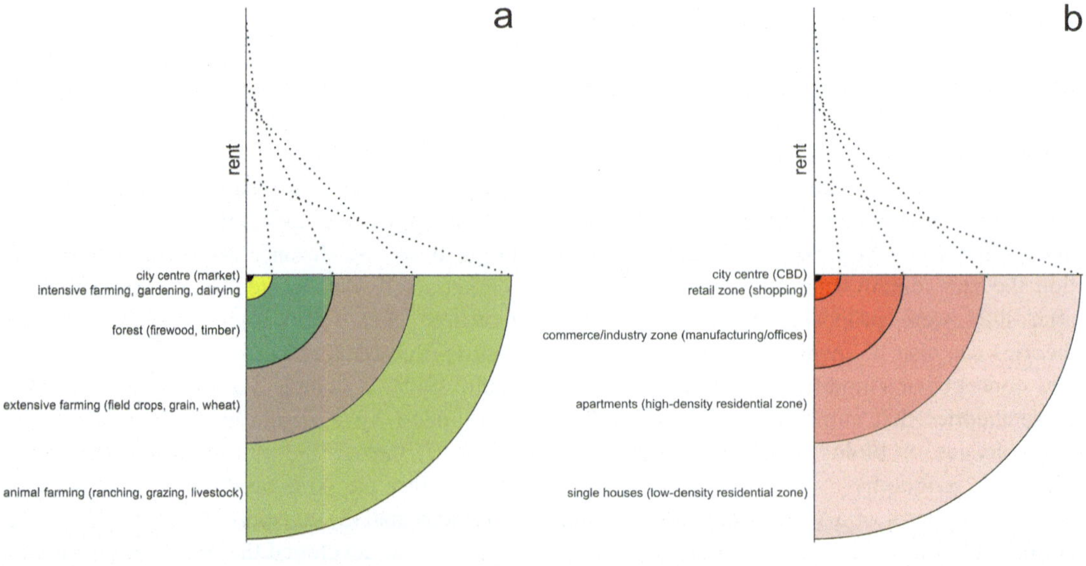

Fig. 7.2 Von Thünen location model (**a**), and modified Alonso's version (**b**). (Source: Author)

employs assumptions such as rational decision-makers endowed with perfect information, perfect competition and a flat surface, to keep intervening forces constant. Although Weber was aware of cultural factors that might influence prices, he did not consider them in theory. It is interesting that Weber was trying to develop a theory too complex, so he needed to ask for help from mathematician Georg Pick, who assisted Albert Einstein in formulating relativity (Barnes 2013b). Figure 7.3 illustrates the complexity of the whole concept and shows that the theory was not only conceptual or "philosophical".

Generally, Weber assumed to place a factory between sources (raw material) and market (agglomeration) considering wages (i.e. costs of the labour force) and transportation costs (of both raw material and final product). In other words, a spatial equilibrium of a factory has to be found (Murray 2009); see Fig. 7.4. Transportation costs play an essential role in his model because we need to consider weight loss or gain of the product. It is favourable to place a factory closer to a source of, e.g. iron ore deposits as the material itself are processed in the factory for example into metal plates; thus the input material loses the weight. At the same time, the new product will bring some added value. It is then profitable to transport the final product on longer distances (rather than to transport input raw material). In the case of weight gaining, e.g. when a factory produces (army) tanks which are more expensive

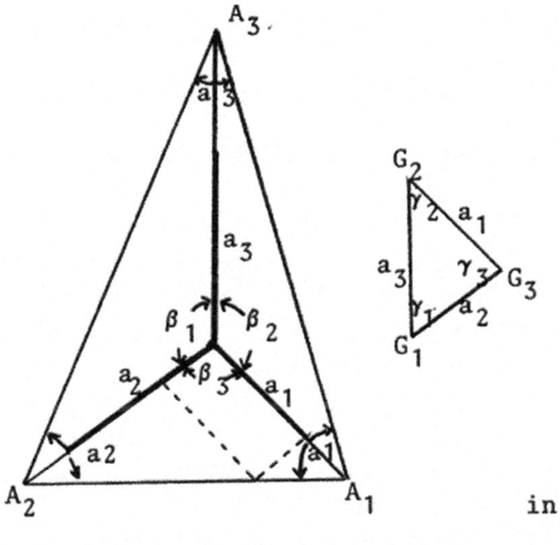

in Weber app.Pick

We obtain in this manner "the triangle of weights" $G_1G_2G_3$ whose angles γ_1, γ_2 and γ_3 are the supplements of the angles β_1, β_2, β_3 and whose sides are determined by a_1, a_2 and a_3 by virtue of the theorem of the parallelogram of forces.

P_o is determined when the position from which the lines connecting it to A_1, A_2 and A_3 form the supplementary angles of γ_1, γ_2 and γ_3 is known: that is, the position of P_o from which A_2A_3 is seen subtending the angle β_1, A_3A_1, the angle β_2 and A_1A_2, the angle β_3.

Now, the angles $\widehat{A_2P_oA_1}$ and $\widehat{A_1P_oA_3}$ will be respectively equal to β_3 and β_2 when P_o is at the intersection point of the arcs passing through A_1A_2 and A_1A_3 (see Figure 2).

Fig. 7.3 Mathematical foundations of Weber's triangle. (Example from Ponsard 1983)

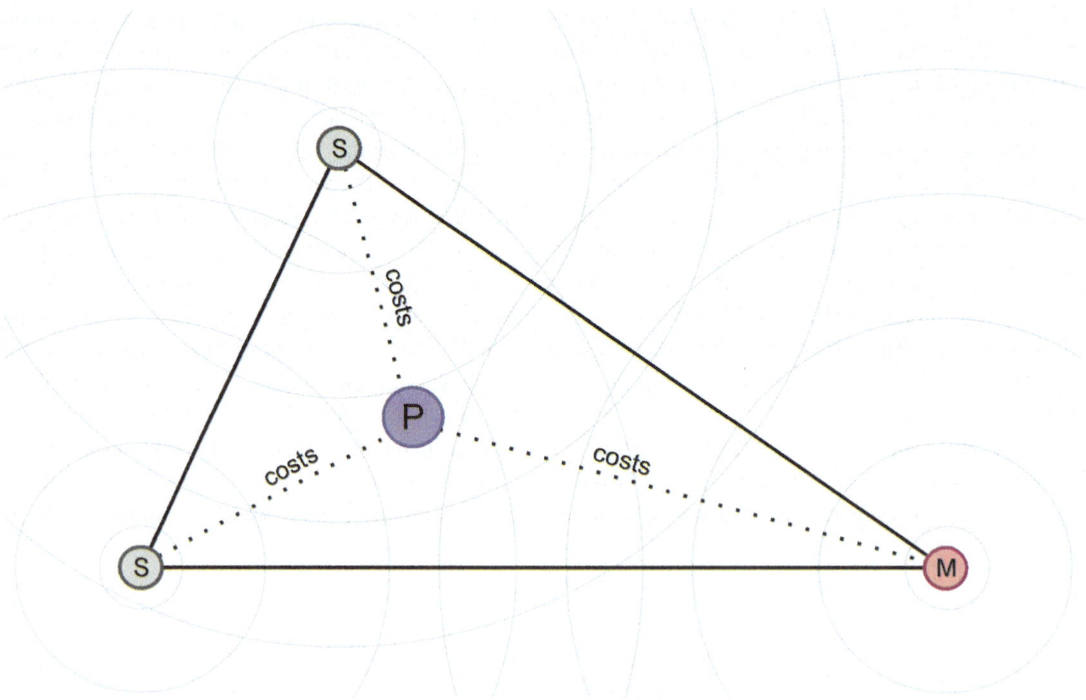

Fig. 7.4 Weber's triangle, where P is a factory (production), M stands for the market, and S represents sources. (Source: Author. Adopted from Weber 1929)

to transport, it is better to find a location of the factory closer to the market. Apparently, in a real-life situation, it is not that straightforward – we need to take multiple inputs into account (and not even counting with heterogeneity/variety of the inputs), and the transportation costs have been slowly losing its significance with the rapid development of transportation industry. However, as Aoyama et al. (2011) notes, Weber's work remains important as many principles of his location theory largely still hold today and broadly explain why firms locate their operations where they do. More about Weber's location theory can be found in Aoyama et al. (2011), Leyshon et al. (2011), or Ponsard (1983).

7.2.3 Christaller Theory of Central Places

Central place theory is probably one of the best-known theory on urban settlements in geography and related field. Paradoxically, neither in geography as a discipline nor in his professional life, Walter Christaller became a recognised person for this theory. First, according to Leyshon et al. (2011), Christaller's work was never truly appreciated among German geographers, that time very much concerned with idiographic and chronological analysis. Second, while Christaller's theory of central places was popular amongst American geographer (especially during the quantitative revolution in the 1960s), in his home country he was reintroduced during 1960s and 1970s through English-speaking textbooks, Christaller was never appreciated in geography during his own time (Leyshon et al. 2011). It might also be caused by his "controversial" political engagement – during Nazi era, he joined the National Socialist Party (and authored a work where he applied the theory on reorganisation of Poland), after the war he joined communist party ending in social democratic party in 1959 (Leyshon et al. 2011). Nevertheless, this has nothing to do with his theoretical contribution to (urban) geography and economy.

According to Johnston (2013), Christaller in his work (Christaller 1933) about central places describes a theoretical statement of the size and distribution of settlements within an urban system in which marketing (especially retailing) is the predominant urban function. In other words, the theory describes a relationship between central places (cities and towns) and hinterlands they served (Murray 2009). Christaller identified two main concepts (Johnston 2013) – (1) the range of good (maximal distance consumers are willing to travel for it), and (2) the threshold for good (minimum volume of necessary sales to maintain selling that good). A clear example is given by Murray (2009): A good, say clothing, is produced and made available in a city, as an example. The demand for this good will be a function of its price and the travel cost for a consumer to purchase it. Thus, retailers locate their businesses to be as near their customers as possible, and at the same time, customers visit the nearest available centre (Johnston, 2013). From a customer perspective, this results in minimum spending on travel and maximum on services and goods themselves.

Assuming a uniform plane with equally distributed population, the principles as mentioned above (range and threshold) cause to form a hexagonal structure from a "normal" settlement structure (Fig. 7.5a), where each hexagon is treated as a "hinterland" (or market area) of a central place. Moreover, this hexagonal grid representing central places is organised further into different levels of hierarchy (Fig. 7.5b). Higher-level cells (e.g. larger cities) offer more services and goods, and lower-level cells (medium and small cities/towns) offer less variety of such goods and services and focus on more regularly used goods and services (e.g. bakery) (Murray 2009). It is evident that larger cities also offer such regular goods and services but is not economically viable for consumers to travel such distance (from their lower-level settlement). On the contrary, typical examples of higher-level services and goods that are worth to travel for are governmental functions, higher education, health services, insurance services or big sports venues. In general, these are mainly quaternary or quinary services.

In addition, Christaller identifies three basic principles based on nesting logic of the hinterlands (hexagons) (Johnston, 2013):

1. Marketing principle – minimises the number of settlements so that each is at the junction of three hexagons at the next hierarchical level up (Fig. 7.5c),
2. transport principle – minimises the length of roads joining adjacent places. Each settlement is located centrally on the boundary line between the hexagons of two places in the next hierarchical level (Fig. 7.5d),
3. administrative principle – every single lower-level settlement and its hinterland is nested within the higher-level hinterland (Fig. 7.5e).

On top of it, we also have two options for how to depiction of the Christaller's central places – using hinterlands mode, and routes mode (more in Johnston 2013). To sum up central place theory, Leyshon et al. (2011) quote Christaller that he wanted to build "theory of location… to correspond with Thünen's agricultural production and Weber's theory of location of industry... derived deductively, by pure reasoning."

7.2.4 Lösch Location Theory

The fourth classical location theory comes from August Lösch and his work The Economics of Location from 1940 (Lösch 1940). He built his theory on Von Thünen's and Weber's models and also merged these ideas with Christaller's findings. Lösch tried to include agriculture and production locations to form a general equilibrium framework of the spatial economy (Leyshon et al. 2011). According to Aoyama et al. (2011) and Leyshon et al. (2011), Lösch sets the optimal location as places where the difference between total revenue and the total cost is the largest, while at the same time each producer maximises its market area. Similarly to Christaller, he ended up with hexagons representing that market area. The size of hexagons will vary across industries dependent on the range and threshold of industry-specific products, and this abstract space divides

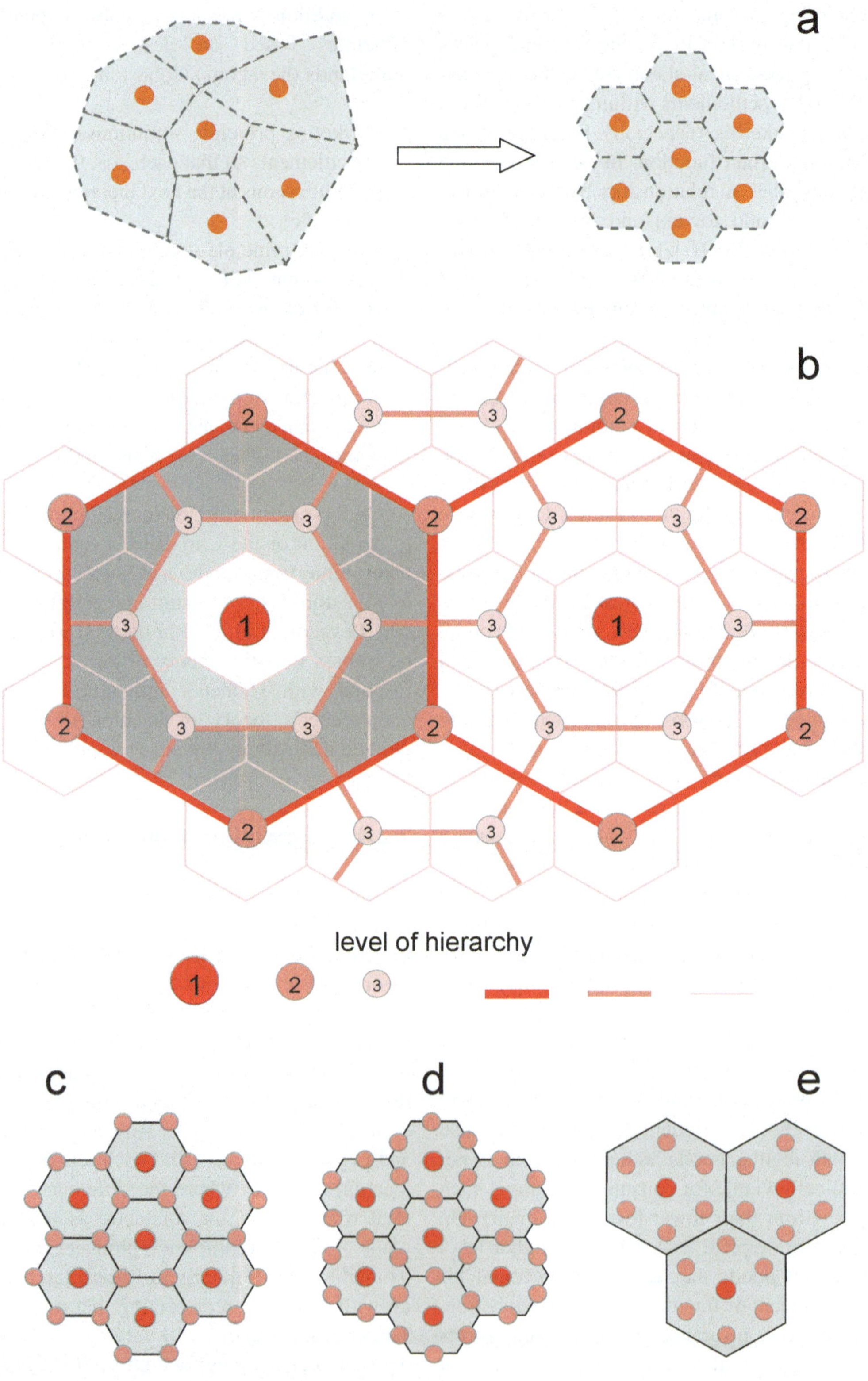

Fig. 7.5 Christaller's central place theory. (Sources: Author)

regions into activity-rich (core) and activity poor (periphery) (Leyshon et al. 2011). As Leyshon et al. (2011) mentions, Lösch's ideal landscape has the following characteristics:

- The greatest number of locations coincide,
- a maximum of purchases can be made locally,
- the sums of distances and consequently shipments between industrial locations are least,
- and that transportation costs and transportation lines are minimised.

In Lösch's model, he assumed that the market areas as overlaid one over another, which implies a certain type of competition amongst producers. For various goods and services, the model uses grids (market areas) with different size, while these grids are gradually overlapped into a regular shape – hexagons. Figure 7.6a demonstrates the

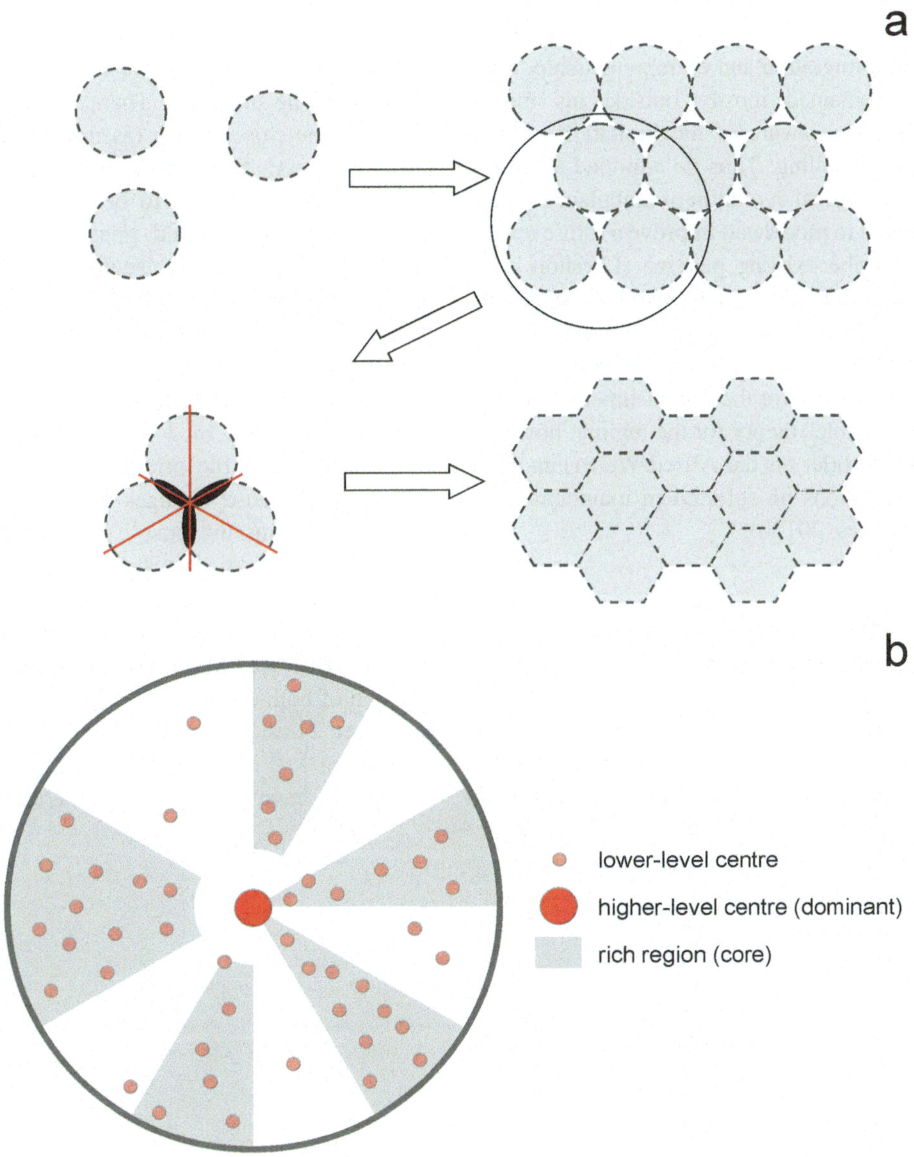

Fig. 7.6 Example of Lösch's location theory. (Source: Author)

process of how market areas fill the space gradually to form a new regular structure based on overlaps. According to Lösch's theory, central places do not necessarily have to offer all the goods and services as at the hierarchical lower-lever places (on the contrary to Christaller's theory). As a result, there are special areas (regions) formed with/by the different structure of production and demand. These regions form radial arrays going out from one dominant centre (Fig. 7.6b).

As noted by Aoyama et al. (2011), Lösch recognised the presence of various important factors that shape regional growth, such as regulations, migration and entrepreneurship, these elements remained mostly outside his model. Lösch was also aware of the limitation of such (artificial) modelling. Thus he admitted a need to study the historical evolution of real places as well as he desired to model and improve a future reality, not to describe existing patterns (Leyshon et al. 2011). A sad history about the life of August Lösch depicts the historical and geographical context, in which he lived and died. He was pushed by Nazi party to work for the establishment by developing a "suitable" theory for the regime; however, he despised Hitler (as did Alfred Weber) and died from deprivations he suffered in maintaining his values (Barnes, 2013b).

7.2.5 Other Location Theories and Concepts

In the last part of this chapter, spatial economic models and theories that do not belong to "big four" concepts will be briefly mentioned, with Hotelling spatial competition model to be described broadly. Starting with quite known spatial competition introduced by Harold Hotelling, American mathematician and statistician, in 1929 in his "Stability of Competition" (Hotelling, 1929). Hotelling draws a model where on a line of a certain length (e.g. street), two sellers A and B offer a homogeneous product, with zero production cost. They might be located at certain distances from the ends of this line. Customers are supposed to be evenly distributed along the line, and each customer consumes exactly a single unit of this commodity, irrespective of its price (which is also assumed to be uniform). Then, a customer will buy from the closest seller (assumed linear transportation costs). The question was – where is the best position for sellers to be placed along the line (street). One should guess that they will divide the street evenly, standing in ¼ of the line length; thus, each seller covers half of the street. But as Murray (2009) noted, this would lead to a situation that if one seller moves towards the other, then he could gain additional demand purely by locational advantage (Fig. 7.7a). Also, the pricing of a product will (in real life) cause changes in market share – e.g. by reducing the cost of the product, one can increase the market share (Murray, 2009). However, this factor is to be set uniformly in the model. With the use of complex mathematical formulas, Hotelling concluded that the best position of both sellers is exactly in the middle of the street. Furthermore, they should be oriented back to back (Fig. 7.7b). This ensures that both sellers will have exactly half of the total market.

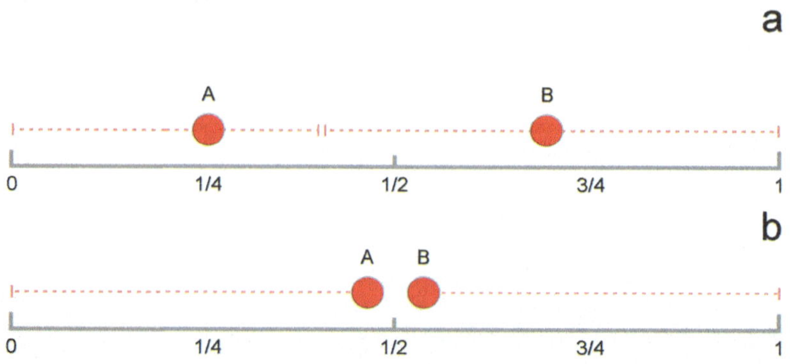

Fig. 7.7 A simplified example of Hotelling model of spatial competition. (Source: Authors)

Besides the spatial composition of two sellers, Hotelling also defined the principle of minimum differentiation. In short, he assumed that there is a tendency of sellers to make the new product very much like the old, applying some small changes which will seem an improvement to as many buyers as possible (Hotelling 1929). However, this was questioned by d'Aspermont et al. (1979) by giving a counterexample to Hotelling's conclusions, i.e. there is a tendency for both sellers to maximise their differentiation, and one should intuitively expect that product differentiation must be an important component of the oligopolistic competition. Although the Hotelling model seems to be quite artificial and not really to correspond with reality, we can find interesting examples of its principles, i.e. sellers with similar products tend to be co-located. This is typical for shopping malls, where product categories are usually concentrated or spatially organised in order to be adjacent. We cannot simply assume that shopping mall managers apply the Hotelling principle as regards placing new shops, but the analogy with the Hotelling model is interestingly precise. However, right reasons could be more prosaic - shops are concentrated in certain places in order to help customers to orient themselves in a mall, their placement might be driven by technical conditions of a mall's construction, marketing and customer behaviour can be a reason, manager wants to guarantee the fairness of shops placement, or it is simply beneficial for all clustered shops (which is the closest to what Hotelling identified). On a higher level of a hierarchy, on an example of supermarkets, there also exist a distinctive spatial agglomeration of such commercial units. In Fig. 7.8, there is an example of two supermarkets placed just across the river.

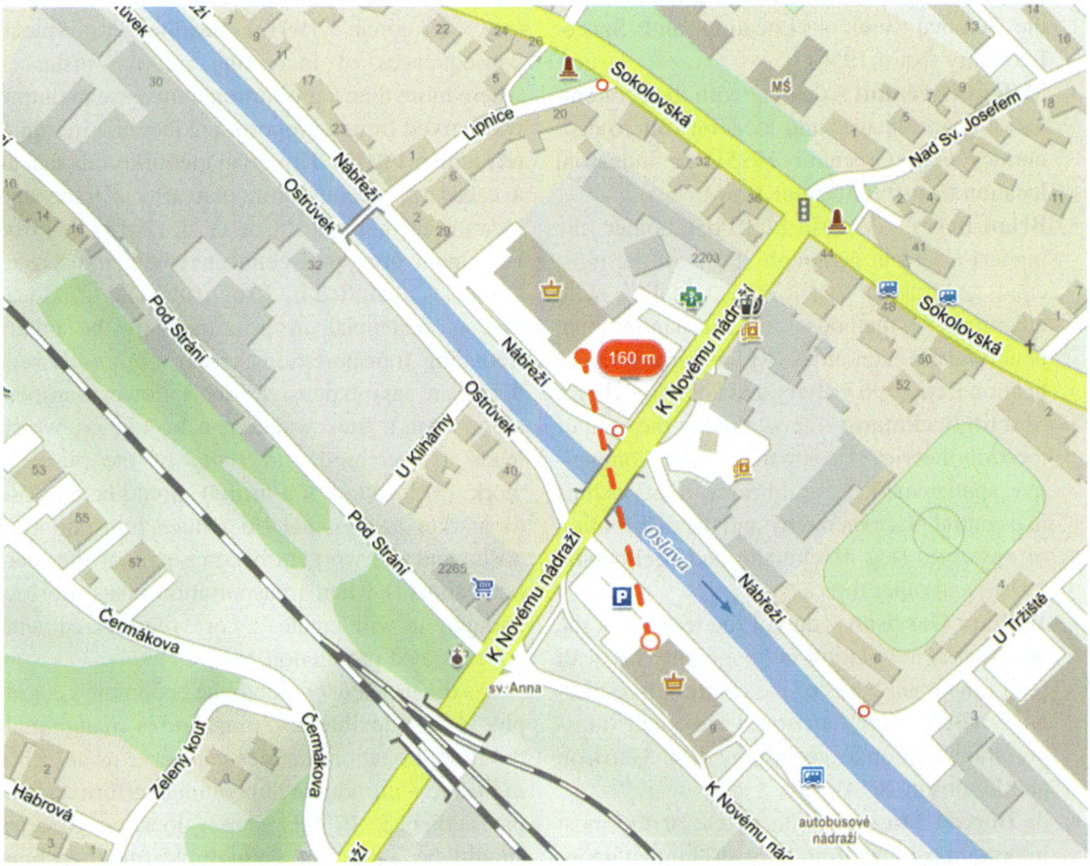

Fig. 7.8 Two supermarkets under the same brand co-located in a neighbourhood of the small-sized town of Velké Meziříčí, Czechia. (Source: Mapy.cz)

Using the Hotelling model, we can assume that one supermarket serves customers on one side of a river, and the other serves the second side. Both of them are profiting from this spatial composition. On top of it, one supermarket chain was acquired by the other, so nowadays (in 2018) these supermarkets belong to the same chain while their distance is around 160 m.

There exists several extension of classical location theories, some of them dealing with a particular problem, some of them tries to investigate its real applicability. However, most of these works are connected with the quantitative revolution in science (generally) after World War II. Inspired by Murray (2009), well-recognised names with research agenda about location theories are (listed by alphabetical order, not by importance):

- **Walter Isard** – one of the fathers of the "regional science" movement, and author of the seminal book – Location and Space-Economy (Isard 1956),
- **Melvin Greenhut** – dealing with the economics of space, spatial and location (micro)economics (e.g. Greenhut 1995), or industrial location theory (Greenhut 1956),
- **Brian Berry** – Vautrin Lud Prize holder (the highest award in geography); one of the most cited geographer at the time, dealing with quantitative methods and urban/regional planning and economics (e.g. remarkable publication Berry 1967; Berry and Garrison 1958),
- **Martin Beckmann** – he worked in the field of location theories (Beckmann, 1968), competitive spatial market equilibrium, relationships in a central place hierarchy and the distribution of city sizes (e.g. Beckmann and McPherson 1970), and other topics,
- **William Garrison** – one of the leaders of the quantitative revolution in geography, he dealt with functional bases of the central place hierarchy (Berry and Garrison 1958), and transportation economic geography (e.g. Garrison and Levinson 2014),
- or **Doreen Massey** – she is one of the most important persons from (British) human geography school, with public engagement (she regularly contributes to media debates) and social theory. Her main contributions cover spatial differentiation, uneven development, and historical and geographical change (e.g. Massey 1995, 2005).

As Murray (2009) further mentions, also works dealing with predictions with formalisation and mathematical abstractions must not be omitted – the work from **Alan Wilson** (spatial interaction), **Leon Cooper** (multiple facility location extending Weber's model), **Louis Hakimi** (transportation networks, optimal location associated with networks), and **Charles ReVelle** (coverage modelling). Also Nobel Prize holder, **Paul Krugman**, must be mentioned as he re-introduced the importance of spatial component (geography) of the economy into the field. He established the so-called "New Economic Geography". Aoyama et al. (2011) note that Krugman combined trade and allocation theory (e.g. Krugman 1991a) leading to spatial consequences of increasing returns, instances where more than a proportional increase in output is observed from a proportional increase in inputs (Krugman 1991b). Krugman identifies this fact as a cause of specialisation that affects trade and location of industries. Coe et al. (2013) describe four main processes connected with transportation costs identified by Krugman: (1) a firm location is beneficial in proximity to the others (reducing transportations costs), (2) goods produced are less expensive (due to lower transportation costs), thus wages can be higher, which leads to increased attractivity of the place to work, (3) location of a firm(s) should be as close to market as possible (to reduce transportation costs, and the workers become part of the market), and (4) when transportation costs are low enough, it will create a core region/area with concentrated production.

As the previously noted, the economic geography is very flexible discipline and is changing all the time. Location theories can serve as an usher in theatre – introducing upcoming performance to the audience. In this sense, location theories should be a starting point in learning/teaching economic geography; however, it is hard to follow complex mathematical concepts, abstractions and formulas. One should be at least aware of location theories existence and should be able to take them into account when dealing with spatial aspects of the economy (especially when dealing with manufactures, firms, and industries together with a socio-demographic aspect of certain geographical extents, let's say a region).

References

Alonso, W. (1964). *Location and land use – Toward a general theory of land rent*. Cambridge: Harvard University Press. 206 p.

Aoyama, Y., Murphy, J. T., & Hanson, S. (2011). *Key concepts in economic geography*. London: Sage. ISBN:978-1-84787-894-6. 278 p.

Barnes, T. J. (2000). Inventing Anglo-American economic geography, 1889–1960. In E. Sheppard & T. Barnes (Eds.), *A companion to economic geography*. New York: Blackwell.

Barnes, T. J. (2009). Economic geography. In R. Kitchin & N. Thrift (Eds.), *International encyclopedia of human geography*. Boston: Elsevier.

Barnes, T. J. (2011). The quantitative revolution and economic geography. In A. Leyshon, R. Lee, L. McDowell, & P. Sunley (Eds.), *The SAGE handbook of economic geography*. London: Sage. ISBN:978-1-84860-114-7. 411 p.

Barnes, T. J. (2013a). Economic geography. In D. Gregory, R. Johnston, G. Pratt, M. Watts, & S. Whatmore (Eds.), *The dictionary of human geography*. Chichester: Wiley.

Barnes, T. J. (2013b). Location theory. In D. Gregory, R. Johnston, G. Pratt, M. Watts, & S. Whatmore (Eds.), *The dictionary of human geography*. Chichester: Wiley.

Beckmann, M. (1968). *Location theory*. New York: Random House. 132 p.

Beckmann, M. J., & McPherson, J. C. (1970). City size distribution in a central place hierarchy: An alternative approach. *Journal of Regional Science, 10*(1), 25–33.

Berry, B. J. L. (1967). *Geography of market centers and retail distribution*. Englewood Cliffs: Prentice Hall.

Berry, B. J., & Garrison, W. L. (1958). The functional bases of the central place hierarchy. *Economic Geography, 34*(2), 145–154.

Castree, N., Kitchin, R., & Rogers, A. (2013). Economic geography. In *A dictionary of human geography*. Oxford: Oxford University Press. Retrieved 21 January, 2019, available from: https://researchguides.dartmouth.edu/human_geography/economic.

Christaller, W. (1933). Die zentralen Orte in Süddeutschland: Eine ökonomisch-geographische Untersuchung über die Gesetzmässigkeit der Verbreitung und Entwicklung der Siedlungen mit städtischen Funktionen (Mit 7 Fig. im Text und 5 Kartenbeilagen). Gustav Fischer. Jena: Gustav Fischer.

Coe, N., Kelly, P., & Yeung, H. (2013). *Economic geography: A contemporary introduction* (2nd ed.). Hoboken: Wiley-Blackwell. ISBN:978-0-470-94338-0. 541 p.

d'Aspremont, C., Gabszewicz, J. J., & Thisse, J. F. (1979). On Hotelling's stability in competition. *Econometrica: Journal of the Econometric Society, 47*, 1145–1150.

Daněk, P. (2013). *Geografické myšlení: úvod do teoretických přístupů* [Geographical thinking: An introduction to theoretical approaches]. Masarykova univerzita. 171 p. https://doi.org/10.5817/CZ.MUNI.M210-6694-2013

Frank, A. G. (1966). The development of underdevelopment. *Monthly Review, 18*(4), 17–31.

Garrison, W. L., & Levinson, D. M. (2014). *The transportation experience: Policy, planning, and deployment*. Oxford: Oxford University Press.

Gibson-Graham, J. K. (2006). *The end of capitalism (as we knew it): A feminist critique of political economy* (1st ed.). Minneapolis: University of Minnesota Press. 348 p.

Greenhut, M. L. (1956). *Plant location in theory and in practice; the economics of space*. Chapel Hill: University of North Carolina Press. 338 p.

Greenhut, M. L. (1995). *Spatial microeconomics. Theoretical underpinnings and applications*. Aldershot: Edward Elgar Pub. 561 p.

Harvey, D. (1969). *Explanation in geography*. London: Edward Arnold. 521 p.

Harvey, D. (1972). Revolutionary and counter revolutionary theory in geography and the problem of ghetto formation. *Antipode, 4*(2), 1–13.

Harvey, D. (1974). What kind of geography for what kind of public policy? *Transactions of the Institute of British Geographers*, 18–24.

Hotelling, H. (1929). Stability in competition. *The Economic Journal, 39*(153), 41–57.

Isard, W. (1956). *Location and space economy – A general theory relating to industrial location, market areas, land use, trade, and urban structure*. Cambridge, MA: MIT Press.

Johnston, R. (2013). Central place theory. In D. Gregory, R. Johnston, G. Pratt, M. Watts, & S. Whatmore (Eds.), *The dictionary of human geography*. Chichester: Wiley.

Kimutai, K. (2017). *What is economic geography? Worldatlas – Economics*. Retrieved 21 January, 2019, available from https://www.worldatlas.com/articles/what-is-economic-geography.html

Kitchin, R., & Thrift, N. (2009). *International encyclopedia of human geography*. Boston: Elsevier.

Krugman, P. (1991a). *Geography and trade*. Cambridge, MA: MIT Press.

Krugman, P. (1991b). Increasing returns and economic geography. *Journal of Political Economy, 99*(3), 483–499.

Leyshon, A., Lee, R., McDowell, L., & Sunley, P. (2011). *The SAGE handbook of economic geography*. London: Sage. ISBN:978-1-84860-114-7. 411 p.

Lösch, A. (1940). Die räumliche ordnung der wirtschaft: eine untersuchung über standort, wirtschaftsgebiete und internationalen handel. G. Fischer.

Market Business News. (2019). What is economic geography? Definition and meaning. *Market Business News*. Retrieved 21 January 2019, available from https://marketbusinessnews.com/financial-glossary/economic-geography/

Maryáš, J., & Vystoupil, J. (2004). *Ekonomická geografie* [Economic Geography]. 1. vyd. Brno: Masarykova univerzita.

Massey, D. (1995). *Spatial divisions of labour: Social structures and the geography of production*. Basingstoke: Macmillan International Higher Education.

Massey, D. (2005). *For space*. London: Sage. 232 p.

Merriam-Webster Dictionary. (2019). Economic geography. *Merriam-Webster Dictionary*. Retrieved 21 January 2019, available from: https://www.merriam-webster.com/dictionary/economic%20geography

Murray, A. T. (2009). Location theory. In R. Kitchin & N. Thrift (Eds.), *International encyclopedia of human geography*. Boston: Elsevier.

Ponsard, C. (1983). *History of spatial economic theory*. Berlin: Springer. 231 p.

Scott, A. J. (2000). Economic geography: The great half-century. *Cambridge Journal of Economics, 24*(4), 483–504.

Smith, D. (2013). Factors of production. In D. Gregory, R. Johnston, G. Pratt, M. Watts, & S. Whatmore (Eds.), *The dictionary of human geography*. Chichester: Wiley.

Thrift, N., & Olds, K. (1996). Refiguring the economic in economic geography. *Progress in Human Geography, 20*(3), 311–337.

Wallerstein, I. (1974). The modern world system I. Capitalist agriculture and the origins of the European world economy in the sixteenth century. With a new prologue. Berkeley/Los Angeles/London 2011.

Weber, A. (1929). *Theory of the Location of Industries*. University of Chicago Press. Translated by C.J. Friedrich from Weber, A. (1909). *Über den Standort der Industrien*.

Part II

Techniques of Data Visualisation

Non-spatial Visualisation

8

Karel Macků

Abstract
An enormous amount of various data is produced every day. With proper data visualisation, an information hidden in the data can be easily and quickly revealed. It is necessary to create a communication channel that could quickly and efficiently transfer the information from the data to the user. By using visual elements like charts, graphs, and maps, data visualisation is an accessible way to see and understand trends, outliers, and patterns in data. This chapter offers an overview of relevant data visualisations divided into thematic categories and supported by examples.

Keywords
Visualisation · Data · Chart · Information

In the world today, we encounter enormous amounts of data every day. To convert data into useful information, data must be presented to the user in a way that allows interpreting, analysing and applying the gained information (Yau 2011). It is necessary to create a communication channel that could quickly and efficiently transfer the information from the data to the user – this can be done with data visualisation. Tableau Software, a company offering a software platform for interactive data presentation, briefly and comprehensively talks about data visualisation: "Data visualisation refers to the graphical representation of information and data. By using visual elements like charts, graphs, and maps, data visualisation is an accessible way to see and understand trends, outliers, and patterns in data" (Tableau Software 2018). According to (Friedman 2008, p. 1) the "main goal of data visualisation is to communicate information clearly and effectively through graphical means. To convey ideas effectively, both aesthetic form and functionality need to go hand in hand, providing insights into a rather sparse and complex data set by communicating its key-aspects more intuitively".

Visualisation is an important step in the whole process of data analysis. Legendary statistician John Tukey often mentions visualisation in the context of using visualisation to find meaning in the data (Tukey 1977). Despite his statistical focus, he believed that a graphic presentation of information plays an immense role. A proper visualisation based on source data can help to understand the data, improve decision making and provide a more objective preview of the problem represented by data (Yau 2013). A graphic can also reveal hidden patterns and relationships.

Visualisation methods have gone far beyond traditional data presentation with simple charts and graphs. Modern trends approach data visualisation as both a science and an art. Of course, certain standards of correctness (e.g. by

K. Macků (✉)
Department of Geoinformatics, Palacký University Olomouc, Olomouc, Czech Republic
e-mail: karel.macku@upol.cz

V. Pászto et al. (eds.), *Spationomy*, https://doi.org/10.1007/978-3-030-26626-4_8

choosing a method according to the characteristics of the data) are still kept, but there is an effort to make the result interesting and catchy to attract the reader's attention. Sophisticated data visualisation and infographics methods offer a variety of exciting charts and diagrams. The advantage of technologies today is the possibility of presenting outputs in the form of online interactive web tools, which makes the processing of the information, that author attempts to communicate, even more intuitive and attractive.

In this chapter, a non-spatial data and its visualisation are discussed. Non-spatial data plays an undeniable role in the field of economics and business intelligence. For that reason, an overview of most common and powerful possibilities how to visualise it will be presented on the following pages.

8.1 Software

Nowadays, a variety of software is easily available, knowledge of some of them is a part of general digital literacy. Almost everyone, who somehow uses a computer, is able to create any visualisation using some of the available software. Most of the computer users are skilled with Microsoft Office Excel – software that doesn't need to be presented (or its opensource alternatives Libre Office/Open Office). Working in these tools is relatively convenient and straightforward, as underlying data and graphical tools are integrated into one user environment, and the whole process is very intuitive. However, this approach does not always offer proper or high-quality graphical outputs and supports the user's tendency to blindly insert data into the provided graphics templates without deeper thinking. By this approach, data loses its ability to interpret the story that is stored in it (Nussbaumer Knaflic 2015). Another point is the technical maturity of the output. In the world of modern technologies, where most of the information is distributed online, it is much more professional to produce outputs that offer a degree of interactivity and support simple distribution in the digital environment. Interactivity allows the viewer to engage with your data in ways impossible by static graphs. With an interactive plot, the viewers can zoom into the areas they care about, highlight the data points that are relevant to them and hide the information that is not (Barter 2017). For this reason, some tools will be introduced offering the possibility of creating interesting graphical outputs.

8.1.1 Tableau Software

The company Tableau Software offers a set of tools of the same name designed for exploratory analysis and data visualisation. The product is especially focused on an effective and highly aesthetic level of visualisation, which undoubtedly attracts many customers. The full version of the program is paid, but the version *Public Tableau* is freely available. In this version, a user can work with many formats such as Microsoft Excel, Access, text files, JSON files, databases and also spatial data (several data formats are supported). After loading the data, the user can easily select the attributes they want to visualise and based on the data type, a set of options is automatically offered to create a visualisation. The main idea of the Public Tableau is the interactivity and presentation of the outputs in the online environment so the result can be shared with other users as attractive interactive data visualisation. The tools are, of course, multiplatform, they can be used as a desktop, mobile or online version.

8.1.2 HTML, Javascript and CSS

HTML, Javascript and CSS are the basis of every webpage. With modern technologies represented by HTML 5, advanced data visualisation running native in the internet browser can be done. This solution is probably suitable only for technically-advanced users/developers, who can handle coding with these technologies. There are several libraries designed for building interactive/static web visualisations, for example, Javascript libraries D3.js, Charts.js or FusionCharts. These libraries offer dozens of charts; detailed information can be found on their websites.

8.1.3 R

It is free and open-source statistical and mathematical computing software, primarily focused on data analysis and modelling. Since R has been developed mainly for statistical analysis, it has a solid background for different types of calculations suitable for data analysis. There is a lot of packages, which can extend the functionality of R software with just a simple code command. Thanks to the packages, R is a very mighty tool for data visualisation. Of course, a knowledge of code writing is required (as well as with HTML), which makes R for many people inapplicable. But once this obstacle is overcome, a new world of data handling and visualisation is opened. All graphics can be saved in vector formats, so it is possible to edit and refine the design of the outputs in suitable graphical software, like Adobe Illustrator or Inkscape. Except for traditional static graphic, also interactive outputs can be produced with special R packages. Sometimes, the interactivity is redeemed by complexity in the form of one extra line of code!

8.1.4 Datawrapper

Datawrapper is an online tool for making the interactive charts. It has a very simple interface; a user can upload data from a file or paste the value directly into the field. The tool generates graphics automatically; a user can choose one of the 16 types of visualisation. Several refining steps can be done, like customising of axis, labelling or colour setting. This tool is an ideal solution when one needs a quick, simple interactive visualisation without any programming.

These examples were just a small slice of what nowaday technologies offers. Everyone is comfortable with a different level of challenge, content control and output options so everyone can find their optimal tool for creation of graphical outputs. There is an overview of another tools for visualisation in following table. Of course, this list is not complete, there are dozens of tools in offer (Table 8.1).

8.2 Charts Classification

There might be a confusion in terminology regarding the visualisation of non-spatial data. Usually, words 'chart' or 'graph' are generally used to describe any visual output. For many people, these two terms mean the same, but there is a difference. A chart is a superior term for a group of methods, how to present information. A graph is a particular graphical tool, which shows a mathematical relationship between sets of data (Blaettler 2018). With this approach, a graph is a subcategory of a chart. For this reason, the term chart will be rather used in this chapter, to keep the description of different methods more board.

Table 8.1 An example of visualisation tools

Name	Output	Difficulty	Pricing
Plotly	Interactive/static	Coding required	Free with some paid plans
Highcharts Cloud	Interactive	Easy to handle	Free
D3.js	Interactive	Coding required	Free
Charts.js	Interactive	Coding required	Free
Inforgram	Interactive/static	Easy to handle	Free, extra features paid
RAWGraphs	Interactive/static	Easy to handle	Free
DataHero	Interactive/static	Easy to handle	Paid
Visually	Interactive/static	Easy to handle	Paid
Visme	Interactive/static	Easy to handle	Free
Google charts	Interactive	Easy to handle	Free
vVizualize.me	Static	Easy to handle	Free

Source: Author

Different types of charts will be described in the following chapter. Since there are dozens of possibilities of visualisations, only the most interesting or most commonly used variants will be introduced. For better thematic logic, the individual methods were divided into thematic groups. The inspiration for this system was the book *Visualize This* (Yau 2011) and the website www.datavizproject.com (Ferdio ApS 2017).

8.2.1 Trend Over the Time

Time series are typical data for many phenomena. Things are changing in time, and this change can be easily captured and presented by suitable graphics. Talking about time series, users try to explore the trend in data. Is the value of the phenomena increasing or decreasing? Are there any repetitive cycles?

Temporal data can be divided into discrete and continuous types. The knowledge about this character of data should guide the user in a decision, which kind of graph should be used. For example, a monthly revenue report is an information referenced to a one-time step – a month, so this can be considered as a discrete phenomenon. Then, a simple bar or point graph can be used. The second type is the continuous data. This is kind of information which can be measured at any time of day during any day of the year. A typical example could be a temperature or another meteorological phenomenon; regarding the economic data, we can use stock exchange prices as an example. The structure of data is same for discrete and continuous phenomena, to distinguish the difference, the proper way of visualisation should be used. The most primitive solution is to connect discreetly plotted data with any line.

8.2.1.1 Bar Chart

Bar charts are commonly used, which means the user doesn't need to ‚learn', how to read the graph. The graphic element is a rectangular bar whose length represents the value. The time axis captures time points, which have to be ordered chronologically. Then every bar stands for one discrete time point. Finally, there are many additional ways how to tune the bar graph, e.g. bars can be placed horizontally or vertically or some of the bars can be highlighted by a different colour (e.g. time points when the value is higher than set limit etc.) (Fig. 8.1).

8.2.1.2 Point Chart

Point chart works on same principle as the bar chart does, except for used geometrical element – it is a simple point here. This can sometimes be more suitable since the points do not represent such graphic content and load as bars. Point chart is also known as a scatterplot when non-temporal data is used. It is crucial to properly create an axis representing the value of the phenomenon, as there is no other way to find out the value.

8.2.1.3 Line Chart

The line chart is a type of chart used for continuous data. The basis of the chart is the same as the basis of point chart. The continuity is added by connection of this points with line segments. Then the chart shows how data changes in the time (particular value is stored in the point), and the line segments create a feeling of continuity. It also better points to the trend between time markers (Fig. 8.2).

There is only a minor difference between the line chart and the spline chart. They differ only with the way how the points are linked. While the line graph uses straight line segments, the spline chart plots a fitted curve through each point from

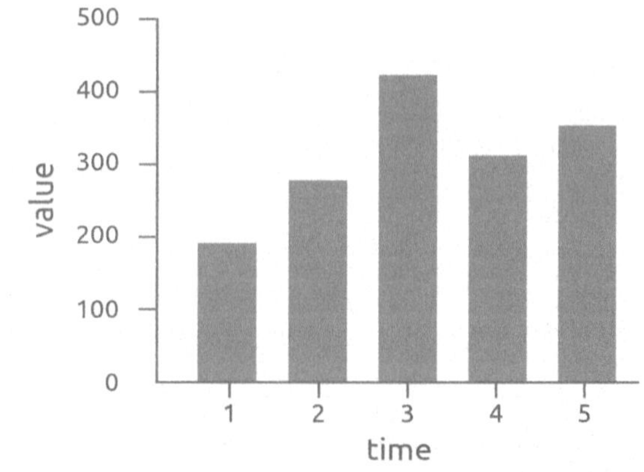

Fig. 8.1 Bar chart. (Source: Author)

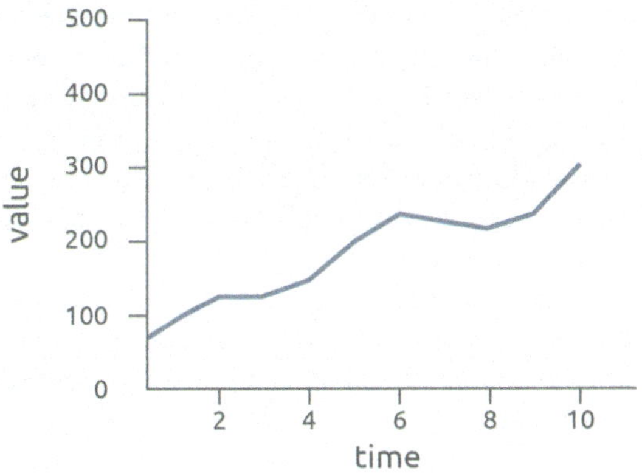

Fig. 8.2 Line chart. (Source: Author)

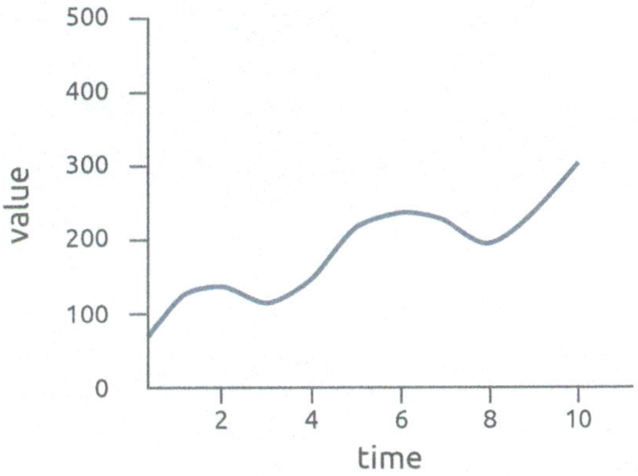

Fig. 8.3 Spline chart. (Source: Author)

the time series. This provides a more smooth and natural course (Fig. 8.3).

An attractive solution for a description of changes between two or several time point is a slope chart. It combines time-approach with multiple observed variables/categories. This helps to see differences in the development of specified categories and also the rate of change in one particular category compared to others (represented by the slope of the connecting line). At the same time, deviations in the general trend can be perfectly observed (Fig. 8.4).

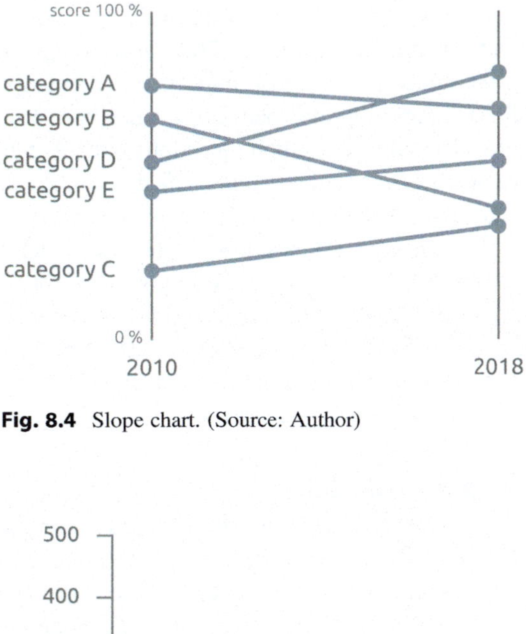

Fig. 8.4 Slope chart. (Source: Author)

8.2.1.4 Step Chart

Last modification of the line chart is a step chart. This one is formed by stepped lines between the time points. It is appropriate to use it in the situation, when the data represents a sudden change in irregular time intervals, for example, a price of any commodity which has been the same for a long time, then in one day the price increased (Fig. 8.5).

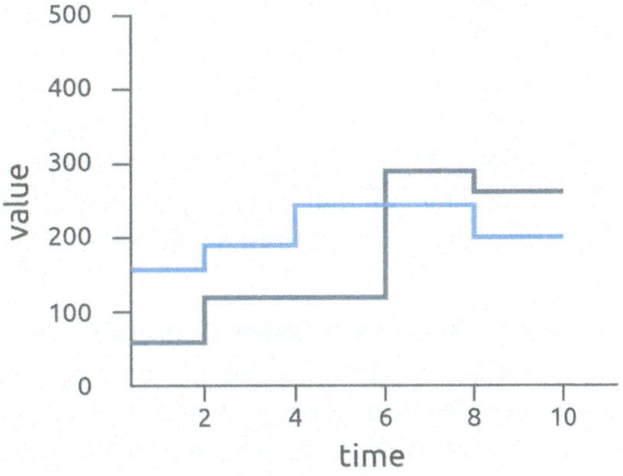

Fig. 8.5 Step chart. (Source: Author)

8.2.1.5 Gantt Chart

This chart visualises via bars duration of several categories in a time series. It illustrates the start and end point of occurrence of any activity/phenomenon. This chart is typically used as a project management tool for a graphical representation of the sequences of activities over time. Tasks or activities, which are parts of the whole project, are displayed in the time context (Fig. 8.6).

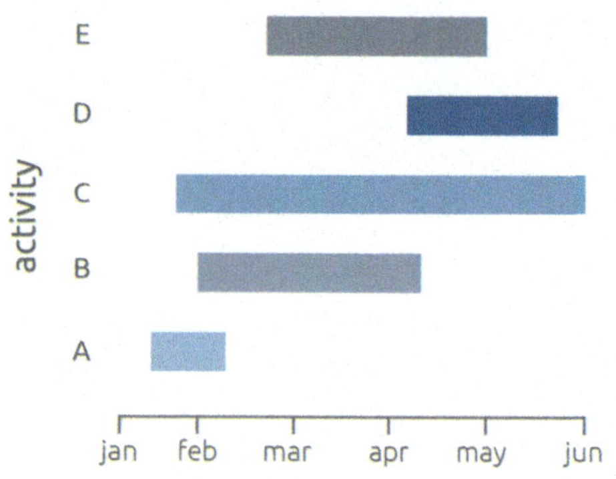

Fig. 8.6 Gantt chart. (Source: Author)

8.2.2 Proportions

Proportion data is grouped by categories/types. Each category represents a possibility, which is part of the certain unit. This distribution of proportions is the most important information for comparing groups between themselves. With proportional visualisation, questions like "Are all of the categories equally represented? Is there any category which dominates?" can be answered.

For this type visualisation, a data needs to have a form of proportions that add up to 1 or 100%. Every part could be stored relatively (as a proportion) and absolutely – total values allow to compare not only proportional part but also total size/amount in different categories.

8.2.2.1 Pie Chart

A pie chart is one of the most often used charts and is typical for an explanation of proportions. The circle which is representing the whole is divided into sectors. The arc length of each segment (or interior central angle, or area) is illustrating the proportion of individual categories. All categories together must form a unit/100%.

8.2.2.2 Doughnut Chart (Fig. 8.7)

The doughnut chart is just a modification of a pie chart, only the blank centre is added. That allows presenting of multiple information at the same time since the inner blank space could be filled with additional related data.

According to some of the resources (Nussbaumer Knaflic 2015), the pie or doughnut chart is an inappropriate way how to visualise proportional data. This is caused by the greater difficulty of perceiving angles or area than distances (which are the key information regarding, e.g. bar charts), it is a common property of human eye perception. In a situation when two or more categories are represented by an approximately same value, it's difficult to decide which one is greater. This issue can be solved by adding labels. Still, several authors recommend using different proportional methods, like a stacked/simple bar charts.

8.2.2.3 Stacked Bar Chart

Instead of pie/doughnut charts, simple bar chart ordered from highest value to least can be used. All bars have the same baseline; the endpoint is easier to compare. Even small differences can be distinguished. The length of bars is recalculated in that way that their sum equals to whole/100%.

A stacked bar chart is a perfect solution for visualising proportion and comparing several classes at the same time. Because of their geometrical representation, they are even more space-saving than pie charts. Stacked bar chart contains multiple values on top of each other, which shows the division of the whole into categories. Concurrently, individual bars represent the different level of categories or even time points. For example – the stacked bar chart can represent sale strategies: every bar signifies a particular strategy (A-E in the Fig. 8.8), different colour shades represent a type of product, and on the y-axis, total sales are displayed.

8.2.2.4 Tree Map/Area Chart

This type of charts uses a structure of rectangles and their area to express the proportion of the whole part. Size of every rectangle represent the metrics. The outer rectangle represents parent categories, and rectangles within the parent are subcategories (Yau 2011). Therefore, primary requirement is that data has to have a tree-based structure (Fig. 8.9).

There is a similar alternative, which doesn't require tree structure in the input dataset. Simple square area chart, also called a waffle chart, uses a regular grid of small cells. If the value of the cell is set, then the proportion is expressed by a number of cells (Fig. 8.10).

Fig. 8.7 Pie and dougnnut chart. (Source: Author)

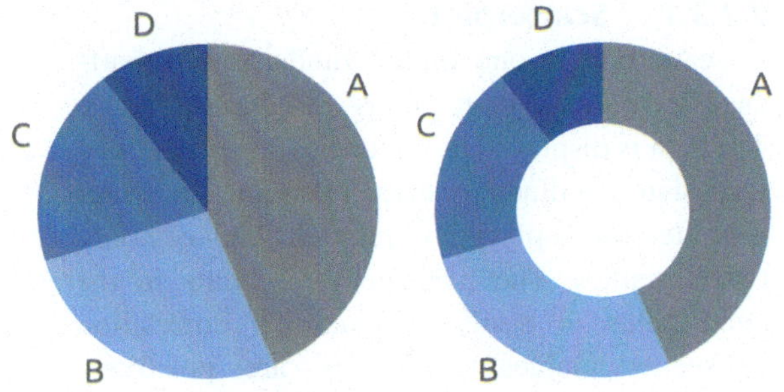

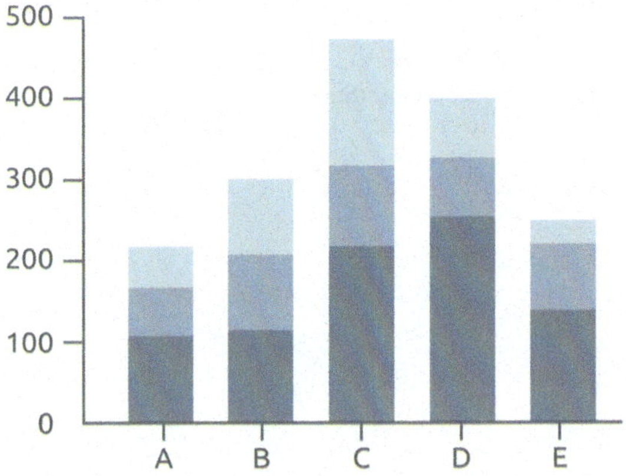

Fig. 8.8 Stacked bar chart. (Source: Author)

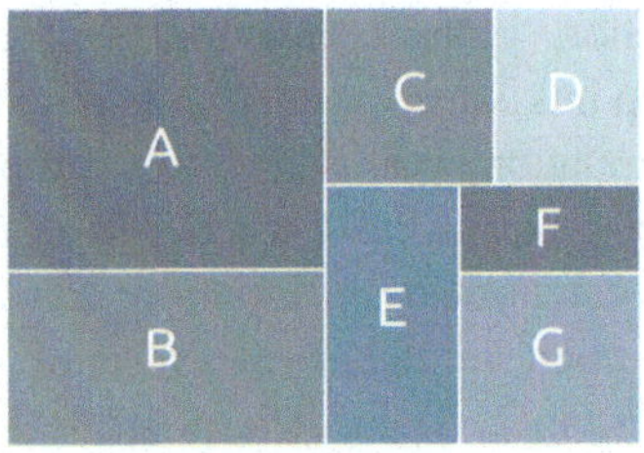

Fig. 8.9 Tree map. (Source: Author)

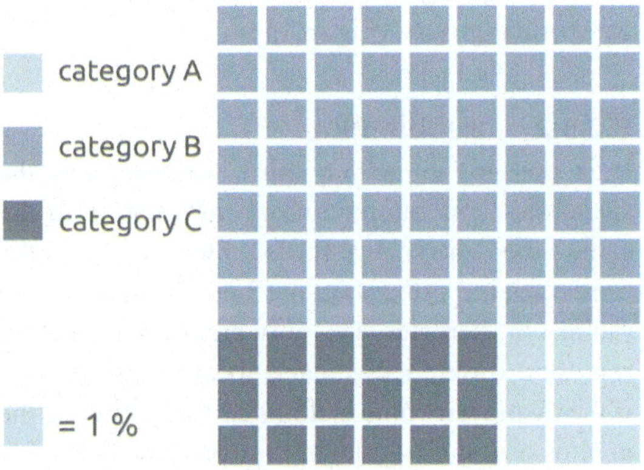

Fig. 8.10 Area chart. (Source: Author)

Regarding the tree map or area charts, there is the same issue with the perception of two-dimensional object as was discussed in pie chart paragraph. In this case, if the area map has a cell-based regular structure, the perception of information can be done correctly by simple counting of cells. Nussbaumer Knaflic (2015, p. 59) describes another situation when area charts are quite helpful: "when visualisation of numbers of vastly different magnitude is needed. The second dimension you get using a square for this (which has both height and width, compared to a bar that has only height or width) allows this to be done in a more compact way than possible with a single dimension".

8.2.3 Relations and Correlation

There are many ways how to quantify relations between several variables/group. A statistical approach provides mathematical tools, such as correlation or regression (if a conditions regarding the characteristics of variables are fulfilled). Sometimes it is much easier just to plot the data to reveal the hidden relations. A correlation simply describes, how two variables change together. Sometimes it is forgotten that correlation doesn't equal causation. Basic correlation of two variables expressed with chart can quickly describe the behaviour of the data, a rate of relation can be estimated, maybe a clustering tendency can be discovered.

8.2.3.1 Scatterplot

A scatterplot is one of the fundamental charts used for plotting of relations and dependencies. The data is displayed as a set of points placed in a Cartesian coordinate system. Therefore, the chart is limited for displaying of relations between only two variables. The placement of points in the chart helps to easily estimate the correlation between variables – if they are positively correlated, points are formed in the line-shaped group, rising with the value of the represented phenomenon. If the correlation is negative, this line group has a decreasing trend. With no significant correlation, points are not grouped in the line-shaped and spread in the field randomly. Another categorical information can be added into the chart in the form of point colour or different shapes. Then, it is possible to observe differences between particular types; they might tend to create a cluster which indicates their similarity, or controversy, points might be overlapping, then there is no clear pattern in the categorical groups (Fig. 8.11).

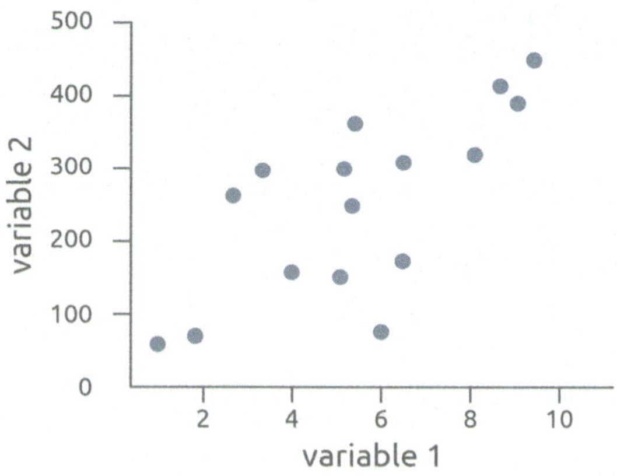

Fig. 8.11 Scatterplot. (Source: Author)

8.2.3.2 Bubble Plot

It is possible to add a third variable into the scatterplot and compare more information at the same time. The size of the bubbles expresses the third variable – the measure here is an area, not radius nor diameter. The only area can accurately represent differences related to original number: if the displayed value is doubled than another, the area of a graphical element must also be double size. If another measure is used (e.g. diameter), the ratio between value and the area of a graphic element wouldn't be the same. Of course, the chart can be modified regarding shapes, squares or triangles could also be used. Attention must be paid to the position of a particular graphics element – the bigger element must not overlap the smaller one, so it would be not visible. This rule should be implemented in software (Fig. 8.12).

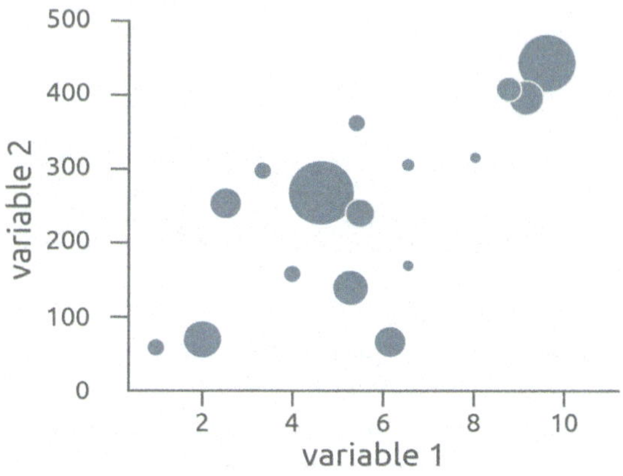

Fig. 8.12 Bubble plot. (Source: Author)

8.2.3.3 Scatterplot Matrix

In case that exploration of more than three variables is required, a scatterplot matrix is a solution. In this case, every possible combination of pairs is plotted by a single scatterplot; subsequently they are all organised into a matrix. This visualisation could be a first step in an exploratory data analysis when the analyst has no clue what is the data about and what is its behaviour. With an increasing number of variables, interpretation of that kind of graphic presentation is more complicated, and the information is still kept on the elementary level of variable pairs (Fig. 8.13).

8.2.4 Differences and Comparison

Comparing a single variable is not a demanding task, the value of every record is displayed by one of the previous-mentioned methods and analysed. Bar charts or simple point charts may well serve

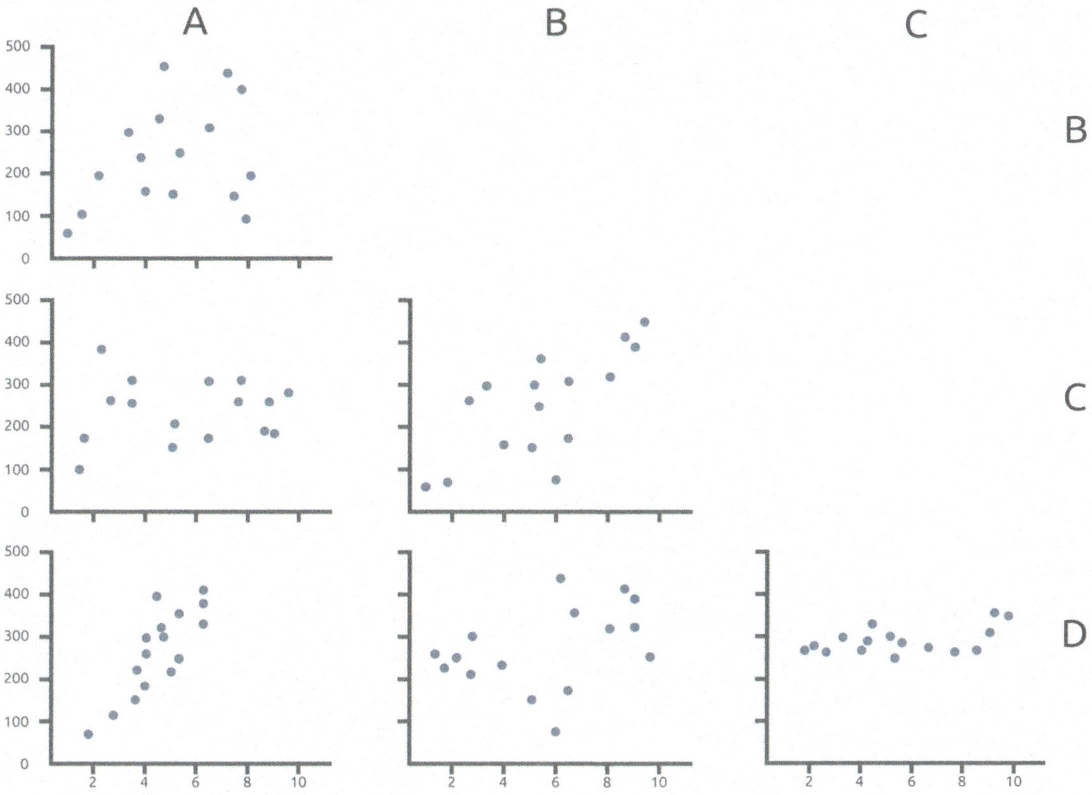

Fig. 8.13 Scatterplot matrix. (Source: Author)

to this task. Considering two or three variables, several charts for this type of visualisation have been introduced in the previous sections. Regarding the data with more variables, known as multidimensional data, different graphical methods have to be used.

8.2.4.1 Heatmap

A heatmap is a simple way how to look at all data at once. Information is displayed in a matrix of regular graphical elements, mostly rectangles or squares. The value for every record's attribute is indicated by colour intensity. The size of the heat map is defined by the number of rows times the number of attributes, heat map has the same number of elements as the input table does. This type of visualisation is not sophisticated for the reading of accurate values in a particular record but provides a great overall view on the complete dataset. Some characteristic patterns in data can be revealed, e.g. tendency to clustering (Fig. 8.14).

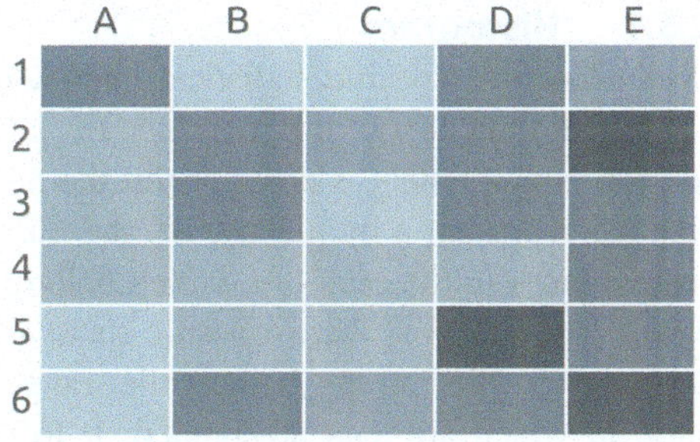

Fig. 8.14 Heatmap. (Source: Author)

8.2.4.2 Paralel Coordinates

Parallel coordinates is another common chart for visualisation of multidimensional data. The number of attributes defines a number of used vertical axes – every single of them represents one variable. It follows that different axes have a different

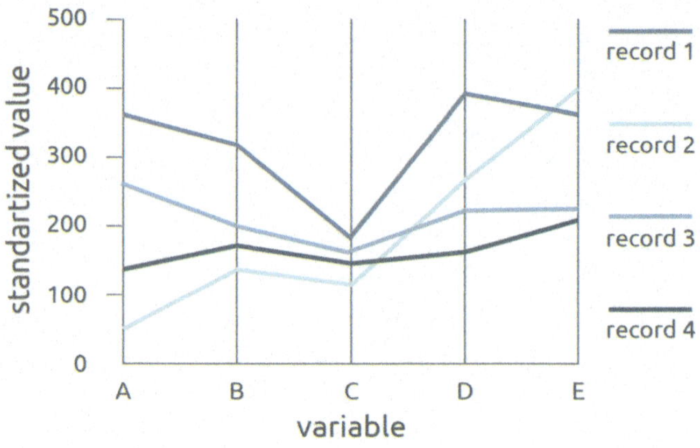

Fig. 8.15 Parallel coordinates. (Source: Author)

scale. To avoid labelling and adding more graphical ballast into the chart, data can be scaled; then all axes have the same scale. Parallel coordinates are suitable for revealing similarities between records. For that reason, labels are not always necessary, it is enough if the plot can define groups of records with a similar pattern or trend on the observed variables (Fig. 8.15).

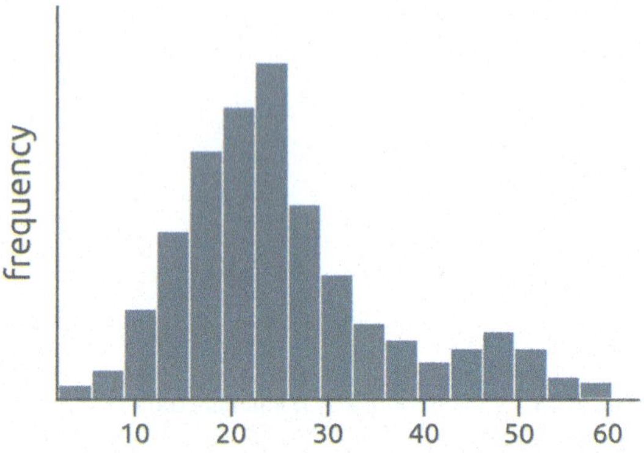

Fig. 8.16 Histogram. (Source: Author)

8.2.5 Statistical Charts

8.2.5.1 Histogram

A histogram is a specific statistical chart which describes the frequency of occurrence of values. The geometric element is a bar again, the height of the bar represent a frequency, i.e. the number of occurrences in the category, which belongs to the bar. The category here doesn't stand for a different type, it rather defines the range of values, in which are data are binded. It follows that both the horizontal and vertical axes are continuous (Fig. 8.16).

8.2.5.2 Distribution Plot

Although the horizontal axis of the histogram is continuous, the distribution is still divided into intervals. If the interval size is not set properly, a lot of information about inter-interval variation is lost. On the other hand, plotting of every single record would make the chart messy and confusing. A compromise between this approaches is a distribution plot, which is able to capture the smaller variation within the distribution and also smoothen the detailed original data. The vertical axis represents the probability of occurrence of value from the sample population. The area under the curve has to be equal to one (or 100%) (Fig. 8.17).

8.2.5.3 Boxplot

A boxplot is an important graphical tool for descriptive statistics. In one picture, it can describe several numeric information – median, first and third quartiles, and minimum and maximum (sometimes minimum and maximum is replaced by the value calculated: mean + − 1.5 ∗ interquartile range of lower/upper quartile). Outliers (values out of this range) are plotted as points. The spacings between parts of boxplots describe how the raw data is dispersed or

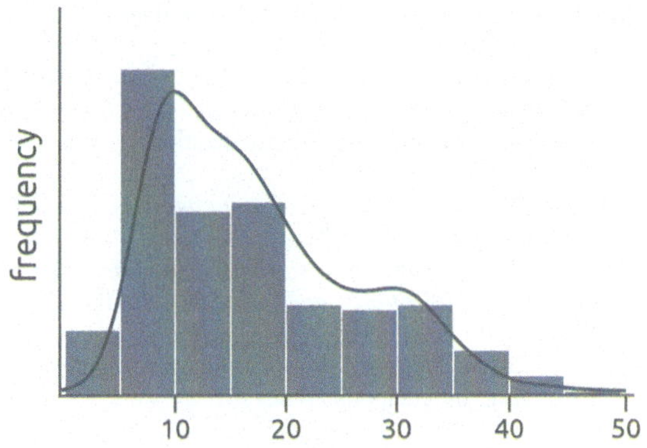

Fig. 8.17 Distribution plot. (Source: Author)

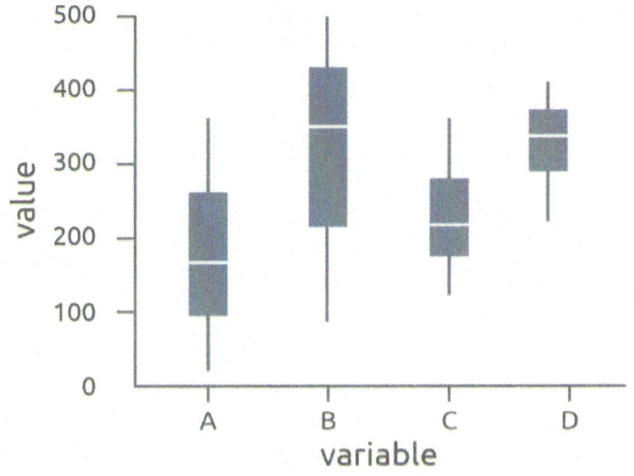

Fig. 8.18 Boxplot. (Source: Author)

condensed. With boxplots, several groups of data can be quickly compared (Fig. 8.18).

8.3 A Good Design

In the beginning, the data analyst has to know the data in detail. Once the analyst understands what kind of information is hidden in the data, what is the data type and character, he can decide which type of chart is the best solution for proper visualisation. Then another step of chart designing follows. The raw default output from the software is not wrong, but usually, it is also not the most attractive result. With an additional improvement of the graphics, information which the author tries to deliver with the chart might be easier to perceive.

It must be always considered, who is the audience, the reader of the created chart, for which purpose is the chart created. By design, the author can manipulate with the way how the chart is read. If it wants to focus on a significant trend, the axis and labels can be de-emphasised with grey colour, and the primary trend line is highlighted. Then, the trend is the information which draws the attention at first. On the other hand, a chart designed with the purpose of reading exact values must have readable and accurate labels of all axes.

Generally, some recommendation can be made. Mostly modest colours should be used. Some of the colour schemes can evoke emotions or feel (e.g. red colours indicate activity that should be addressed; neutral pastel colours means that all features in the chart are equal etc.). Proper labelling should be done – through user might know the context in which is the chart placed, he doesn't know the meaning of every single element of the chart. Therefore, a title of the chart, axes names and value labels and legend with explanations of colours should be a part of the visualisation. Geometrical aspects are also important. Sometimes it's more suitable just to rotate the chart, what makes it much easier to read (e.g. bar chart with long category names – rotation to horizontal is more natural, because it follows the way how we read the common text). A different spatial arrangement of geometrical features can solve the issues with blank space or can fit better into a whole graphic design (text or poster). Transforming the geometrical elements into pseudo-3D and displaying data that way should be avoided (the only exception is plotting a three-dimensional data with a 3D plot). Unfortunately, for example, visualisation of the pie chart in 3D is quite popular. As discussed above, the pie chart is not always a good choice for visualisation of proportional data; the combination with 3D makes it much more difficult to read or compare with others pie charts because the third dimension is problematic for perception. A perspective in 3D visualisation can also be misused for promotion – a segment of pie chart placed in the foreground looks larger than a segment of similar size in the background.

References

Barter, R. (2017). *Interactive visualization in R*. http://www.rebeccabarter.com/blog/2017-04-20-interactive/. Accessed 13 Nov 2018.

Blaettler K (2018). *The difference between charts & graphs. Sciencing*. https://sciencing.com/difference-between-charts-graphs-7385398.html. Accessed 2 Jan 2019.

Ferdio ApS. (2017). *Data viz project*. https://datavizproject.com/

Friedman, V. (2008). Data visualization and infographics. *Smashing Mag*.

Nussbaumer Knaflic, C. (2015). *Storytelling with data*. Hoboken: Wiley.

Tableau Software. (2018). *What is data visualization? A definition, examples, and resources*. https://www.tableau.com/learn/articles/data-visualization. Accessed 30 Oct 2018.

Tukey, J. W. (1977). *Exploratory data analysis*. Reading: Addision-Wesley Publishing.

Yau, N. (2011). *Visualize this: The flowingdata guide to design, visualization, and statistics*. Indianapolis: Wiley.

Yau, N. (2013). *Data points – Visualization that means something*. Indianapolis: Wiley.

Spatial Visualisation

9

Jiří Pánek

Abstract

According to the International Cartographic Association (ICA), Cartography is the discipline dealing with the art, science and technology of making and using maps; while a map is a symbolised representation of geographical reality, representing selected features or characteristics, resulting from the creative effort of its author's execution of choices (ICA 2018). This subchapter is focused on step-by-step procedure on how to create a thematic map, which are the basic cartographic elements and how to select the optimal visualisation method.

Keywords

Cartography · Spatial visualisation · QGIS · Choropleth maps · Thematic maps

9.1 Introduction

For creating a map in a vector-based model one can use points, lines and polygons. All of the features have parameters (Fig. 9.1), that can be changed in order to create unique visual interpretation. In 1967 Jacques Bertin proposed an original set of "retinal variables" in Semiology of Graphics (Bertin 1967):

- Position
- Size
- Shape
- Value (lightness)
- Colour hue
- Orientation
- Texture

The ultimate combination of the variables above allows the author (cartographer) to create a unique visual outcome of his/her work – a map. On the following pages we will have a look on what basic types of maps can be used and how to prepare them in open-source software QGIS (QGIS.com 2018).

The most widely used methods of expression in thematic cartography are choropleth maps. These are maps in which the intensity of the phenomenon is expressed using colour fill or raster/pattern in polygon, converted to a unit of the surface of the observed territory. The other options for spatial visualisation are Cartograms, thematic maps, where variables such as travel time, population, or GNP are substituted for land area or distance. In case of Cartograms geometry or space of the map are distorted. Another visualisation method would be proportional symbols, that would use points or lines to visualise the variable.

J. Pánek (✉)
Department of Development and Environmental Studies, Palacký University Olomouc, Olomouc, Czech Republic
e-mail: jiri.panek@upol.cz

V. Pászto et al. (eds.), *Spationomy*, https://doi.org/10.1007/978-3-030-26626-4_9

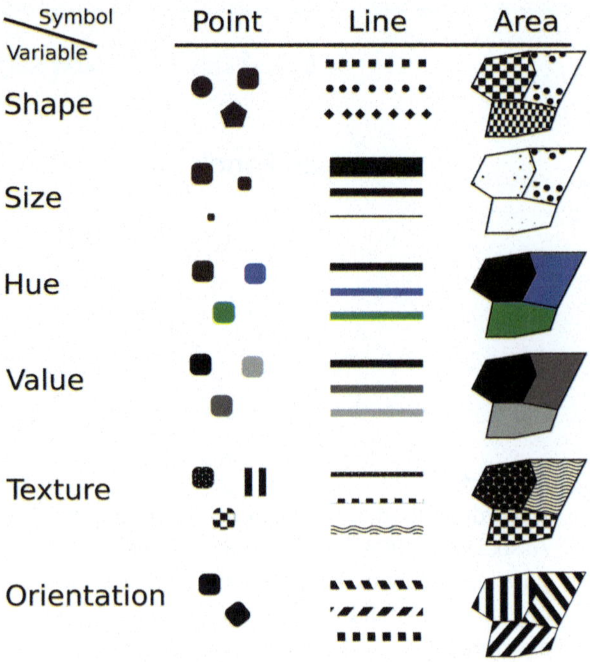

Fig. 9.1 Bertin's visual variables. (Source: Authors)

9.2 Creation of the Choropleth Map

From a strictly cartographic perspective choropleth maps express the selected phenomenon converted into a surface unit – e.g. population density (number of inhabitants/km^2) or density of road network (total length in km/km^2). Geographers, demographers or sociologists however often need to express the intensity of a phenomenon (even if related to a certain surface) converted to a unit of a non-surface character, most frequently in the form of percentage or per thousand share – e.g. proportion of economically active persons in total population of the selected locality (%) or gross birth rate (‰). As mentioned above, choropleth maps express only relative values. It is a cartographic error if data on absolute values of the observed phenomenon is displayed – e.g. number of inhabitants in districts or number of small hydroelectric power stations in catchment basins. In these cases it is not possible to speak of a choropleth map in the geographical or cartographic sense of the word.

In geographical practice, the creation of a choropleth map is performed most frequently in territories which can be delineated by various types of boundaries. This primarily concerns boundaries of an administrative character (e.g. states, regions, cities), but also boundaries delineated on the basis of either selected socio-economic geographical aspects (e.g. commuter belts, urban agglomerations) or physical-geographical aspects (e.g. catchment basins, geomorphological units).

However, it is also possible to encounter the method by which the entire territory is covered by a network of regular cells of identical size, and the data expressing the intensity of the observed phenomenon is then presented for the individual cells of the aforementioned network.

The method of expression of the choropleth maps is frequently used in combination with the proportional symbol method (see Sect. 9.4). This method of presenting data has a great advantage in simultaneous display of relative (choropleth maps) and absolute (proportional symbol) values of the phenomenon of the observed territory, thanks to which the user is able to determine more information from a single map.

9.2.1 Creating the Choropleth Map in QGIS

Choropleth maps are created using the offer Properties (click with the right button on the name of the layer and select Properties). In the window you open the Symbology (Fig. 9.2) menu, where the visualisation of the layer can be defined. A number of options are available here for visualization of data:

- No symbol
- Single symbol
- Categorized
- Graduated
- Rule-Based
- Inverted polygons
- 2.5 D

In order to create the choropleth map "Categorized" or "Graduated" are the most often used options within the symbology menu. The main difference between categorized and graduated options is that categorizes works with numeric as well as text values and assigns each value a unique symbol/colour – hence it is more suitable for qualitative data. On the other hand, graduated scales work with number fields only and are most often used for quantitative values.

9.2.1.1 Graduated Choropleth Map

In creating the example of density of population map graduated option is selected. The attribute from which the choropleth map is created is selected in the column offer. The program will automatically sort data into categories (equal interval is a default option) and allocate a colours

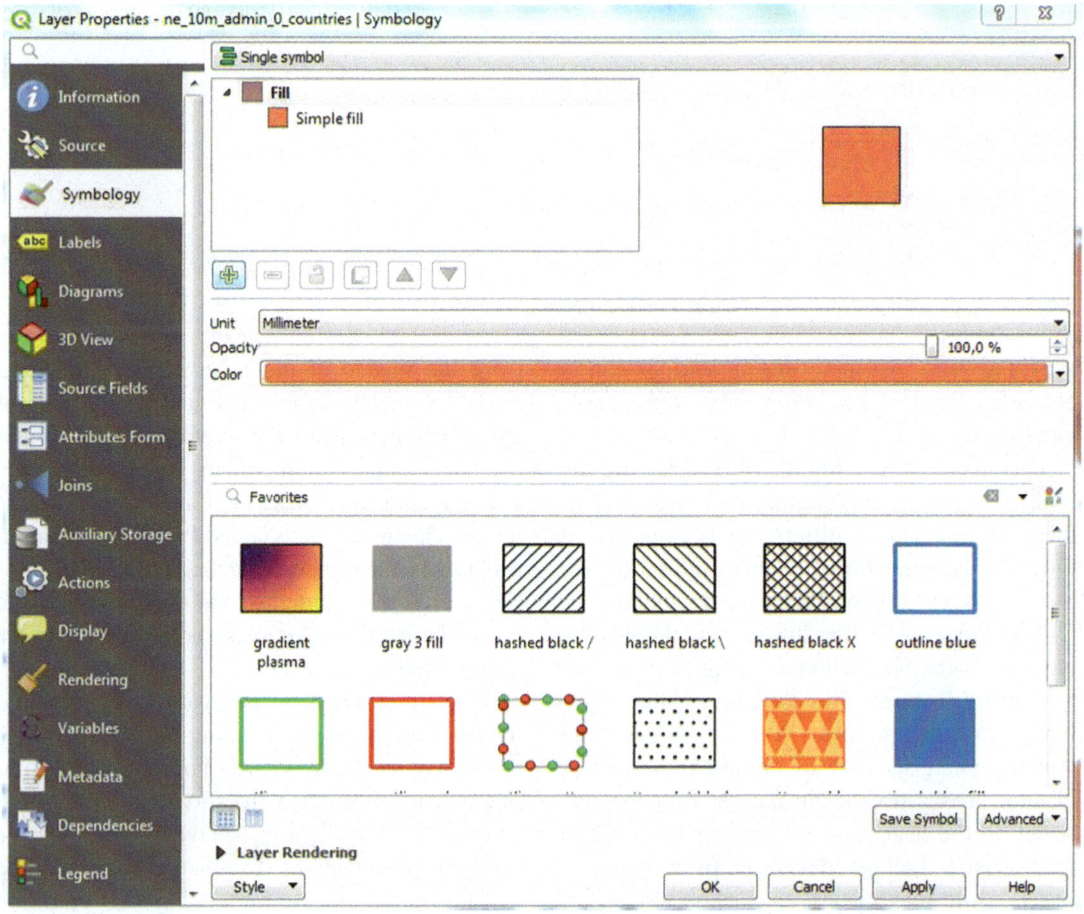

Fig. 9.2 Symbology menu in QGIS. (Source: Authors)

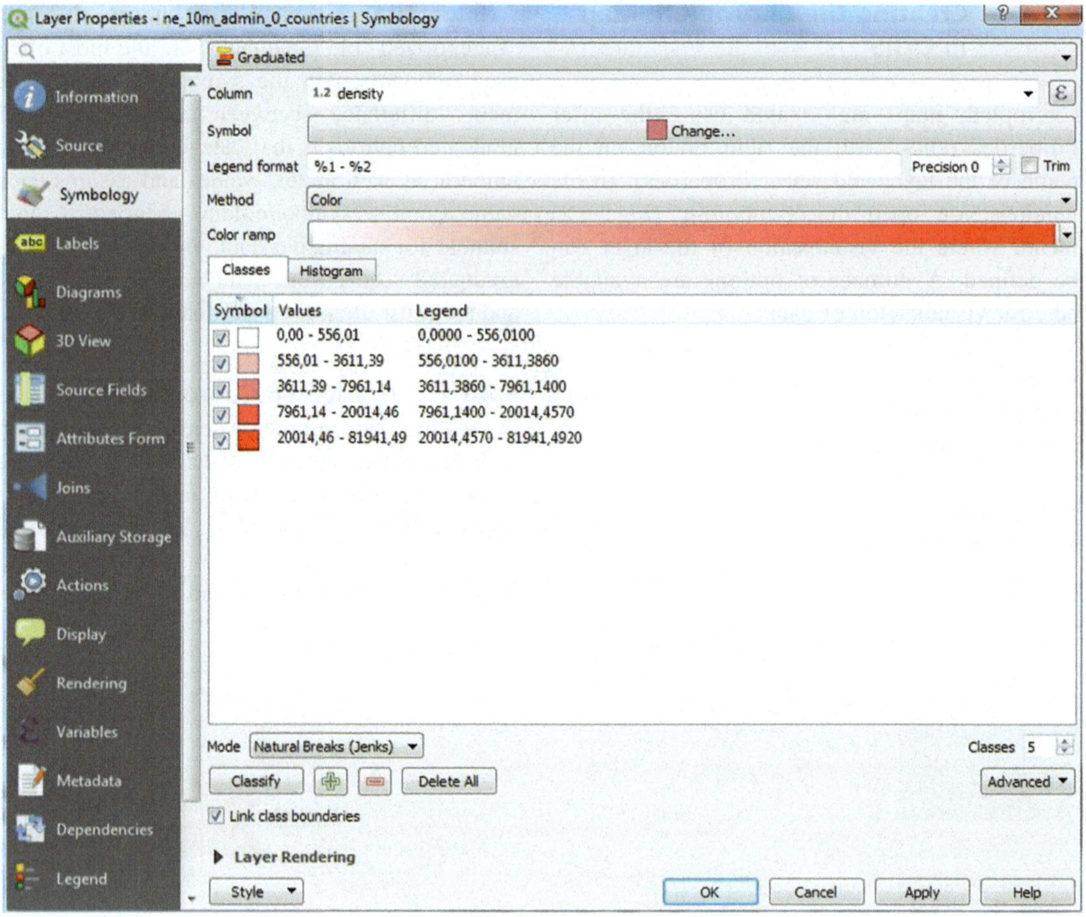

Fig. 9.3 Basic settings within the graduated option. (Source: Authors)

to it (Fig. 9.3). All these pre-set settings can be changed freely according to the user's requirements.

One has to be careful while selecting the "Mode" of data distribution into the bins, as there are tremendous differences between Equal interval, Quantile (Equal count), Natural breaks, Standard deviation and Pretty breaks modes.

Equal Interval divides the dataset into bins that have the same size without taking into account how many features is in each bin. Quantile has bins with equal number of cases in each bin, without taking the size of the bin into account. Natural breaks divide the dataset into the bins based on the histogram by seeking to minimize each class's average deviation from the class mean, while maximizing each class's deviation from the means of the other groups. Standard deviation makes equal intervals from the input data's distance from the average value. If you use an input column which has enough data below the mean to make more than one category, it will. The last mode is called Pretty breaks and it is based on the statistical package R's pretty algorithm. It is a bit complex, but the 'pretty' in the name means it creates class boundaries that are round numbers.

Adjustments to the classification of values into intervals are made by selecting the offer *Classify* or *Histogram*. In the section *Values* one can simply rewrite the Break Values to the required limit values. The number of displayed classes can be set in the offer *Classes*. These characteristics, like the histogram, serve for setting the optimum intervals. All settings are performed after clicking the button *OK* or *Apply*.

The description of the interval which appears in the legend can easily be adjusted by clicking on the required value in the *Legend* field. The range of colours used can be adjusted using the offer *Colour Ramp*, which contains a large number of predefined colour ranges. After clicking on the *Apply* button the changes are made in the data field.

Creation of Choropleth maps is subject to fundamental cartographic rules. Amongst the most important are: the scope of intervals should be logical (e.g. linear growth, exponential, decimal categories etc.); the number of elements in each interval of the cartogram should be approximately the same; use the smallest possible number of intervals (if a layer has 30 elements it is not suitable to use 10 intervals) and use corresponding colours (it is not suitable to use black, the selected colour should correspond to the map topic). For further information one can read (Field 2018) (Fig. 9.4).

9.2.1.2 Categorized Choropleth Map

When working with qualitative values, such as country names, one cannot use graduated colour ramps, but unique random colour schemes. While creating the political map of a given region cartography follows the Five colour theorem or the Four colour theorem. Both of them state, that a given plane separated into regions, such as a political map of the counties of a state, the regions may be coloured using no more than four/five colours in such a way that no two adjacent regions receive the same colour. While the five colour theorem was proved already in the 1800s, the four colour theorem was proved in 1976 by Kenneth Appel and Wolfgang Haken, but only after many false proofs and counterexamples. It was the first major theorem to be proved using a computer. Initially, their proof was not accepted by all mathematicians because the computer-assisted proof was infeasible for a human to check by hand. Since then the proof has gained wide acceptance, although some doubters remain. (Robertson et al. 1996).

By default, QGIS cannot work with the theorems mentioned above and if the user decides to create a political map for example of Africa, s/he will end up with 54 states coloured by 54 different colours (see Fig. 9.5).

In order to apply the colour theorems mentioned above the user has to prepare the data by

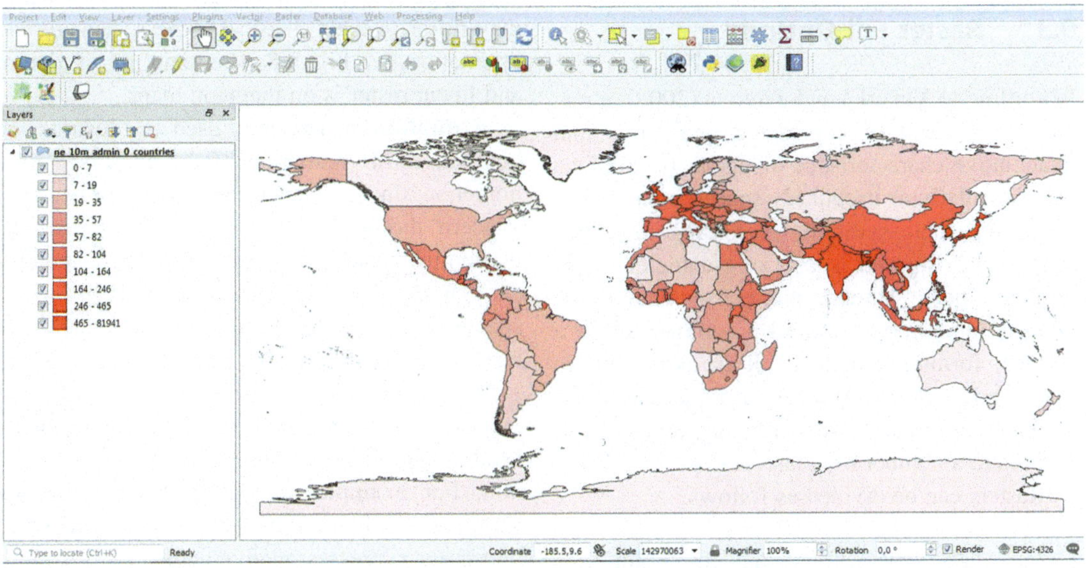

Fig. 9.4 Example of population density visualisation via Choropleth map using Quantiles with ten bins. (Source: Authors)

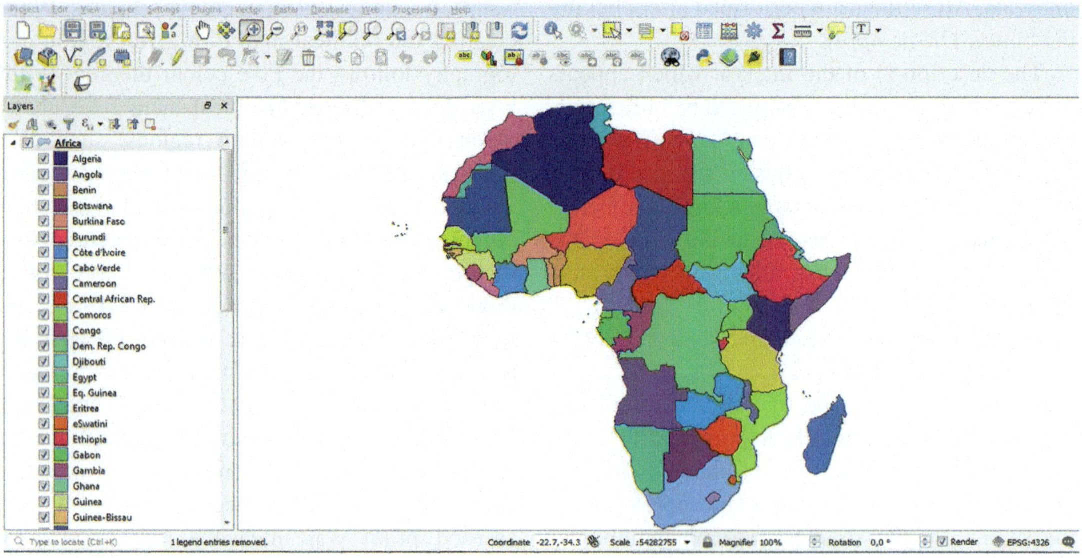

Fig. 9.5 Example of 54 coloured political map of Africa. (Source: Authors)

a Topological Colouring tool in the Cartography Toolbox (Fig. 9.6a), where a certain number of colours can be set (Fig. 9.6b). This toolbox will create a new column within the attribute table, which will contain a numeric value (1–4 for four colour theorem, 1–5 for five colour theorem), etc. (Fig. 9.7).

9.3 Raster Fill Options

According to Veverka and Zimová (2008), a raster is a method of expressing qualitative and quantitative characteristics of planar phenomena using regularly or irregularly spaced point or linear cartographic symbols. According to Čapek et al. (1992), a raster is a set of graphic elements (points, lines, letters, numerals) which are repeated and spaced around a certain part of a surface, forming a pattern. It is used both in black and white maps, where it replaces colour, and in colour maps, in which it supplements colour where a number of areas overlap.

Rasters can be divided as follows:

- according to geometrical character: point, line;
- according to spacing of symbols: regular, irregular;
- according to version: colour, black and white;
- according to use: qualitative, quantitative.

A raster formed by patterns or points (point raster) is called a pattern raster, and may differ in its shape, density, dimensions and layout. A linear raster is distinguished by the concurrent layout of lines (sometimes indicated as hatches) leading in one more directions (crossing). Lines may differ in their shape, thickness, density and orientation, see Fig. 9.8. The main use of a point and linear raster is on thematic maps.

A **qualitative raster** is used to illustrate the qualitative differentiation of the expressed phenomenon, thus its categories. It may for example concern the type composition of a forest, the predominant nationality composition, a geological map etc. Point and linear symbols in regular spacing are most frequently used for qualitative rasters, exceptionally also with irregular spacing.

For quantitative differentiation of the expressed phenomenon a **quantitative raster** is used, i.e. the degree of intensity of the phenomenon. For example population density, hectare yields etc. Unlike a qualitative raster, a quantitative raster expresses relative values.

The QGIS program offers several options for the visualisation of data into maps using rasters.

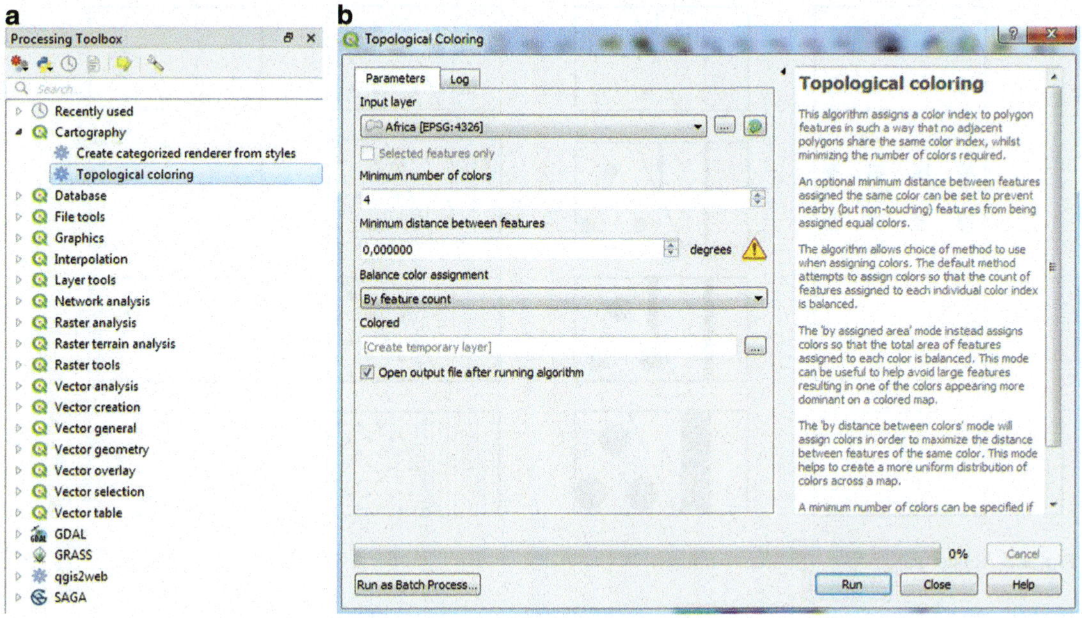

Fig. 9.6 (**a**) Topological colouring tool in the Cartography toolbox, (**b**) Setting of the Topological colouring tool. (Source: Authors)

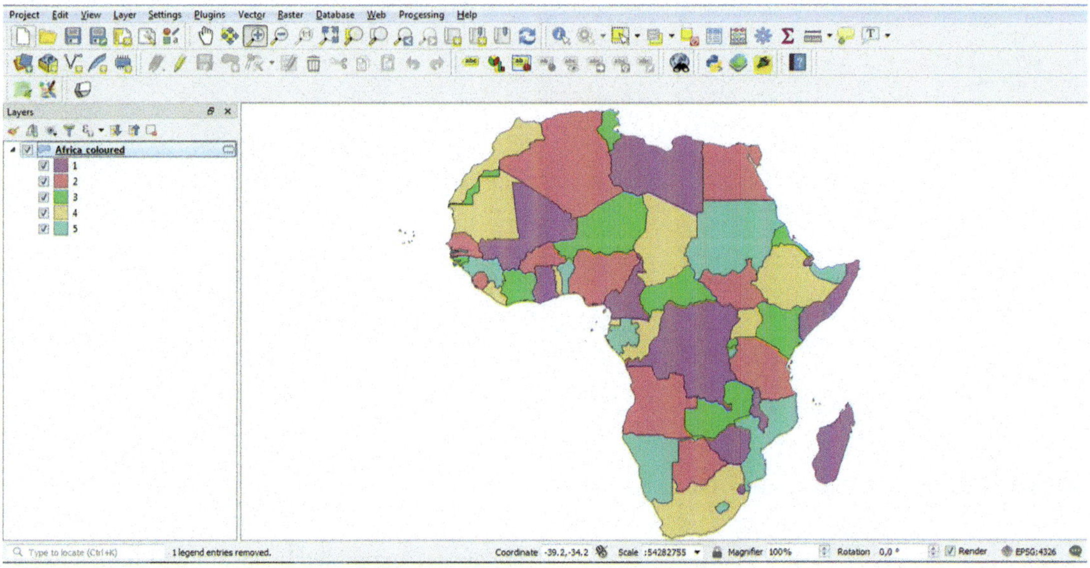

Fig. 9.7 Political map of Africa with using only five colours. (Source: Authors)

A simple example of a qualitative raster may be a different land use visualisation (Fig. 9.9). Unlike in other GIS software creating a raster fill is not a default option, but the raster has to be created by single line or point elements and by combining them, changing the size, orientation, offset, etc. (see Fig. 9.10).

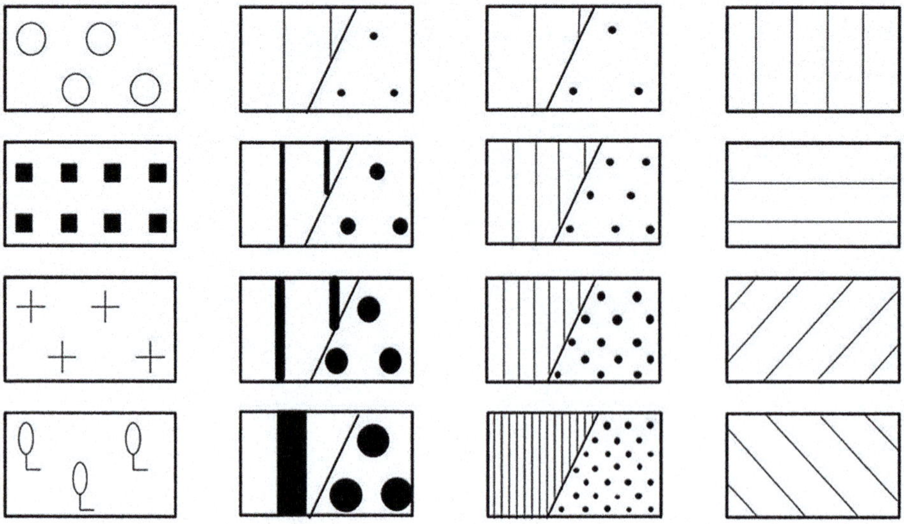

Fig. 9.8 Raster parameters. (Source: Authors)

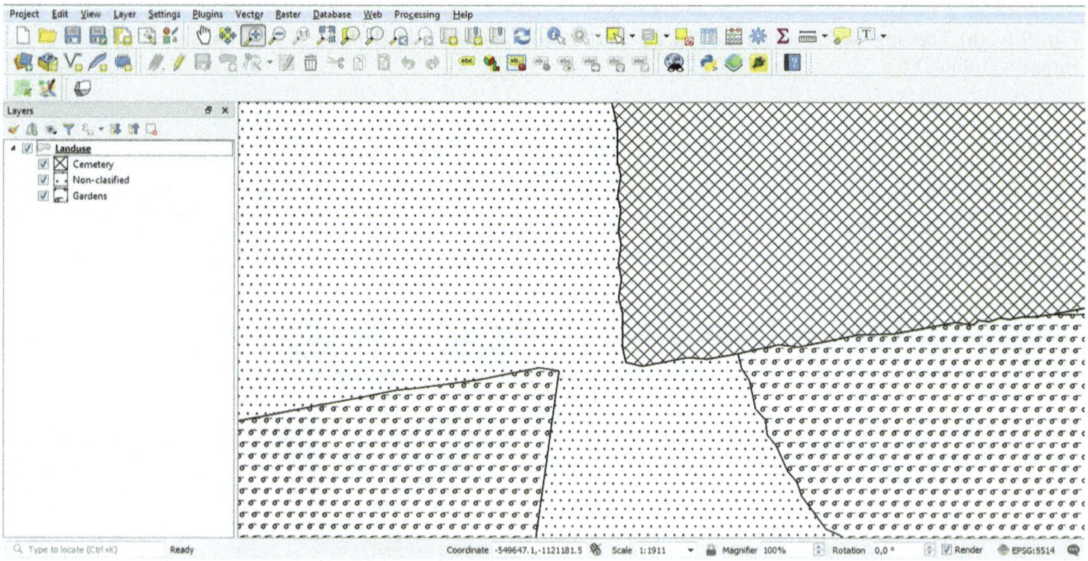

Fig. 9.9 Example of qualitative raster. (Source: Authors)

9.4 Proportional Symbols

According to Voženílek and Kaňok (2011), proportional symbols are used on maps with partial territorial units into which statistical data (absolute values), mostly of a geographical character, is illustrated by means of diagrams. Unlike choropleth maps, values in proportional symbols are always expressed in absolute form. An exception is the structure of a certain character, where the values are mostly stated in percentages (Čapek et al. 1992). Čapek et al. (1992), and Voženílek and Kaňok (2011) divide proportional symbols into three types: point, linear and planar (or planar structural).

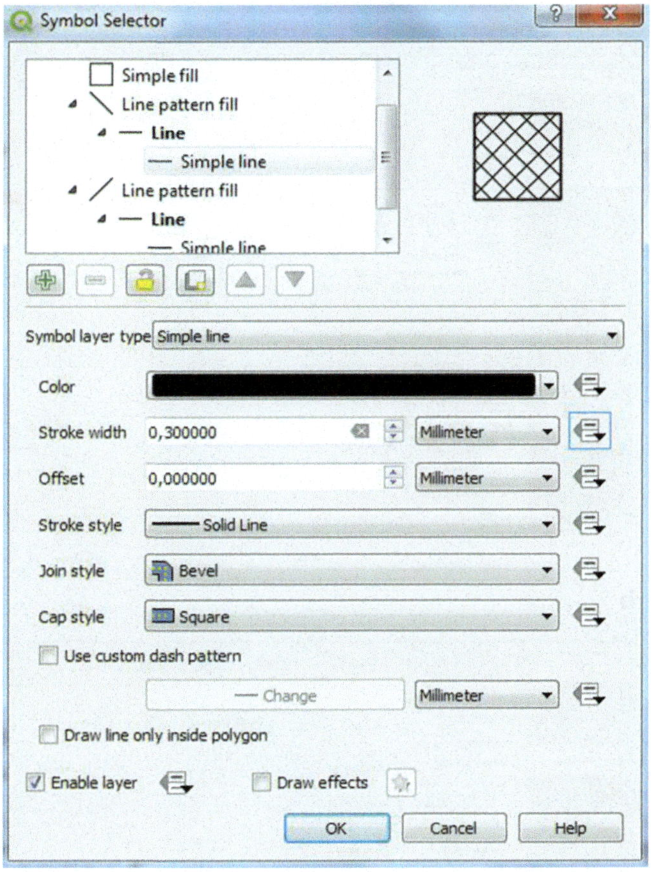

Fig. 9.10 Example of creating raster via symbol selector. (Source: Authors)

9.5 Method of Graduate Symbols

Simple graduated symbology is dealt within the QGIS program by the offer Symbology (similarly as raster above). The symbols can be graduated by colour (Fig. 9.11a) or by size (Fig. 9.11b). One can easily see, that for a larger number of features, like cities on a whole continent, colour is a better solution than size. On the other hand, when working with fewer features, the size option can be more optimal.

9.6 Using Charts to Visualise Proportions

Proportional symbols visualise relations between two or more counts, such as age distribution in the population as a percentage of total population. To display the proportional relationships in QGIS is via the menu Diagrams in Layer Properties. Among others, Pie Charts (Fig. 9.12a) and Histograms (bar/column chart – Fig. 9.12b, c), are available options. Like the Proportional Symbol map, the Pie Chart map plots a single symbol usually at the centroid of each geometry. Each chart can vary in size, representing a certain value – such as population, while the division of the chart will represent the age distribution within the population. Similarly to Pie Charts, Histograms can visualise proportional values. Histograms can be oriented as a bar or a column graph.

9.7 Cartograms

A cartogram is a specific type of map, where the theme (such as travel time, wealth, or population with HIV/AIDS) is substituted for land area or

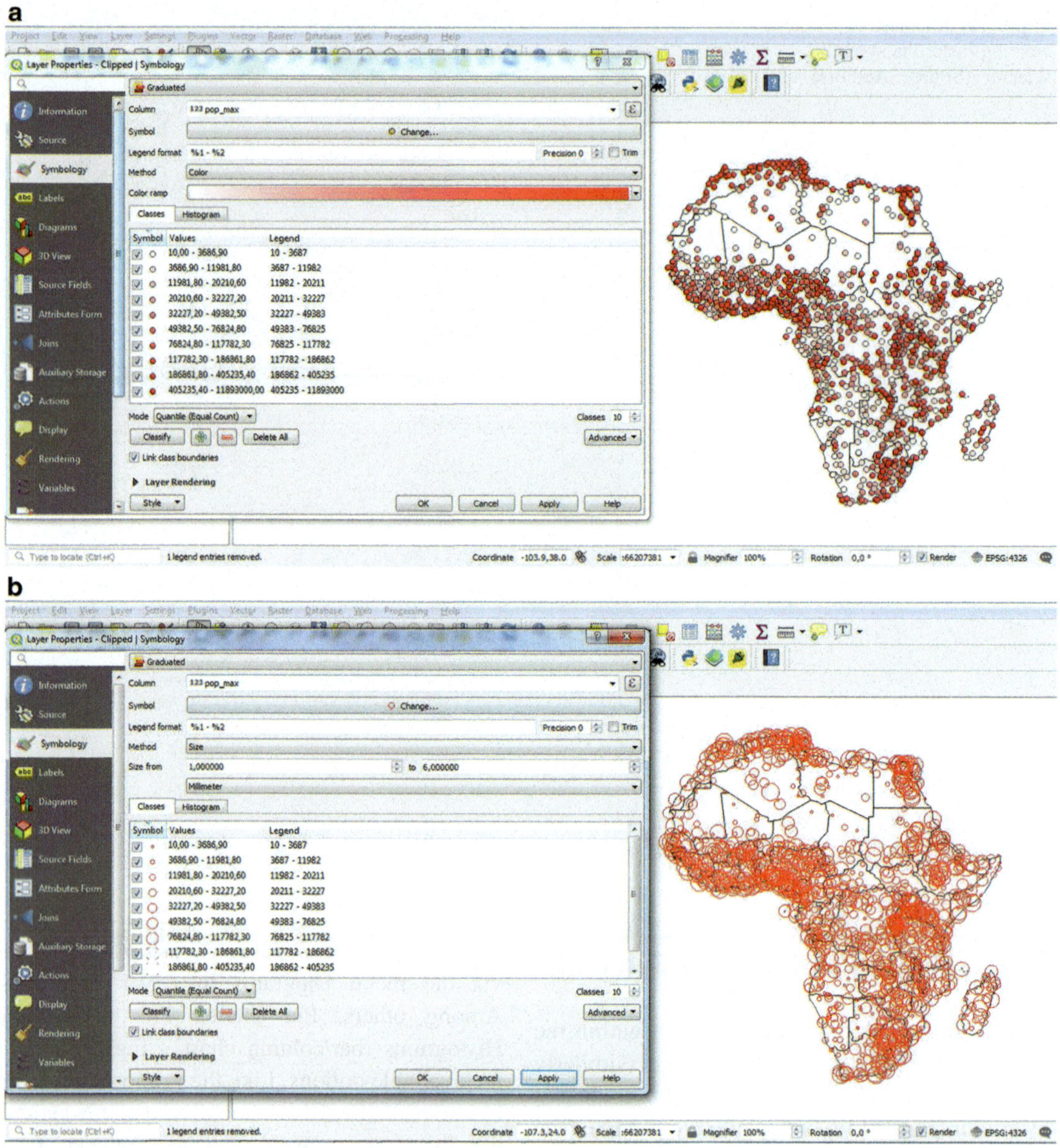

Fig. 9.11 (**a**) Visualisation by colour. (**b**) Visualisation by size (source: Authors)

distance. The geometry is distorted in order to convey the information of this alternate variable.

9.8 Map Composition

Map composition as the layout of the basic structural map elements is a result of the *carto*graphic creativity of the author. The visual presentation (Fig. 9.13) should follow some cartographic rules and its purpose, scale, map sheet format etc.

There are five main elements of a map layout, that should be present in every map:

- Name/Title
- Scale
- Legend
- Map field
- Imprint belong

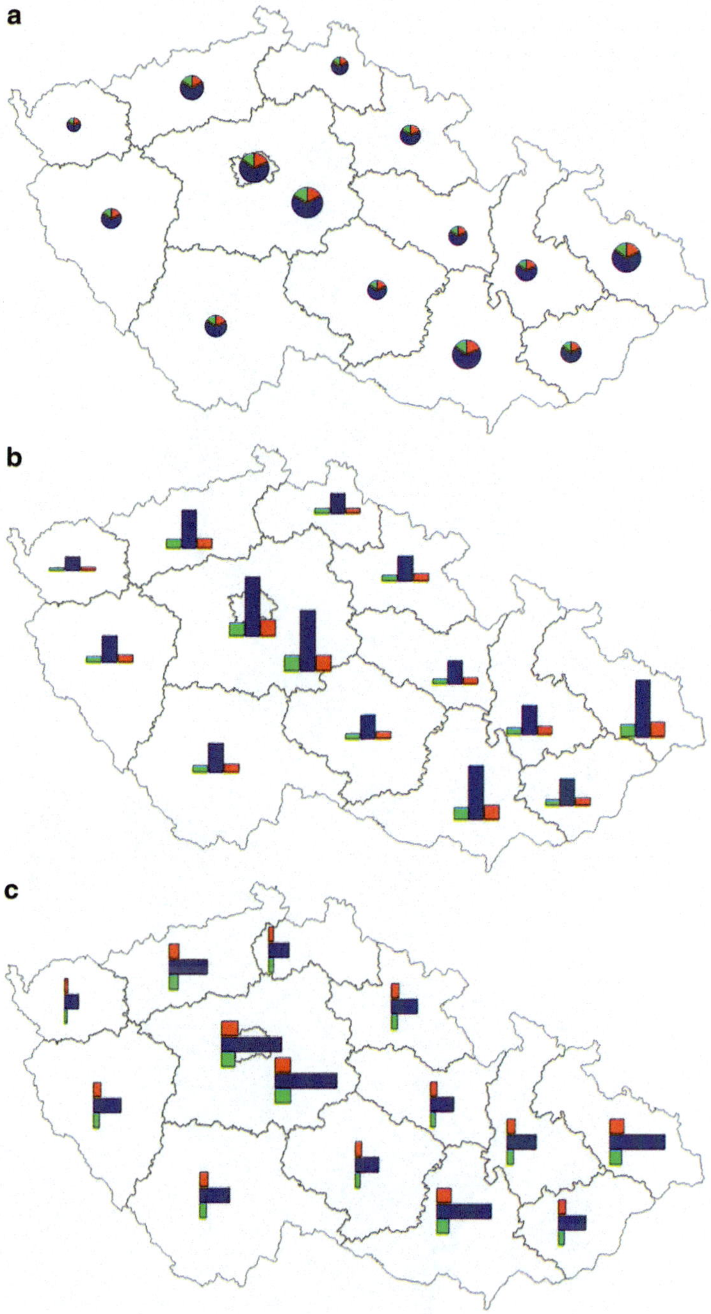

Fig. 9.12 Visualisation of same data as (**a**) Pie Chart, (**b**) Column Chart, (**c**) Bar Chart. (Source: Authors)

Additional map elements, such as a North arrow or Coordinates/Grid can be present as well.

The name of the map/Title represents the subject (WHAT), an area (WHERE) and a time specification (WHEN) of the mapped event. It is usually placed at the top of the map and often centred and written large enough to be readable from the distance. The word "map" is usually not used in the name as it is clear that it is a map.

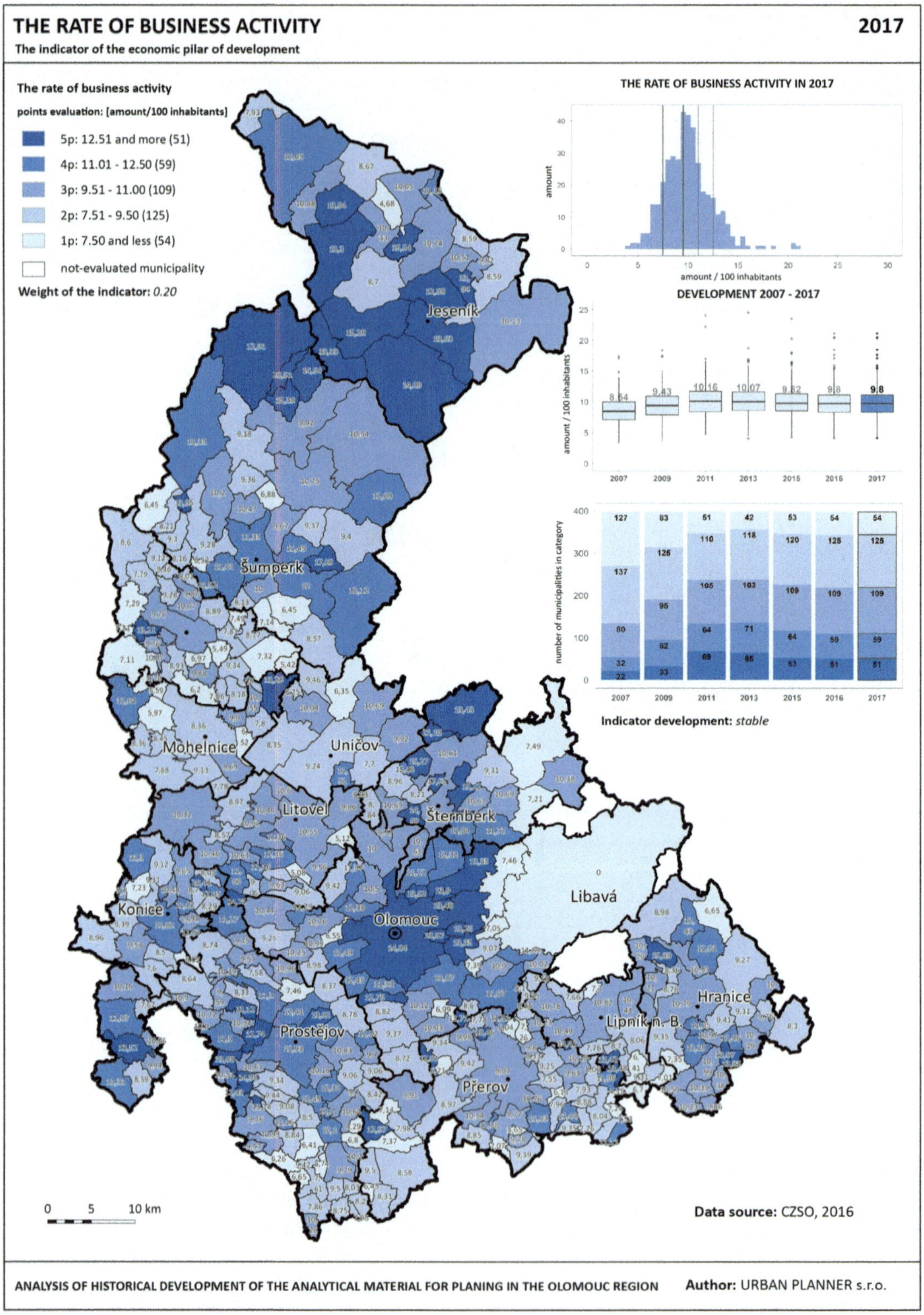

Fig. 9.13 Example of map with all obligatory components. (Source: Authors)

Scale is a graphical and numeral ratio between distance on a map and the corresponding real distance on the ground. Graphical scale is more preferable as it maintains proportional compared to the numeral scale while copying (reducing or enlarging).

Creating a **legend** belongs to the most difficult tasks in the map creation and the following rules must be obeyed during the process:

- Completeness of the legend – every symbol from the map has to be in the legend and vice versa,
- Independence of the legend – every dissimilar object must also have a different symbol, it is also not possible that two different symbols would be assigned to the same object,
- Organisation of the legend – similar symbols are put into groups and the most important sets of elements are placed the highest,
- Structure of the legend – proportion, structure and colours in the legend have to be the same as used in the map,
- Comprehensiveness of the legend – the legend is simple, combined or complex combined according to the map character.

The map field is an area representing the map content itself which is limited by an inner map frame. It can have any shape or it is rectangular. It is either geometrically limited by a regular frame map (rectangular, square, etc.), or the frame border is an area (country, island) in case of individual territories.

The imprint contains a résumé of the information connected with the map creation – such as author, date of creation, data sources, publisher, etc.

Supplemental elements can include also logos, graphs, charts, figures, diagrams, text fields, smaller maps, etc.

References

Bertin, J. (1967). Semiology of graphics: Diagrams, networks, maps (Translation of "Sémiologie Graphique. Les Diagrammes, les Réseaux, les Cartes (1967)" by William J. Berg). Madison, Wisconsin.

Čapek, R., Mikšovský, M., & Mucha, L. (1992). *Geografická kartografie*. Praha: SPN.

Field, K. (2018). *Cartography : A compendium of design thinking for mapmakers*.

ICA. (2018). *Definitions*. https://icaci.org/mission/. Accessed 27 Nov 2018.

QGIS.com. (2018). *QGIS*. http://qgis.com/

Robertson, N., Sanders, D. P., Seymour, P., & Thomas, R. (1996). Efficiently four-coloring planar graphs. In *Proceedings of the twenty-eighth annual ACM symposium on theory of computing – STOC '96* (pp. 571–575). New York: ACM Press.

Veverka, B., & Zimová, R. (2008). *Topografická a tematická kartografie. České vysoké učení technické v Praze*. Praha: Stavební fakulta.

Voženílek, V., & Kaňok, J. (2011). *Metody tématické kartografie: vizualizace prostorových jevů*. Olomouc: Univerzita Palckého v Olomouci.

Online Visualisation 10

Jiří Pánek and Jaroslav Burian

Abstract
This chapter deals with online tools used for spatial data visualisation. Currently, there exist many tools and software that can be used for this purpose, both commercial (e.g. ArcGIS Online) and non-commercial (e.g. MapBox, Leaflet or CartoDB). Within this chapter, we are focusing on the most used software solutions. The text serves the basic overview of existing solutions, that can be used with no or minimal programming skills.

Keywords
Visualisation · Online · Web · GIS · Maps

10.1 Introduction

The use of online mapping and spatial search has become ubiquitous, with hundreds of millions of desktop and smartphone users regularly accessing mapping services (Smith 2016). There is various numbers of platforms, that allow creating online maps without the need of coding knowledge. This section will firstly focus on the commercial online GIS applications from Esri called ArcGIS Online (https://www.arcgis.com/home/index.html), Collector for ArcGIS and platform Story Maps (http://storymaps.arcgis.com/en/). Furthermore, other platforms such as Leaflet, MapBox, or CartoDB will be briefly described as well.

10.2 ArcGIS Online

ArcGIS online a cloud-based mapping and analysis platform from Esri (Esri 2018a), that allows users to access the workflow-specific apps, maps and data from around the globe, and tools for being mobile in the field (Pánek and Glass 2018). In order to use ArcGIS online platform, one has to have a login – it can be either ArcGIS Public Account or Enterprise login (Esri Account). The Public Account is free, but has some limitations, such as users with Public Account cannot publish Feature Services from ArcGIS Desktop to a Public Account. It is possible to upload shape files that are available as layers (so long as not greater than 1000 features) and total storage limit is 2GB. On the other hand, the Enterprise login is less restricted, but usually available only to institutions, who pay for it. The data and maps are stored in a cloud, hence can be access from anywhere and anytime. The interface of ArcGIS online (Fig. 10.1) resembles other web mapping platforms, and allows users to upload offline data in a form:

- Shapefile (ZIP archive containing all shapefile files)

J. Pánek (✉)
Department of Development and Environmental Studies, Palacký University Olomouc, Olomouc, Czech Republic
e-mail: jiri.panek@upol.cz

J. Burian
Department of Geoinformatics, Palacký University Olomouc, Olomouc, Czech Republic
e-mail: jaroslav.burian@upol.cz

V. Pászto et al. (eds.), *Spationomy*, https://doi.org/10.1007/978-3-030-26626-4_10

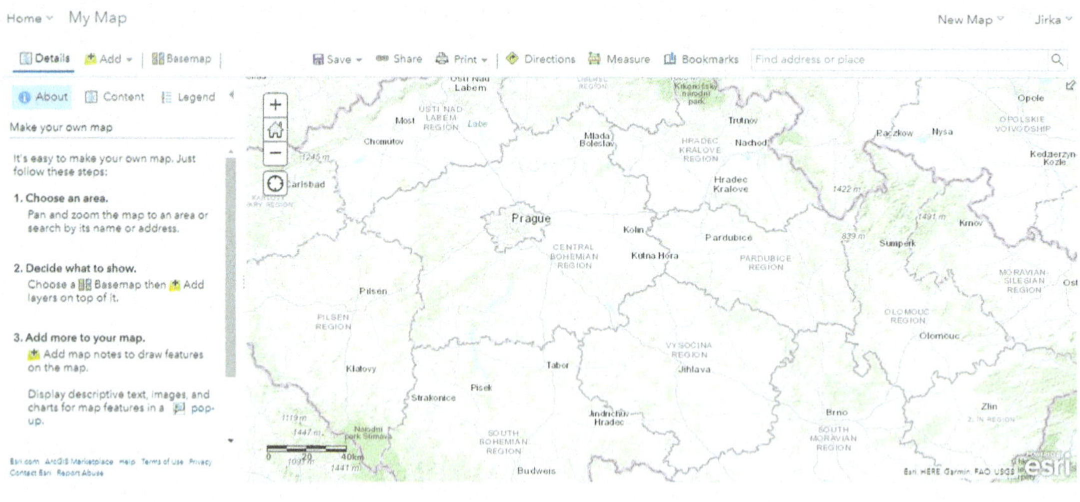

Fig. 10.1 Interface of ArcGIS Online. (Source: Authors)

- CSV or TXT files with optional address, place or coordinate locations (comma, semi-colon or tab delimited)
- GPX (GPS Exchange Format)
- GeoJSON (open standard format for simple geographical features).

Furthermore, online data can be linked as ArcGIS Online layers, ArcGIS Server web services, OGC WMS/WMTS/WFS, Tile layers, KML files or GeoRSS files.

ArcGIS online is not just a tool for creating online maps, it gauges users with options to make web apps or 3D web scenes – with none or limited coding skills.

10.3 Collector for ArcGIS

Another Esri application usable for spationomy research, or field research in general, is Collector for ArcGIS – a stand-alone Esri mobile application available for use with iOS, Android, and Windows. Collector for ArcGIS is an application for designing field survey apps. Unfortunately, the user needs an ArcGIS organizational account to make full use of the application. Previous studies note the utility of Collector for educational applications, including geography classrooms (Peirce 2016; Kolvoord et al. 2017; Pánek and Glass 2018) and field trips (Cho and Kang 2017), typical topics for Collector deployment include also mapping ecosystem services (Edsall et al. 2015) and cadastral mapping (Mourafetis et al. 2015; Apostolopoulos et al. 2016). Collector is used less frequently for social and cultural applications, despite its possibilities for aggregating data on social and physical phenomena from multiple active field researchers. Similar to Collector capabilities is a Crowdsource Story Map template (Esri 2018b), that is unfortunately no more an active project and wont be developed in the future (ArcGIS Blog 2018).

There also exists free and open source alternative to Collector for ArcGIS – Open Data Kit (ODK) which is available at https://opendatakit.org/. ODK allows to design a survey for data collection and later collect spatial and non-spatial data in the field. Open-source alternative to Collector can be QField (https://www.qfield.org/) – data collection application for Android devices directly linked with QGIS.

10.4 Esri Story Maps

Esri Story Maps let you combine authoritative maps with narrative text, images, and multimedia content. They make it easy to harness the power of maps and geography to tell your story. Using StoryMaps for presenting study materials is becoming popular in recent years (Kerski 2015), in fields like history

(Abrate et al. 2013; Coleman et al. 2015), but also migration (Kerski 2013) or protections of ecosystems (Crocker et al. 2015; Fox 2016).

First of all user needs to sign-in by using ArcGIS account either the institutional or public (free). One can also use Facebook or Google to log-in, nevertheless the public accounts have some limitations – especially in storage and some in functionality as well.

There are several templates one can use for creating his/her own online map presentation. In this section we will focus on two of the easier templates. The first template is "Shortlist" (see Fig. 10.2) and this template presents a set of photos or videos along with captions, linked to an interactive map. It's ideal for walking tours or any sequence of places one would like users to follow in order. The only thing author needs is to link photos from online repository (in case of free account) or upload photos (in case of paid account). If the pictures are geocoded – they have GPS coordinate included in the picture information, for example from smartphone, they will automatically appear on the map. In case they do not contain this information, one can easily place them manually on the map, by clicking at the picture and then clicking on the map, where the picture was taken. Once all the pictures are located, it is possible to edit captions, texts, labels, etc. Final version of the map is saved in the cloud – no need to download anything and is available online anytime.

The second map template presented in this section is "Story Map Journal", that is a bit more sophisticated and allows creating an in-depth narrative organized into sections presented in a scrolling side panel. As users scroll through the sections in your Map Journal they see the content associated with each section, such as a map, 3D scene, image, video, etc. Each "page" of the journal is built from two parts – stage (very often a map) and side panel (usually a text). Maps that are in the stage area can be pre-created, or you can create them during the process. In the side panel one can insert any text and link words in the text to webpages, photos, videos or even to a maps (Fig. 10.3). Nevertheless, Esri announced new version of StoryMaps, that will not support any of the "old" templates showed in this chapter.

10.5 Google Fusion Tables

Google Fusion Tables is a cloud-based service for data management and integration (Gonzalez et al. 2010). There can be found many examples of use

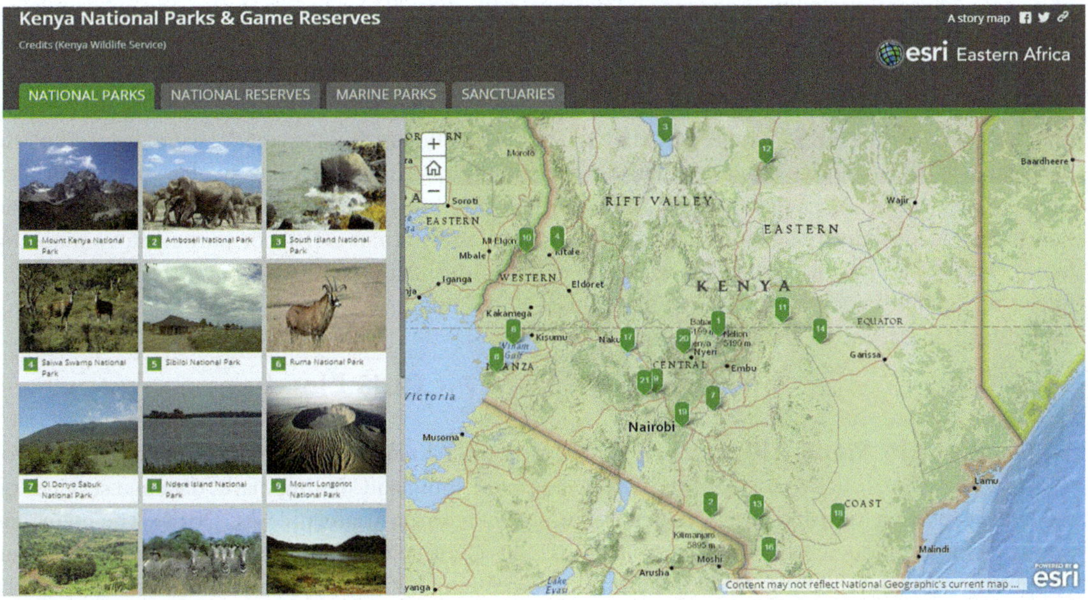

Fig. 10.2 Example of the Shortlist template. (Source: Esri Eastern Africa 2018)

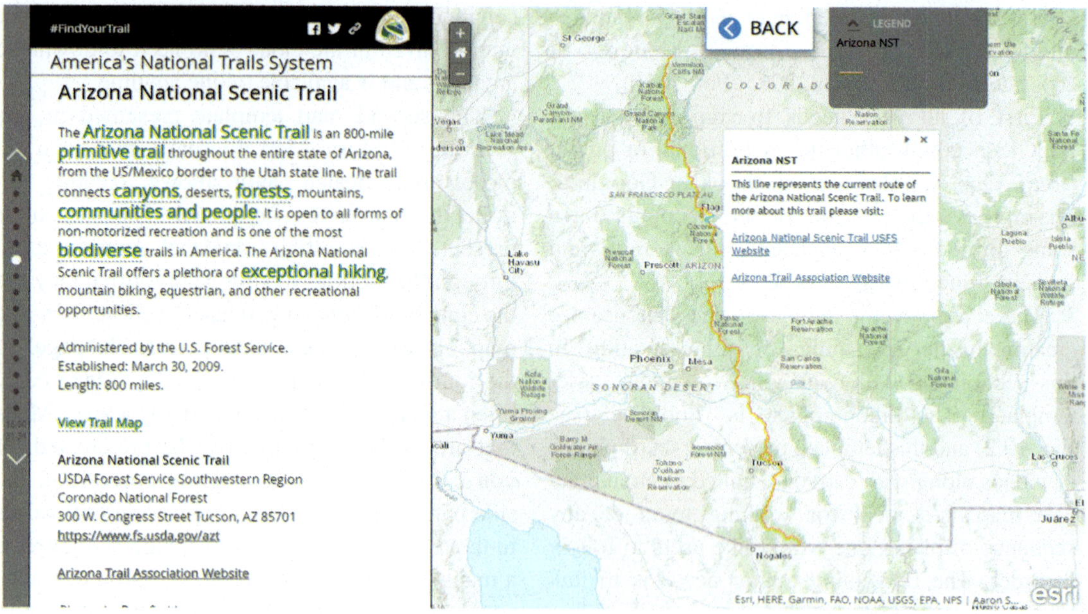

Fig. 10.3 Example of the Map Journal template. (Source: US National Park Service 2018)

Fusion Tables in the research (Bowie et al. 2014; Signore 2016). Fusion Tables enables to upload tabular data (spreadsheets, CSV, KML) and visualise them in many ways (pie charts, bar charts, lineplots, scatterplots, timelines, and maps). Visualisation is based on Google Maps (street, satellite, and terrain) that also allows to geocode the input data without coordinates (e.g. geocoding of postal addresses into point in the map). Fusion Tables also supports the rendering of heat maps (maps of the density of features). Data and all results can be shared and also exported in many graphic, tabular and GIS formats (KML). All users have 1 GB of storage quota for their tables. There exist also API that allows external developers to design applications using Fusion tables (Fig. 10.4). Fusion Tables are available at https://fusiontables.google.com, but will not be available after December 2019.

10.6 Google Maps API

Google Maps is a web mapping service developed by Google. Google Maps also offers an API (Application programming interface), that can be used for development custom web mapping applications based on Google Maps (including different map types, Street View, geocoding services or route planning tools). API is based on the JavaScript and allows to create maps with your own content and imagery for display on web pages and mobile devices. There exist many online tutorials how to start to use Google Maps API, that allows to use it also for programming beginners. Google Maps API is a payed service, but Google offers $200 monthly credit that is good enough for creation of many basic map applications. Google maps have been used in many research fields like health studies (Boulos 2005), logistics (Fu et al. 2010), GPS navigation (Li and Zhijian 2010), or participatory mapping (Boroushaki et al. 2010) (Fig. 10.5).

10.7 QGIS Cloud

QGIS Cloud is a powerful Web-GIS platform for publishing and sharing maps, data and services on the internet (Sourcepole AG 2018). The main idea of QGIS cloud is a direct connection between desktop QGIS application and online QGIS

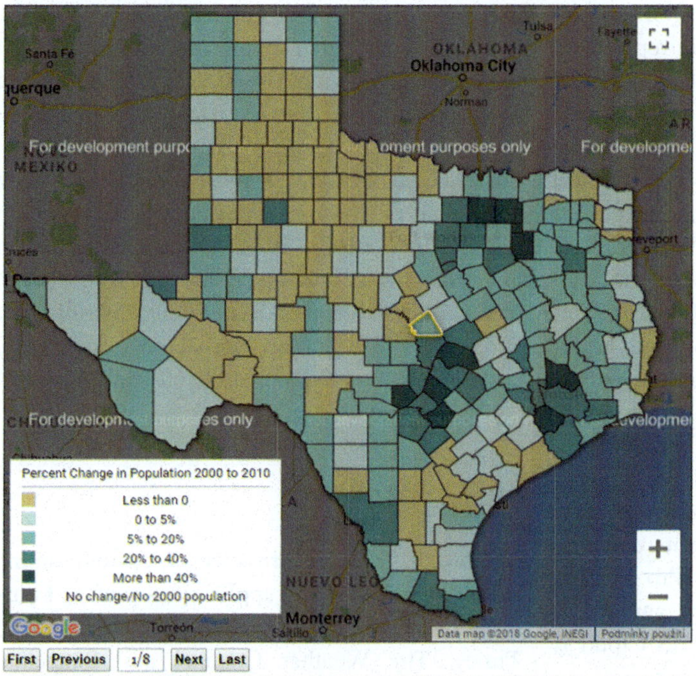

Lampasas County			
Race/Ethnicity	2000 Pop.	2010 Pop.	Change
All races	17,762	19,677	10.8%
White	15,409	16,792	9.0%
Black	550	632	14.9%
Asian	134	201	50.0%
American Indian	124	177	42.7%
Pacific Islander	11	39	254.5%
Other	1,153	1,216	5.5%
Two or more races	381	620	62.7%
*Hispanic origin	2,677	3,443	28.6%

County	All races (2010)	All races (2000)	All races (Change)	Hispanic (2010)	Hispanic (2000)	Hispanic (Change)
Anderson County	58,458	55,109	6.1%	9,287	6,705	38.5%
Andrews County	14,786	13,004	13.7%	7,195	5,202	38.3%
Angelina County	86,771	80,130	8.3%	17,145	11,496	49.1%
Aransas County	23,158	22,497	2.9%	5,690	4,571	24.5%
Archer County	9,054	8,854	2.3%	675	431	56.6%

Fig. 10.4 Example of use of Google Fusion Tables for Census data visualisation. (Source: Murphy and Stiles 2012)

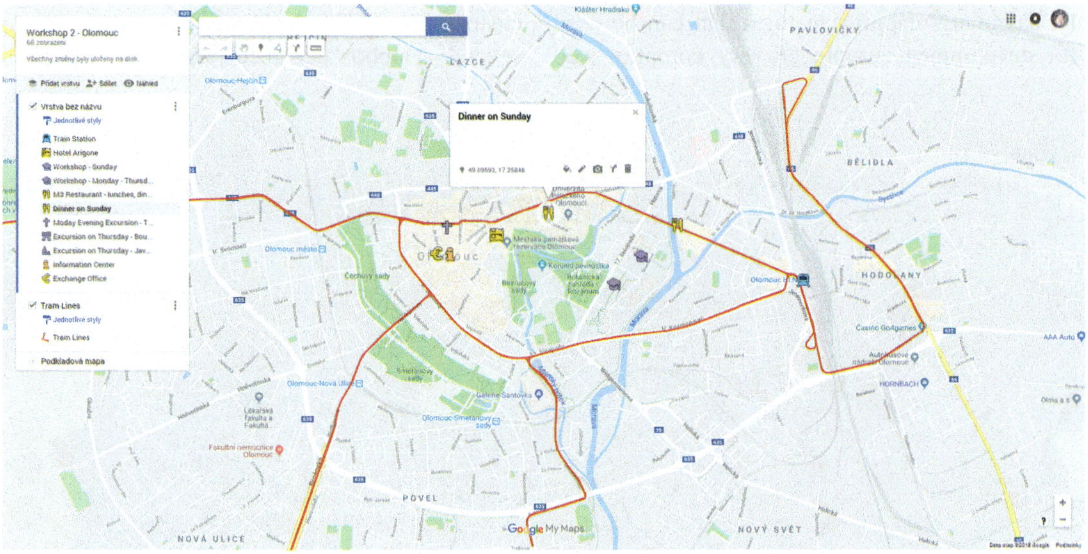

Fig. 10.5 Example of Google Maps API used for private purposes – conference organisation. (Source: Authors)

cloud. This is realised through the QGIS Cloud plugin from the official QGIS plugin repository. Thanks to this connection, the application can be prepared as a map in QGIS environment and transferred to the web at the end. Maps can be shared over OGC (Open Geospatial Consortium) compliant web services. Maps can be displayed via WMS or downloaded via WFS. With WFS-T, you can edit your data directly over the web service. There exist also the Mobile Client integrated in QGIS Cloud. If the data should not be publicly accessible, QGIS Cloud pro allows you to restrict access by protecting the resources with a password. A free account offers unlimited public maps (only non-commercial/non-government use) and one PostGIS 2.0 database (max. 50 MB total, max 10 concurrent database connections). Payed QGIS Cloud Pro version offers more databases, more space and many additional functions (Fig. 10.6). QGIS Cloud is available at https://qgiscloud.com/.

10.8 Leaflet

Leaflet is the leading open-source JavaScript library for mobile-friendly interactive maps. Leaflet has many mapping features that can be used for development simple or very complex web mapping applications. Leaflet is designed with simplicity, performance and usability in mind. It works efficiently across all major desktop and mobile platforms, can be extended with lots of plugins, has a beautiful, easy to use and well-documented API and a simple, readable source code that is a joy to contribute to (Agafonkin 2017) (Fig. 10.7). It is used for the main OpenStreetMap website map, as well as on many other websites like Flickr, Washington Post, The Wall Street Journal or Geocaching.com. Leaflet is available at https://leafletjs.com/.

10.9 Mapbox

Mapbox is a large provider of custom online maps for websites and applications such as Foursquare, Lonely Planet, Facebook, the Financial Times, The Weather Channel and Snapchat. Mapbox is the creator of, or a significant contributor to some open source mapping libraries and applications, including the MBTiles specification, the TileMill cartography IDE, the Leaflet JavaScript library, and the CartoCSS map styling language and parser (Mapbox 2018). The Mapbox Maps SDK allows advanced map customisation. The developer can choose among several Mapbox-designed styles or design a

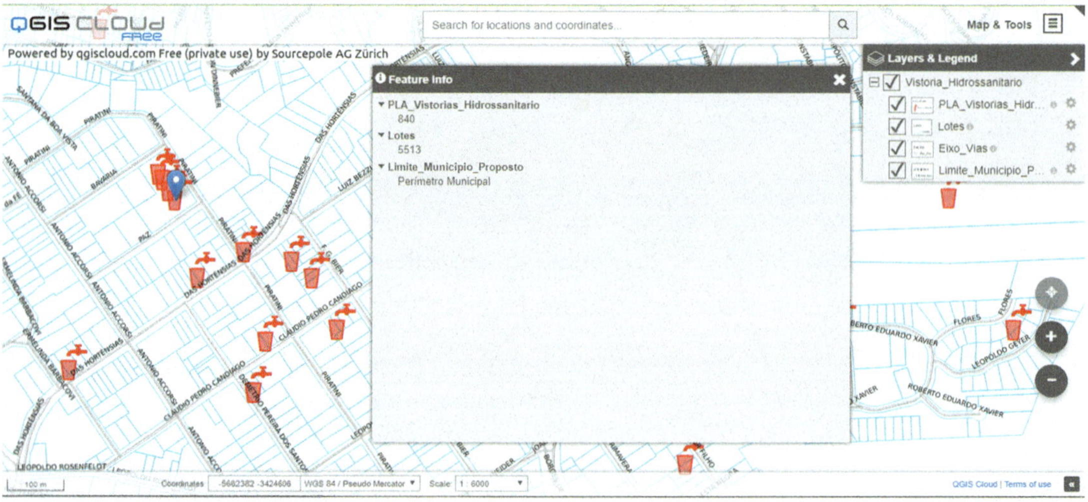

Fig. 10.6 Example of QGIS Cloud application. (Source Sourcepole AG 2018)

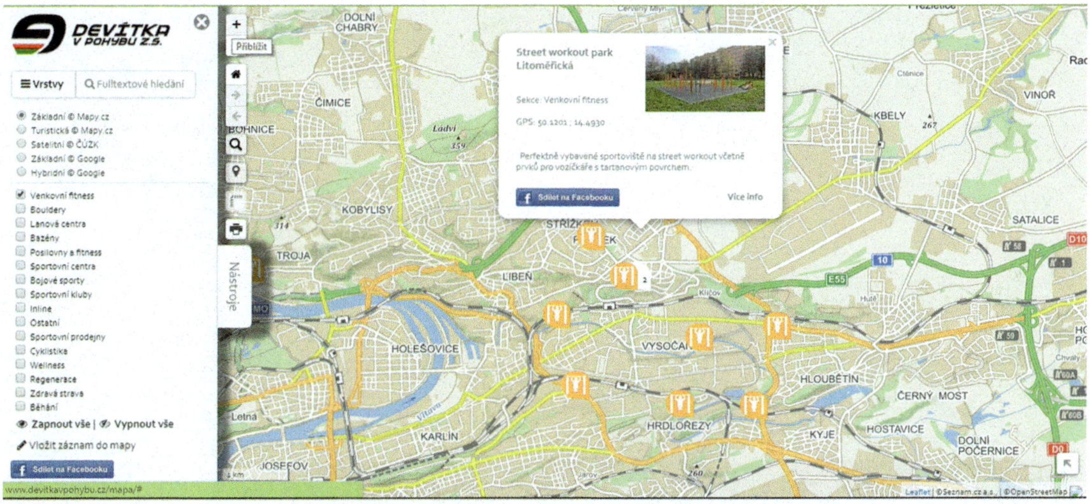

Fig. 10.7 Example of QGIS Cloud application. (Source: Nétek 2016)

custom style in the graphical style editor of Mapbox Studio (Mapbox GL 2018). Mapbox is a payed service, but if you will not exceed some limits (map views, page requests), you can use it for free (Fig. 10.8).

10.10 Carto

CARTO (formerly CartoDB) is a Software as a Service (SaaS) cloud computing platform that provides GIS and web mapping tools for display in a web browser (Wikipedia 2018). Use of CARTO for data analysis and visualization that do not require previous GIS or development experience. CARTO users can use the company's free platform or deploy their own instance of the open source software. For smaller amount of data, CARTO is offered as free service. CARTO is built on PostGIS and PostgreSQL. The tool uses JavaScript extensively in the front end web application, back end Node.js based APIs, and for client libraries (CartoDB 2011) (Fig. 10.9).

10.11 OpenLayers

OpenLayers is an open-source JavaScript library for creation of dynamic web maps. It can display map tiles, vector data and markers loaded from any source. OpenLayers has been developed to further the use of geographic information of all kinds (OpenLayers 2018). It is completely free, Open Source JavaScript, released under the 2-clause BSD License (also known as the FreeBSD). OpenLayers supports GeoRSS, KML (Keyhole Markup Language), Geography Markup Language (GML), GeoJSON and map data from any source using OGC-standards as Web Map Service (WMS) or Web Feature Service (WFS). API is provided for building custom web map applications (Fig. 10.10).

10.12 Advanced Mapping Tools

Except the tools mentioned above there exist many advanced web mapping software or environment that allows to create very advanced web map applications with very specific tools. Currently, there are two leading open source platforms – MapServer (https://mapserver.org/) and GeoServer (http://geoserver.org/). These are used mostly by GIS professionals and requires more complex programming skills. MapServer and GeoServer represents classical mapping server with a huge variety of GIS functions for data management, analysis and visualisation. In comparison with simple mapping tools, map servers have many benefits. They offer advanced

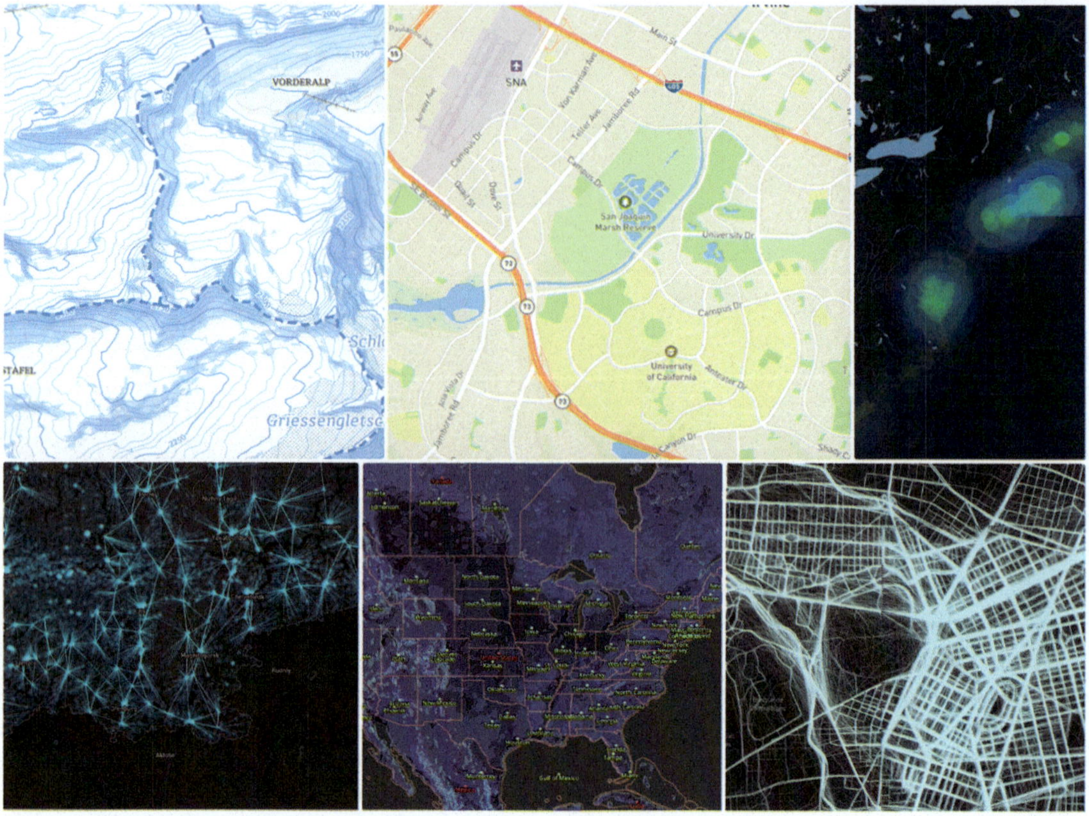

Fig. 10.8 Examples of Mapbox customisation. (Source: Mapbox 2018)

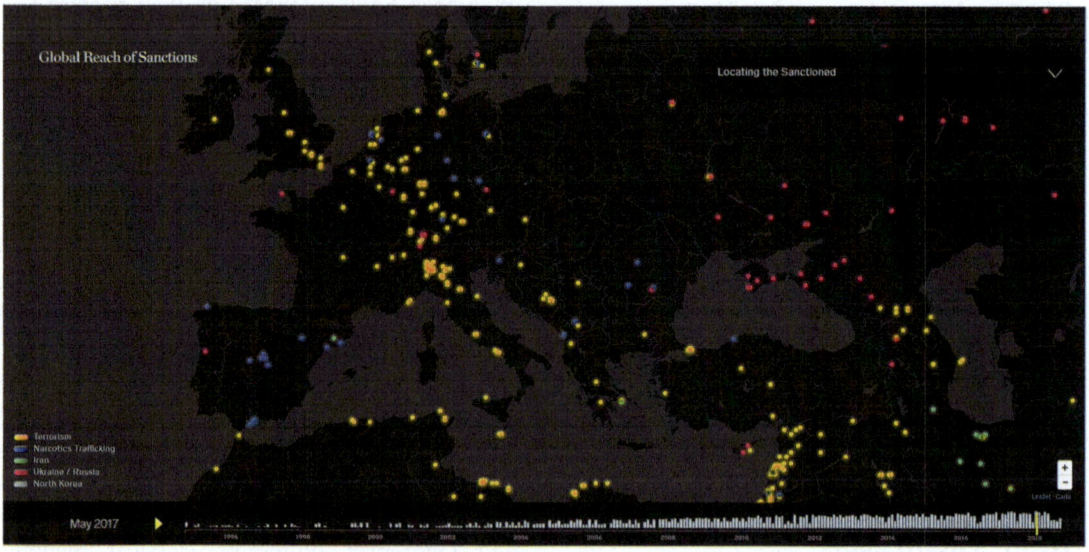

Fig. 10.9 Example of use of CARTO for sanctions visualisation. (Source: EnigmaPublic 2018)

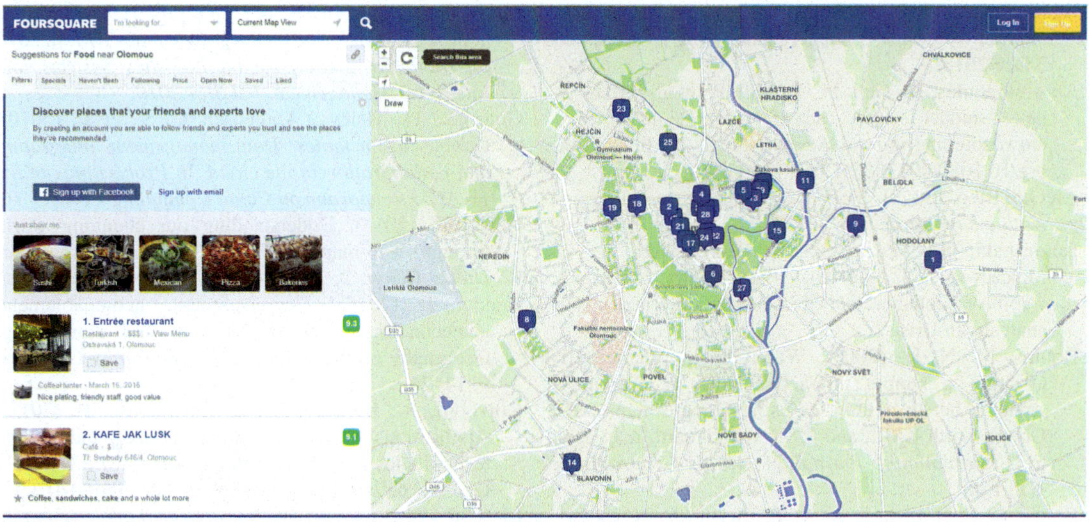

Fig. 10.10 Example of use of OpenLayers. (Source: Foursquare 2018)

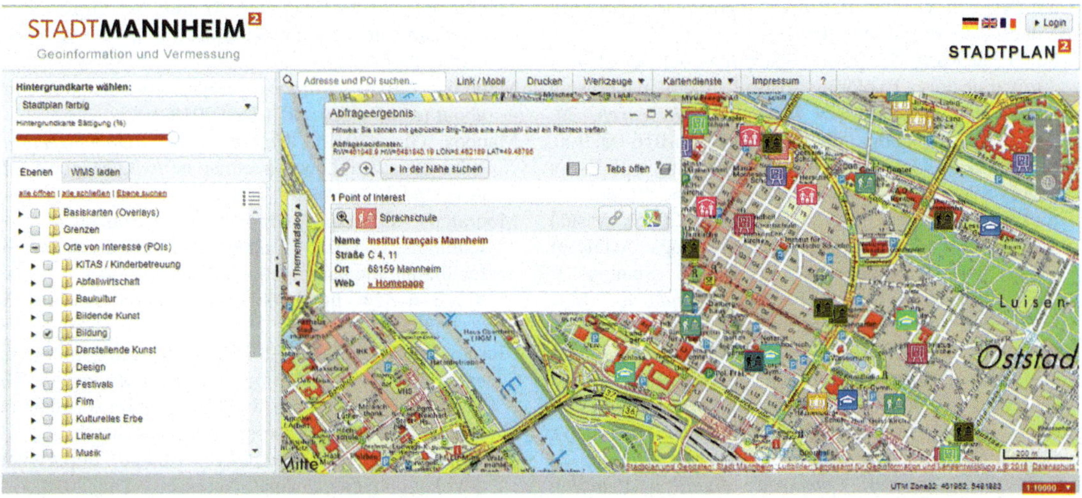

Fig. 10.11 Example of MapServer Application. (Source: Mannheim 2018)

cartographic options like displaying data dynamically, cartographic projections, professional symbol styling, different spatial data formats support, or publishing mapping services (e.g. WMS and WFS) (Fig. 10.11).

References

Abrate, M., Bacciu, C., Hast, A., et al. (2013). GeoMemories—A platform for visualizing historical, environmental and geospatial changes in the Italian landscape. *ISPRS Int J Geo-Information, 2*, 432–455.

Agafonkin, V. (2017). *Leaflet – A JavaScript library for interactive maps*. https://leafletjs.com/. Accessed 3 Dec 2018.

Apostolopoulos, K., Geli, M., Petrelli, P., et al. (2016). A new model for cadastral surveying using crowdsourcing. *Survey Review, 50*, 1–12. https://doi.org/10.1080/00396265.2016.1253522.

ArcGIS Blog. (2018). *Story map crowdsource moves to mature support in June 2018*.

Boroushaki, S., Malczewski, J., Boroushaki, S., & Malczewski, J. (2010). Participatory GIS: A web-based collaborative GIS and multicriteria decision analysis. In Urisa J.

Boulos, M. N. K. (2005). Web GIS in practice III: Creating a simple interactive map of England's strategic health authorities using Google maps API, Google earth KML, and MSN virtual earth map control. *International Journal of Health Geographics, 4*(1),1–8.

Bowie, G. D., Millward, A. A., & Bhagat, N. N. (2014). Interactive mapping of urban tree benefits using Google fusion tables and API technologies. *Urban Forestry & Urban Greening, 13*, 742–755. https://doi.org/10.1016/j.ufug.2014.06.002.

CartoDB. (2011). *CartoDB code repository*. carto.com/docs. Accessed 4 Dec 2018.

Cho, N., & Kang, Y. (2017). Space-time density of field trip trajectory: Exploring spatio-temporal patterns in movement data. *Spatial Information Research, 25*, 141–150. https://doi.org/10.1007/s41324-016-0079-x.

Coleman, B., et al. (2015). History storytelling with Esri's story maps application. *Teaching History, 49*, 60.

Crocker, S., Walters, B., & Morin, R. (2015). Visual analysis of forest health using story maps: A tale of two forest insect pests. In S. M. Stanton & G. Christensen (Eds.), *Pushing boundaries: New directions in inventory techniques and applications: Forest inventory and analysis (FIA) symposium 2015* (p. 315). Portland: Department of Agriculture, Forest Service, Pacific Northwest Research Station.

Edsall, R. M., Barbour, L., & Hoffman, J. (2015). *Complementary methods for citizen mapping of ecosystem services: Comparing digital and analog representations* (pp. 295–307). Cham: Springer.

EnigmaPublic. (2018). *Sanctions tracker*. https://labs.enigma.com/sanctions-tracker/. Accessed 4 Dec 2018.

Esri. (2018a). *What is ArcGIS online?*

Esri. (2018b). *Story Map Crowdsource (Mature)*.

Esri Eastern Africa. (2018). *Kenya National Parks & Game Reserves*.

Foursquare. (2018). *Foursquare*. https://foursquare.com. Accessed 4 Dec 2018.

Fox, C. (2016). *The value of story mapping for coastal managers*. University of Rhode Island.

Fu, C., Wang, Y., Xu, Y., & Li, Q. (2010). *The logistics network system based on the google maps API*. In *2010 International Conference on Logistics Systems and Intelligent Management, ICLSIM 2010*.

Gonzalez, H., Halevy, A., Jensen, C. S., et al. (2010). *Google fusion tables: Data management, integration and collaboration in the cloud*. In *Proceedings of the 1st ACM symposium on Cloud Computing – SoCC '10*.

Kerski, J. (2013). Understanding our changing world through web-mapping based investigations. *J Res Didact Geogr, 2*, 11–26.

Kerski, J. (2015). Geo-awareness, geo-enablement, geotechnologies, citizen science, and storytelling: Geography on the world stage. *Geography Compass, 9*, 14–26.

Kolvoord, R., Keranen, K., & Rittenhouse, P. (2017). Applications of location-based services and mobile technologies in K-12 classrooms. *ISPRS Int J Geo-Inf, 6*, 209. https://doi.org/10.3390/ijgi6070209.

Li, H., & Zhijian, L. (2010). *The study and implementation of mobile GPS navigation system based on Google maps*. In *Proceedings of ICCIA 2010–2010 international conference on computer and information application*.

Mannheim. (2018). *Stadtplan Mannheim*. https://www.gis-mannheim.de/mannheim/index.php. Accessed 4 Dec 2018.

Mapbox. (2018). *Mapbox*. https://www.mapbox.com/about/team/. Accessed 4 Dec 2018.

Mapbox GL. (2018). *Mapbox GL – OpenStreetMap Wiki*. https://wiki.openstreetmap.org/wiki/Mapbox_GL. Accessed 4 Dec 2018.

Mourafetis, G., Apostolopoulos, K., Potsiou, C., & Ioannidis, C. (2015). Enhancing cadastral surveys by facilitating the participation of owners. *Survey Review, 47*, 316–324. https://doi.org/10.1179/1752270615Y.0000000009.

Murphy, R., & Stiles, M. (2012). *Census 2010 interactive map: Texas population by race, Hispanic origin*. The Texas Tribune.. https://www.texastribune.org/library/data/census-2010/. Accessed 3 Dec 2018.

Nétek, R. (2016). *Devítka v pohybu – mapa*.. http://www.devitkavpohybu.cz/mapa/. Accessed 3 Dec 2018.

OpenLayers. (2018). *OpenLayers*.. https://openlayers.org/. Accessed 4 Dec 2018.

Pánek, J., & Glass, M. (2018). Gaining a mobile sense of place with collector for ArcGIS. *Journal of Geography in Higher Education*, 1–14. https://doi.org/10.1080/03098265.2018.1515190.

Peirce, S. (2016). Making learning Mobile: Using Mobile technologies to bring GIS into the geography classroom. *Teach Innov Proj, 6*, 1.

Signore, A. (2016). Mapping and sharing agro-biodiversity using open data kit and Google fusion

tables. *Computers and Electronics in Agriculture, 127*, 87–91. https://doi.org/10.1016/j.compag.2016.06.006.
Smith, D. A. (2016). Online interactive thematic mapping: Applications and techniques for socio-economic research. *Computers, Environment and Urban Systems, 57*, 106–117.
Sourcepole AG. (2018). *QGIS cloud.*. https://qgiscloud.com/. Accessed 3 Dec 2018.
US National Park Service. (2018). *America's National Trails System.*
Wikipedia. (2018). *CartoDB.*. https://en.wikipedia.org/wiki/CartoDB. Accessed 4 Dec 2018.

Part III

Spatial Exploration of Economic Data

Introduction to Spatial Exploration of Economic Data 11

Vít Pászto

Abstract

In the introductory chapter, firstly, the summary of how this chapter evolved is provided as well as the organisation of the whole Part III of this book is described. Then, the main body serves as an overview of how the spatial exploration of economic data and respective methods of interdisciplinary analytics can be approached. Based on the authors experiences up to this date, five levels or stages of an analytical approach to spatially analyse economic data are defined. At the lowest level, the author focuses on a "simple" visualisation of data, following by the level of merging (multivariate) statistics of economic data with their spatial component. In the middle, as a third level, spatial statistics – as an implicit use of statistics in the spatial analysis – is mentioned. As the fourth stage, a workflow of the previous ones is depicted, which leads to the final and most advanced level of spatial-economic modelling. This chapter strives to define a universal workflow in economic data analysis. The conceptual framework introduced in this chapter is based on 3 years of interdisciplinary cooperation within the Spationomy project.

Keywords

Geovisual analysis · Spatial statistics · Exploratory analysis · Spatial modelling

11.1 Introduction

This chapter has been formerly work-titled as a "methods of interdisciplinary analytics", which turned out to be a rather ambitious plan as it might take a whole book to write about methods. In this book, we talk about a fusion of several distinct fields – geoinformatics/geomatics, geography, spatial analysis, geovisual tools and (geo) visualisation on one side, on the other, we refer to economy, business, business informatics, economic data and quantitative methods to work with them, and also about a management. In the Spationomy project, to simplify the mixture of disciplines, we call the former disciplines and people (staff and students) simply as "geo" part; keeping the same logic, the latter (disciplines and people) had a label as "eco". As the "eco" label may be confused with "ecology", we lately re-branded the label to "business" part. Anyway, each one above mentioned has a broad theoretical framing, old concepts, methodologies, and contemporary issues to deal with. It will be almost impossible to capture every aspect of these disciplines in on coherent text, and it was not intended at all. However, in previous parts of the book (Parts I and II), we provide a comprehensive overview of the subjects' bases. The Parts

V. Pászto (✉)
Department of Informatics and Applied Mathematics, Moravian Business College Olomouc, Olomouc, Czech Republic

Department of Geoinformatics, Palacký University Olomouc, Olomouc, Czech Republic
e-mail: vit.paszto@gmail.com

V. Pászto et al. (eds.), *Spationomy*, https://doi.org/10.1007/978-3-030-26626-4_11

I and II are meant to be an optimal start for those interested in the spatial economic topics.

Part III is dedicated to examples and case studies on how the “geo” and “business” part can be used jointly. Following chapters illustrate a few application of how common knowledge from geomatics, geography, economy and business informatics could be used in practice and research. Firstly, in Chap. 12, an interesting fusion of geospatial tools (mainly for presenting data) with purely business and managerial needs in a water management company is described in detail. This fusion is an example of how both parts (“geo” and “economic”) can be utilised in real-life situations. On the other hand, the case study in Chap. 13, shows an artificial site selection of a fictitious furniture store. This example is a typical process task for a new branch, store, or any facility allocation. In Chap. 14, the issue of demographical development was merged with spatial planning in cities. It represents a standard research paper approach to study given phenomenon, with a unique combination of the main, very actual topic (population ageing), and its spatial pattern in the studied region. Lastly, in Chap. 15, another example of a scientific study is provided. The study aims to capture the spatial implications of a European CO_2 emission trading system over the ten years. Cartographic methods and geovisual analysis are used to (spatially) explore basic environmental and economic data referring to pollution allowances market. The set of overviewing introduction and four different case studies demonstrates how methods of interdisciplinary analytics can be deployed in both real-life situation and research. It brings new knowledge and opens possibilities for novel approaches in the new joint field of the “spationomy”.

Now let’s get back to the “methods of interdisciplinary analytics” or “(spatial) exploration of economic data”. During the 3 years of the Spationomy project, we learnt how beneficial it is to combines ideas, methods, approaches, and topics among (and from) already mentioned disciplines. From various experiences – learning and teaching material creation, joint scientific paper preparations, brainstorming and discussions – that we had a chance to have in the Spationomy team, and also from the interaction with other stakeholders, we feel the need to propose an optimal workflow of the spatial and economic data analytics. It is also a practice of the author of this chapter to advice students to follow the next five steps towards a successful bachelor or master thesis. In the following pages, a way to approach the data analysis is presented from the simple to more advanced and complex application of spatial and statistical methods. Introduced five levels reflect the authors’ experiences that were proven in practice during the project. However, it is not a dogma that could not be changed, modified or adjusted to the reader’s needs.

11.1.1 Level 1 – (Geo)Visual Analysis

The geovisual analysis represents the first step in the exploration of the data (spatial, economic, business and any other types). The general objective of data visualisation is to transform textual or numerical information into the form of its graphical representation. Whether it is the picture, scheme, chart, graph, workflow, infographics, map, interactive application, 3D graphics or something else, it focuses on a transfer of information to the reader. The visualisation also serves as a tool for data exploration. We can perform simple (and effective at the same time) exploratory analysis, e.g. to find extreme values, outliers, in a data. By depicting a boxplot, scatterplot, or just linear chart (see Part II, Chap. 8), we can immediately see such outliers, which could be hardly detected when “looking” at the numbers. Indeed, the experienced data analyst can find out anomalies in a raw dataset, or when we have a small data sample, it is easy to capture outliers. However, in the case of big data or other highly heterogeneous data, the hidden pattern could be revealed with considerable difficulties, if at all. With the use of visualisation techniques, we can analyse such messy data, we can describe data patterns inside the dataset, uncover and show extreme values, find relationships in data and compare them, and most importantly to communicate the information much more clearly.

Visualisation as the seemingly the simplest level of proposed ways of economic data exploration, however, must also follow some rules and recommendations (for further reference, see Part II). Otherwise, visualisation tools might be misused, which would consequently lead to possible misinterpretation of the graphical representation of data. A great example of the strength and appropriateness of data visualisation is the Anscombe quartet. It is the unique dataset containing four subsets of two-variable data. By calculating mean, variance, the correlation between two variables in each dataset, the same values are returned. Thus, statistically, all four datasets are the same (share identical statistical properties). But if we visualise Anscombe quartet, we achieve something completely different (Fig. 11.1). Of course, it is vital to be familiar with basic statistical properties of analysed data, but we strongly recommend complementing it with any form of a visualisation.

The first step of data analysis should always include visualisation of the data (besides basic statistics). In other words, level one in the data exploration should be that we take the data and visualise them. At this level, we do not modify, filter, select, or aggregate the data before we visualise it. That is why it is the first and rather straightforward way for data analysis. When we talk about geovisualisation, it is nothing more than using maps as a means of the medium to visualise data. In the case of geovisualisation, we

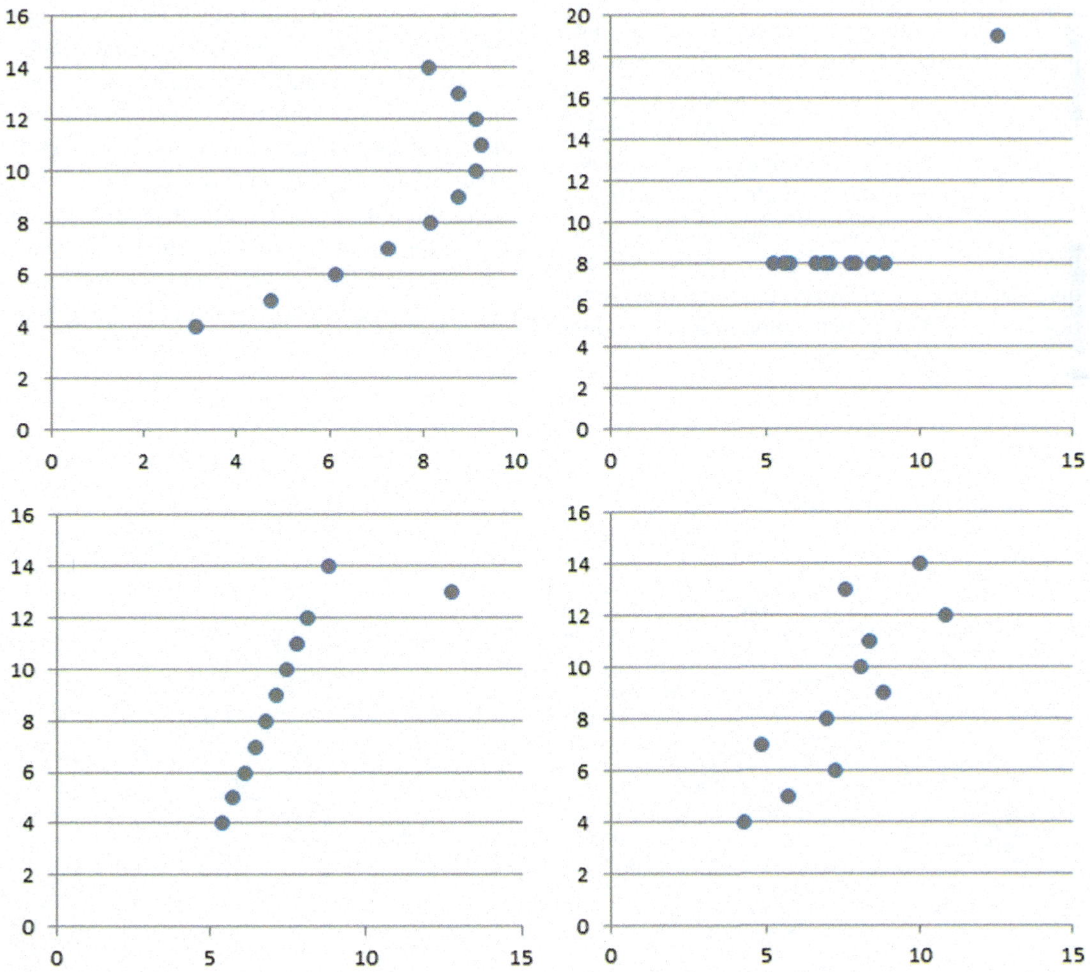

Fig. 11.1 Scatterplots of the Anscombe quartet. (Source: Authors)

have cartography and the rules that we need to follow. At the same time, we should still keep in mind that the information transfer is the primary goal of (geo)visualisation, not blind conformation to the rules. Figure 11.2 serves as an example of geovisualisation, with sample points representing economic subjects displayed on the left (with no attributes reflected), and population density map within administrative units on the right. From both examples in Fig. 11.2, we can easily see otherwise hidden spatial pattern of data.

11.1.2 Level 2 – Statistics, Exploratory Data Analysis and Its (Geo)Visualisation

At this level, all the analytical processes take place outside the GIS environment, or better to say without a spatial component implicitly included in the data. We understand this level to be covering advanced techniques of statistics (e.g. testing hypothesis, multivariate statistics, regression or correlation analysis and such), mathematics. For example, one of the most frequently applied techniques is multivariate statistics. In the field of spatial exploration of economic data, we commonly work with multiple attributes of geographical units (e.g. regions' GDP, income, unemployment rates, demographic structure and others). Unfortunately, there is a somewhat limited number of implemented tools of multivariate statistics directly in the GIS environment. In other words, we can use more advanced settings, or several more variations of such statistical tools outside the GIS environment (or in the environment directly working with a spatial component of data). That is why it is often better to "run away" from GIS to "normal" statistics or mathematics – and multivariate data analysis is the case. In (spatial) data analysis, we commonly use instruments like Correlation analysis, Factorial analysis, Analysis of Variance (ANOVA), Principle Component Analysis, or clustering methods (see Fig. 11.3). These statistical tools help us to reveal relationships in the data variables, reduce the dimension of data that can be then easier handled in GIS or find groups with similar variable values (characteristics). Non-spatial visualisation of such analyses can help us to understand a data better, and it also serves as the support for data interpretation. Since we usually work with data that are geographically referenced (e.g. to the country level, particular geographic

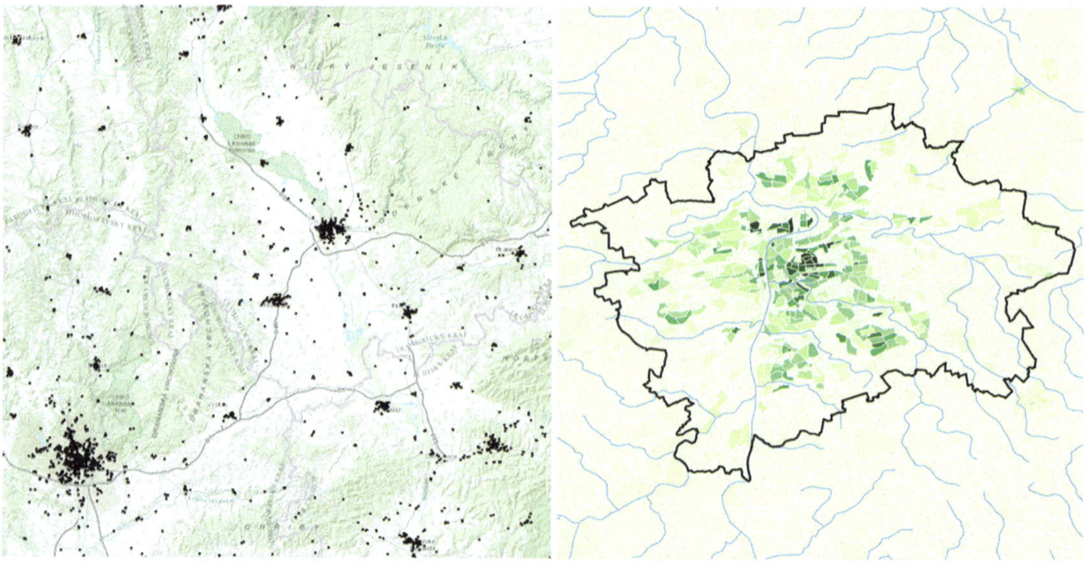

Fig. 11.2 Distribution of economic subjects – sample data (left), population density visualisation in administrative units (right). (Source: Authors)

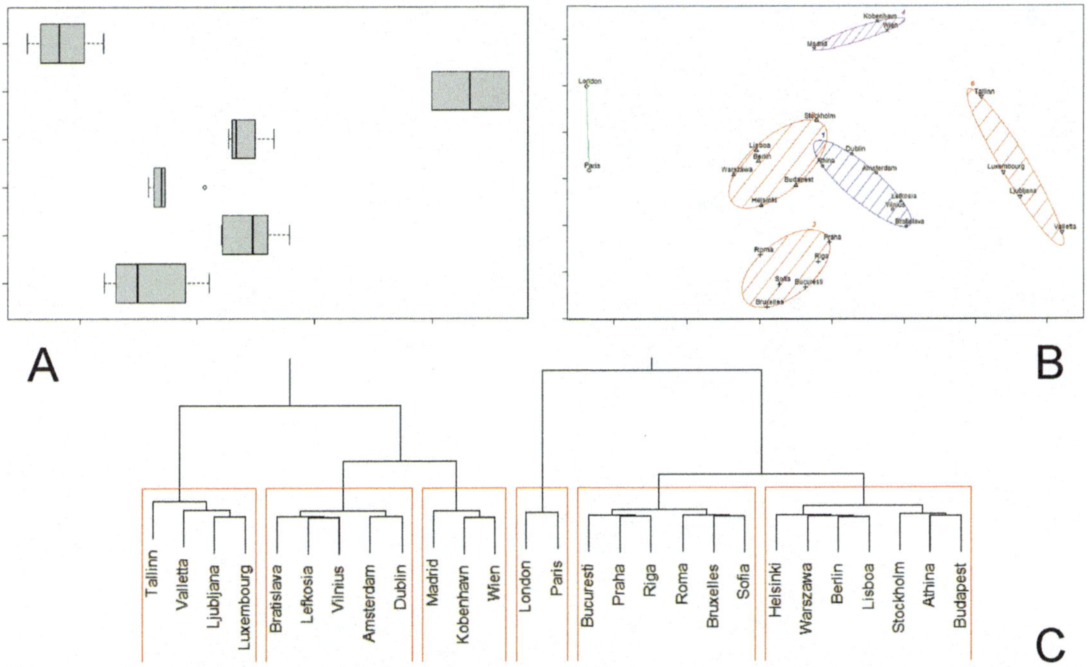

Fig. 11.3 Examples of statistical exploration of data visualised in the form of boxplots (**a**), hierarchical clustering tree (**b**), and in-hierarchical clustering (**c**). (Source: Authors)

region or area, or even to a concrete position – XY geographical coordinates), it is possible to display a data in the form of a map. In these kinds of geovisualisation, however, we need to bear in mind that we depict results of non-spatial analyses spatially. Therefore, no spatial relationships are taken into account during the analysis, and we can "only" observe if data relationships and patterns are also in correspondence in the geographical/spatial context.

11.1.3 Level 3 – Spatial Statistics, Exploratory Spatial Data Analysis and Its (Geo)Visualisation

The level three is referring about types of analyses that use some techniques from the previous part but this time with the spatial component inherently included. Analogically to level two, the methods used for (proper) spatial exploration of data are then labelled as spatial statistics or exploratory spatial data analyses. Examples of such tools include for instance spatial autocorrelation, Morans' I, Local Indicators of Spatial Associations, area local or global statistics, geographically weighted regression and others. Spatial statistics and exploratory data analysis help us to examine and measure the geographic distribution of your data, look for global and local outliers, search for global trends, examine local variation or spatial autocorrelation, analyse geographical patterns, mapping clusters, or find a spatial relationship in data. The main advantage of such methods is that they implicitly include the spatial component in the analysis. For instances, if we deal with a point data, the XY coordinates are taken into account. So in the case of a grouping analysis (Fig. 11.4a), points are grouped based on both their attributes and their position. Another example could be a use of neighbouring characteristics of data, meaning that when performing cluster analysis (Fig. 11.4b), a predefined number surrounding polygons/countries are included in clustering to capture also the spatial configuration of data. Figure 11.4 provides examples of (geo)visualisation of some

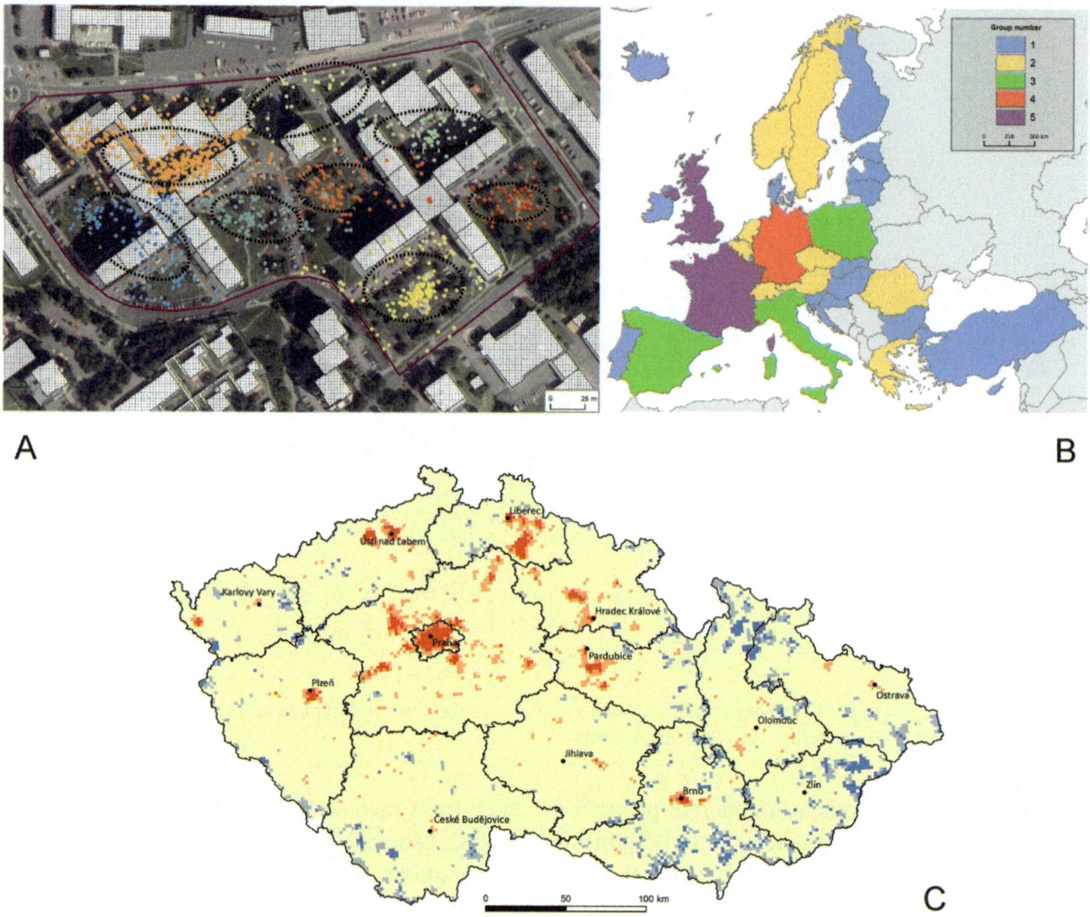

Fig. 11.4 Examples of results from the exploratory spatial data analysis – grouping analysis and standard deviation ellipses (**a**), spatial clustering (**b**), and hot-spot analysis (Getis-Ord Gi∗) (**c**). (Source: Authors)

of the methods – grouping analysis of point patterns with evaluation of their dispersion around their geographical centre using standard deviations ellipses (a), spatial cluster analyses taking neighbourhood proximity measures as a spatial constraint (b), or (c) hot-spot analysis identifying statistically significant spatial clusters of high values (hot spots) and low values (cold spots).

11.1.4 Level 4 – A Combination of Analytical Methods (Level 2 and 3)

Level four represents a fusion of level two and three. Ideally, we should analyse (spatial) data concerning both of their characteristics – non-spatial and spatial. Therefore, we advise to start with a basic and advanced statistics followed by their non-spatial and spatial visualisation (levels one and two), and then to move on to the solely spatial methods (level three). By a combination of such methods, we can grasp the most important properties of (spatial) data and deliver results, interpretations, and practical implications. Sometimes, it is required to repeat particular steps in the joint analysis since some preliminary or intermediate results might influence the further investigation of the data. For example, we perform correlation analysis to identify redundant variables. Remaining data variables are then inputs for a Principal Component Analysis that is later used for a (spatial) cluster analysis and

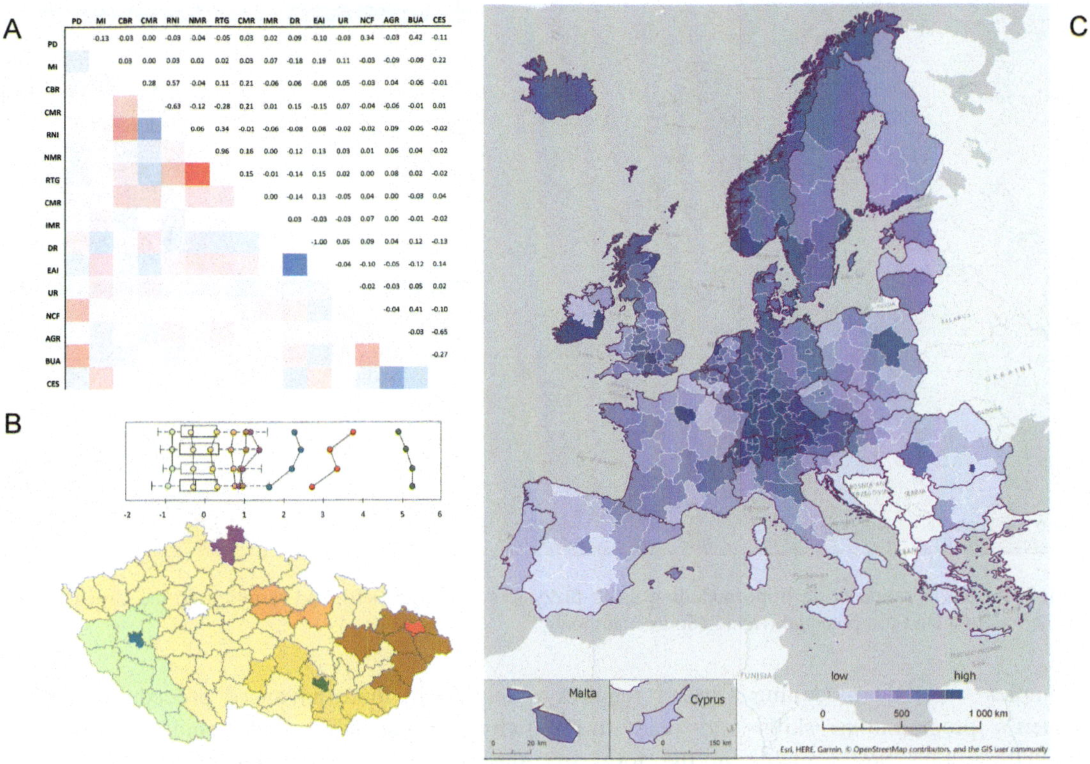

Fig. 11.5 Examples of the combination of analytical methods – correlation matrix (**a**), boxplots with (spatial) clustering (**b**), and visualisation of the first component from PCA (Principal Component Analysis) (**c**). (Source: Authors)

(geo)visualisation. In the end, we might end up with results that need to be validated, modified or adjusted. In this case, we go through the whole cycle again to find the best fit of methods to given data which leads to finer exploratory analysis and understanding of results. This complex process of (spatial) data analysis encompasses a great variety of techniques that are time-consuming, and also demanding as regards the researcher expertise. Thus, interdisciplinary cooperation in forming an expert team is often inevitable. There is no need that one person masters all the skills to perform a complex analytical workflow presented here. To be more specific, examples of individual techniques for (spatial) data analysis is given in Fig. 11.5, where a correlation matrix highlighting variables with low/high values is depicted (a). In Fig. 11.5b, a joint visualisation of spatial clustering in a chart together with boxplot (upper part) and map output (lower) part illustrates an excellent showcase of the combination of methods. Finally, geovisualisation of the first component from a Principal Component Analysis is depicted in Fig. 11.5c.

11.1.5 Level 5 – (Spatial) Modelling

The highest level of data analysis involves techniques connected with modelling approaches. Again, the modelling part of data analysis can take place outside a "geographical" domain, e.g. mathematical, statistical, machine learning, or another computation is performed first. Then, if possible, such modelling results can be visualised by charts or on a map. The second approach includes modelling directly counting with a spatial component of data. However, this could also be done with no use of a geographical information system. But most of the GIS software offers ways to run modelling within its environment. It must be noted, that such modelling within GIS is more or less a sequence of separate analytical tools connected into a model workflow.

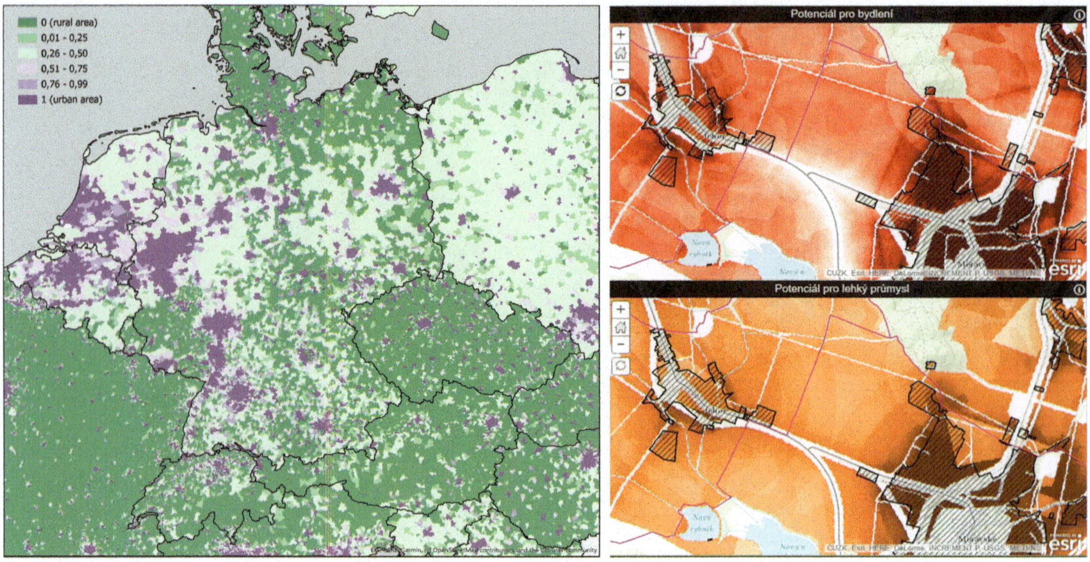

Fig. 11.6 Visualisation made from modelling using fuzzy sets and logic (left), and results from the Urban Planner modelling of a land-use potential (right). (Source: authors)

Complex spatial modelling often requires expert's programming skills and ready-to-go modelling tools are available in the form of plugins or specialised extensions. In Fig. 11.6 (left), there is a geovisualisation of the non-spatial modelling that took place outside GIS by using special fuzzy inference system. On the contrary, Fig. 11.6 (right) shows results from the Urban Planner extension to GIS software used for a land suitability modelling.

11.2 Summary

This chapter addressed an ideal workflow of "methods of interdisciplinary analytics", but we need to note that it is delivered from the author's practical experiences. Nevertheless, when analysing data, it is advised to proceed from the simplest to more advanced procedures. That is why, the first step to understand a data is to use proposed level one techniques, i.e. simple (geo) visualisation. Then, level two and three can be applied if we aim to explore specific characteristics of data or to conduct comprehensive statistical or spatial analysis. Ideally, after this stage, a combination of both should follow – again, with respect to the research goals. Finally, a modelling phase can be the concluding step in the whole workflow. As mentioned in a previous text, to follow such workflow, it usually requires cooperation among several experts. Therefore, as synopsis to the very first sentence in this introductory chapter – "methods of interdisciplinary analytics" for (spatial) data exploration is indeed a good label. In the next chapter of this part of the book, one synthetic/artificial and four real-life examples are presented.

Spatial Informatics in Water Supply Management: The Case of Mariborski Vodovod

12

Danilo Burnac, Bojan Erker, Simona Sternad Zabukovšek, and Samo Bobek

Abstract

More and more companies in different industries realized benefits of GIS and its integration with business informatics. The chapter describes attempts on the area of GIS and business integration in public company for water supply in Maribor, Slovenia. The company Mariborski vodovod is the leader in using GIS in the region. Value-added which the company is achieving for GIS and business informatics integration is explained, and best practice is presented. The company already moved from the implementation of basic functionalities to more advanced functionalities. The case describes how such integration helps the company to run their daily operations better and how management is supported with a better and in-depth view on operations.

Keywords

Water supply management · Enterprise resource planning · ERP · Business intelligence · BI · Geographical information systems · GIS · Water supply SCADA

D. Burnac · B. Erker
Mariborski Vodovod, Public Limited Company, Maribor, Slovenia
e-mail: danilo.burnac@mb-vodovod.si; bojan.erker@mb-vodovod.si

S. S. Zabukovšek (✉) · S. Bobek
Faculty of Economics and Business, University of Maribor, Maribor, Slovenia
e-mail: simona.sternad@um.si; samo.bobek@um.si

12.1 Introduction

Mariborski vodovod is a public company located town Maribor in Slovenia. Maribor is the second biggest town in Slovenia and is a centre of its northeast region. The company is owned by municipalities in the region, and its main activity is the collection, purification and distribution of sufficient quantity and quality of drinking water for more than 166 thousand of customers – users – which is about 11% of the Slovenian population. They have the largest water distribution network in Slovenia. The distribution network consists of over 1615 km of water supply pipeline system and over 260 facilities. They operate ten water sources and some smaller reservoirs which provide the water to their customers. Annually they pump 13,7 million cubic meters of drinking water which supplies 16 municipalities. Their largest water source and pumping station is Vrbanski plato which provide 760 litres per second of fresh water. This location has natural as well as technical protection and is irreplaceable for supplying water.

Their complementary activities are manufacturing, servicing and sale of water meters, construction and assembly work on the water supply network, engineering projects and technical consultancy in the field of water supply.

V. Pászto et al. (eds.), *Spationomy*, https://doi.org/10.1007/978-3-030-26626-4_12

They are also issuing approvals for the project documentation and construction documentation. They are also carrying construction and building works for connecting customers to the public water supply network.

They are a public company, and as such, they operate as a non-profit organisation according to their mission of providing quality drinking water supply. Their strategy is to be a modern and efficient public company. One of their very important goals is to implement all aspects of the modern way of business including environmental friendliness and sustainability and also with a high level of social responsibility. In all business decisions, they include the environmental management issues in accordance with the principles of sustainable development. Their digital transformation orientation and implementation of e-business are focused in: more efficient core operations of water supply, nature prevention, user-friendliness to their customers, and improved connectivity with their stakeholders (owners and partners). With E-procurement in the online ordering processing, they implemented paperless procurement of materials and services. They have implemented e-invoicing so more, and more of their customers and partners are receiving e-invoices to pay them with a »single click«.

Their Facebook profile serves as very effective communication tools for forwarding relevant, up-to-date information regarding water supply to customers and is also offering insight into their operational environment and activities. Their website is mainly used as a communication tool and a platform where the company information is published. Their web site also offers special services for blind people and for people with eye-difficulties providing them information about the latest news and also information about invoice details as voice messages.

They are one of the most successful public companies in Slovenia. They are recipients of the European certificate of business excellence; their annual report is the »Best Annual Report among other organisations in 2012 in Slovenia«. They are the first company to receive the award from the Slovenian Chamber of Communal Economy. By receiving the Family-Friendly Company certificate, they proved that they follow the principle that a successful career and family life are not incompatible. Therefore they implement the solutions and activities to enable the coordination of work and family life of employees. They also received several Slovenian awards for social responsibility »HORUS«. Thus they have proved that they are aware of social responsibility regarding the environment, their employees, their business partners and general society.

They know how important it is to raise the awareness of citizens and the wider public about saving drinking water. Therefore they associate themselves with educational institutions and particularly for the younger generation. They prepare lectures about how to deal efficiently with drinking water and how important it is that the environment near water sources remains clean.

In the area where Mariborski vodovod supplies citizens with drinking water, they placed 33 outdoor drinking fountains where passers-by can refresh themselves on hot days with cold and tasty drinking water. Because animals need water too, we also added drinking fountains for pets to the same of the fountains.

Their drinking water supplies are under constant surveillance. Both, the internal and external microbiological and chemical analyses are carried out, which ensures that safe drinking water reaches the user's tap. With the careful selection of suppliers and installing quality materials, they try to ensure customers satisfaction. Of course, they also conduct regular maintenance of water supply facilities and the water supply network, and they are trying to raise awareness among the general public how to deal with drinking water efficiently.

12.2 Information Systems Infrastructure and Architecture in Mariborski Vodovod

Water supply is among the most important and crucial utilities delivered by countries. EU Directive 114/2008 defines water supply infrastructure as »critical infrastructure« of EU. Legislation in Slovenia follows these guidelines and further defines approaches and procedures regarding critical infrastructure operation and maintenance. This includes requisites for uninterrupted supply of healthy drinking water 365 days and 24 h daily.

Information infrastructure has crucial role In fulfilment of these requirements while the collection, purification and distribution of sufficient quantity and quality of drinking water is possible only in a highly digitalised environment. Information infrastructure implemented and information systems used are considered with the same importance as water supply network while they are integrated. They have to operate as crucial infrastructure all time, and they have to fulfil the requirements for information systems security and safety. Mariborski vodovod is using smart water network layered approach for intelligent management of their system (Fig. 12.1).

Majority of operations in the water supply are conducted outside in open-air space; therefore it is very important that information systems support space issues and enable the use of geolocation and spatial data. Information systems architecture in Mariborski vodovod comprise of three categories of information systems, applications and tools:

- Business information systems – comprised in its core of Enterprise resources planning solutions and other applications for support of business activities of the company
- Geo information systems (GIS) – includes access to cadastre of the territory on which Mariborski vodovod supplies water and applications and tools for processing of spatial data and their visualisation
- Technical (SCADA) information systems – includes applications needed for monitoring and control of pumping and distribution system of water.

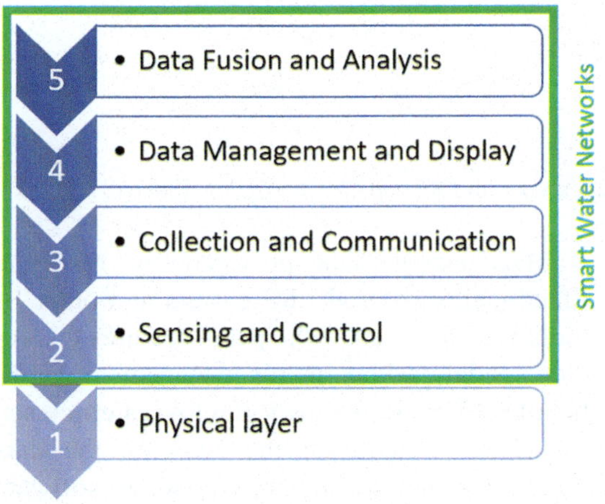

Fig. 12.1 Layers of smart water networks. (Source: Authors)

Key information systems in Mariborski vodovod are (Fig. 12.2): Enterprise resource planning applications, Asset management, Work order management, Customer information systems with billing, GIS, Water loss, Water quality, Control room. Information infrastructure of the company consists of a modern network of servers and workstations which allows running all necessary information systems/applications/tools also in a mobile environment.

12.3 Integration of ERP and GIS

Enterprise resource planning functionality is in Mariborski vodovod achieved by integration of accounting, warehouse management and human resources management application modules with billing, work order management, asset management, water loss management, smart meter reading, and vehicles management.

To implement necessary spatial features and functionalities needed for better quality of conducting their business activities – use of geo-visualisation of ERP data – they have implemented ArcGIS from ESRI. Employees in Mariborski vodovod perceive ArcGIS as a very user-friendly and easy-to-use application. With the integration of ERP functionality and GIS functionality, they added value to support for operational activities (Fig. 12.3) in the sense of enhanced support for employees.

Mariborski vodovod recently extended their ERP applications with DMS system including e-archive which improved accounting processes

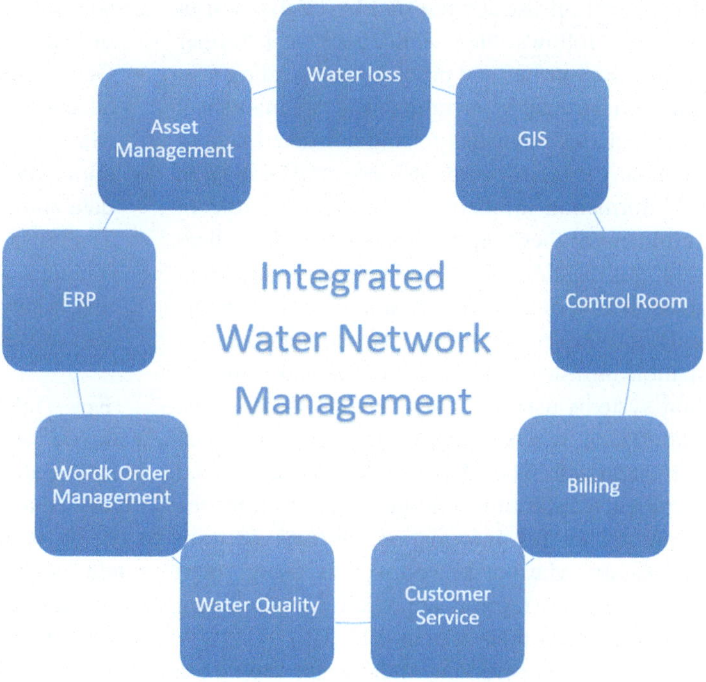

Fig. 12.2 Information systems architecture for integrated water network management. (Source: Authors)

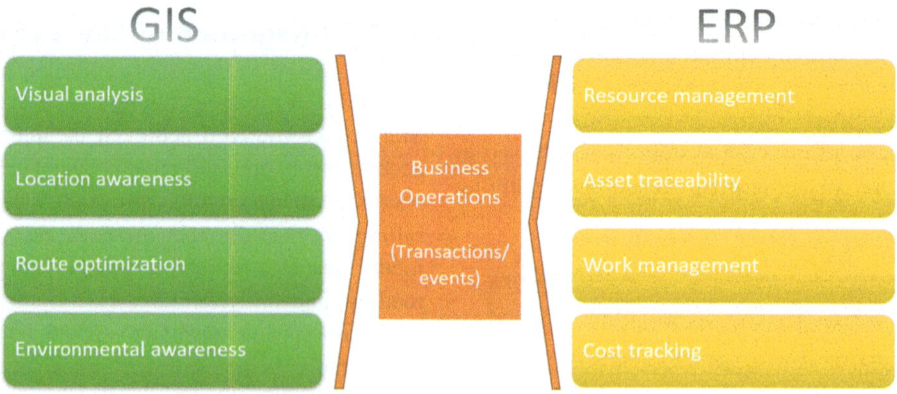

Fig. 12.3 Enhanced support for employees on an operational level with ERP and GIS integration. (Source: Authors)

and accessibility to the documents and their tracking. Currently, they are integrating the DMS system with GIS system to add spatial features. Report on Fig. 12.4 shows how working/route documents of employees assignments are linked with spatial data showing vehicle license plates showed on the map to allow better work force management. The data source for such reports are an ERP system and a DMS system.

Management data provided by reporting functionality of used ERP system and other BI tools are enhanced with spatial features to provide spatial dashboards for management. Dashboard on Fig. 12.5 shows employees efficiency for chosen month on the basis of the data obtained from smart meter reading software and information is visualised on the map. The map shows the number of smart water meter

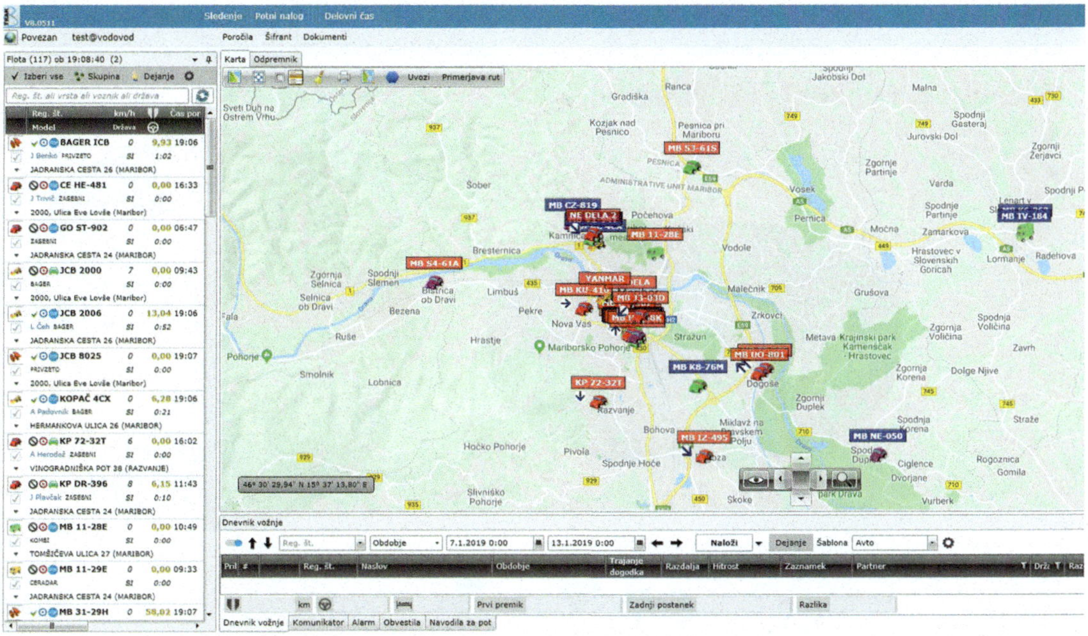

Fig. 12.4 Linking spatial data with the working/route documents of employees assignments with vehicle license plates showed on the map. (Source: Authors)

reads (remote reading). It enables zooming by which detail information about each water meter (reading date and time, volume, reading distance) can be seen.

12.4 Integration of SCADA and GIS

Technical information systems and applications – SCADA - intended to manage water pumping and distribution to the customers. On the basis of the measurements of flow rates, pressures, surfaces of the water in containers that are part of the network system, SCADA maintains the correct pressure in the network, operates pumps in the pump stations and stations for drawing water, and if necessary, it doses needed quantities of disinfectant and ensures uninterrupted supply of drinking water. For the security reasons, this part of the information system is to some extent isolated from other parts of information systems. The maintenance of this part is essential, as in the case of operational difficulties it may lead to immediate disruptions of the water supply (Fig. 12.6).

12.4.1 The Project "Ruptured Pipelines in the Network"

The project was created to visualise locations of the ruptured pipelines in the network and to help the decision makers to decide where to invest and reconstruct water supply network. The basic idea of this project was to make a map of the water supply network that shows the points where the pipelines are broken. It allows detailed insight into the individual data on ruptures, and employees can search needed information by various criteria (time, types of materials, the reasons for a rupture, etc.) and insight into related repair costs. Employees can have a "statistical" view showing the number of events in the area, which is particularly suitable for decision makers. Increased number of ruptures in a smaller area indicates that they have to invest in a specific part of the network. Technology allows employees to have immediate insight into the current situation on the basis of real-time information (Figs. 12.7, 12.8 and 12.9).

The project "Ruptured pipelines in the network" enables better management of water supply network delivering following functionalities

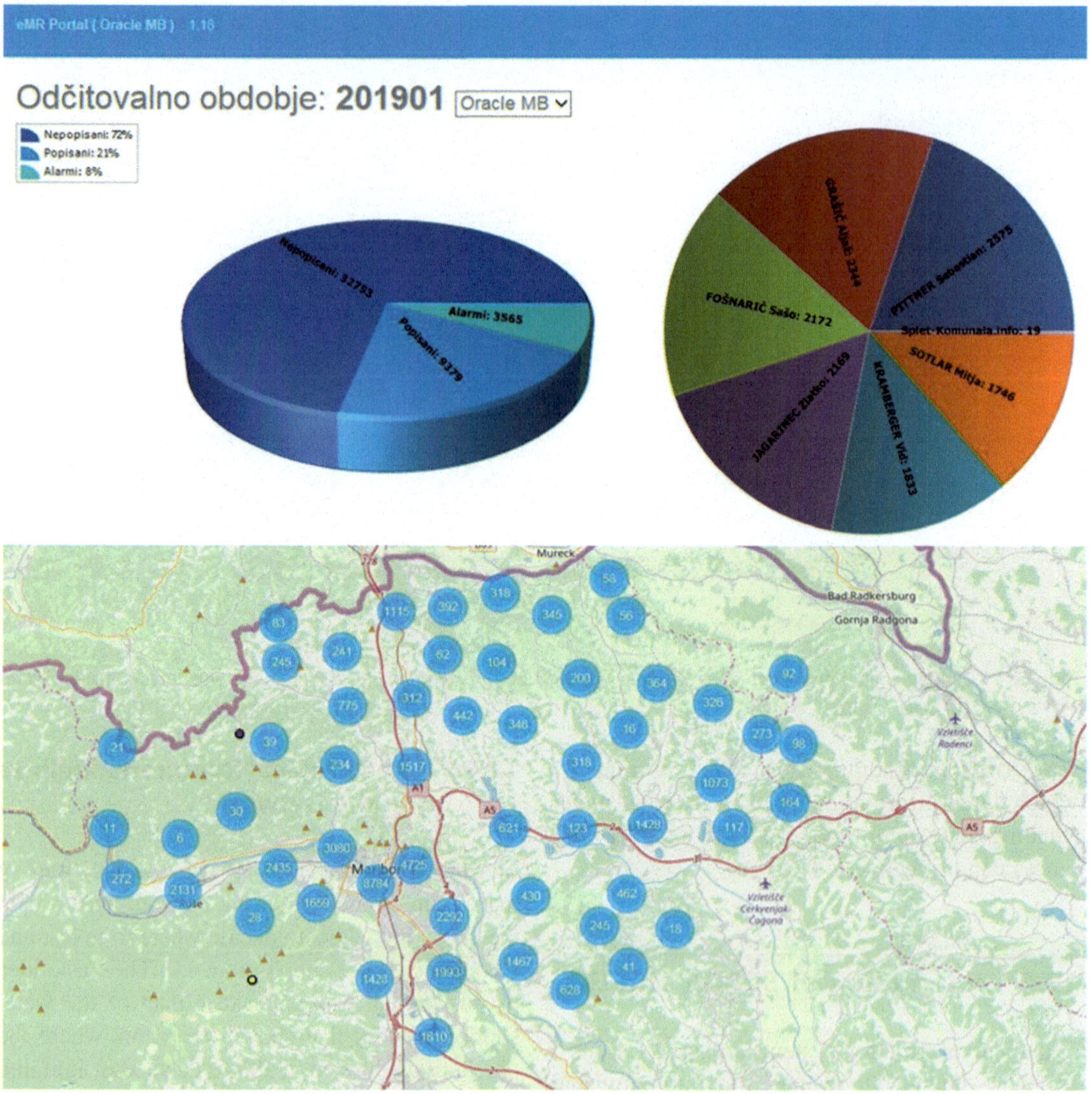

Fig. 12.5 Spatial dashboard is showing smart meter reading data analytics. (Source: Authors)

(Fig. 12.10): monitoring and alerts based on the remote control, reporting for decision making, feedback and automation, evidence-based planning.

12.4.2 Project »Smart Metering and Pay-As-You-Use Billing«

Modern concepts of services and utilities billing emphasise the »pay-as-you-use« models. In some industries such as electricity and telecommunications, this is quite easy because of technology platforms used. For water supply implementation of »pay-as-you-use« concept is more difficult while customers have mechanical water meters which are mechanical. For such devices, it is necessary that water supply company employee visits customer premises and reads the water meter and the data is edited to tablet. This is usually done once a year. During the year customers are billed on the basis of forecasted/calculated use. The problem arises if there is substantial leaking of water for which customer is not aware while the customer gets a huge bill at the end of the period.

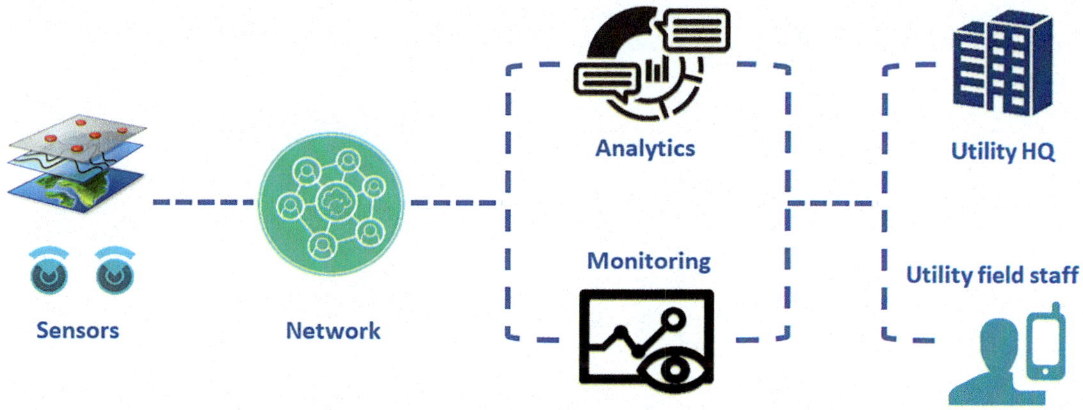

Fig. 12.6 Monitoring of water supply technical systems. (Source: Authors)

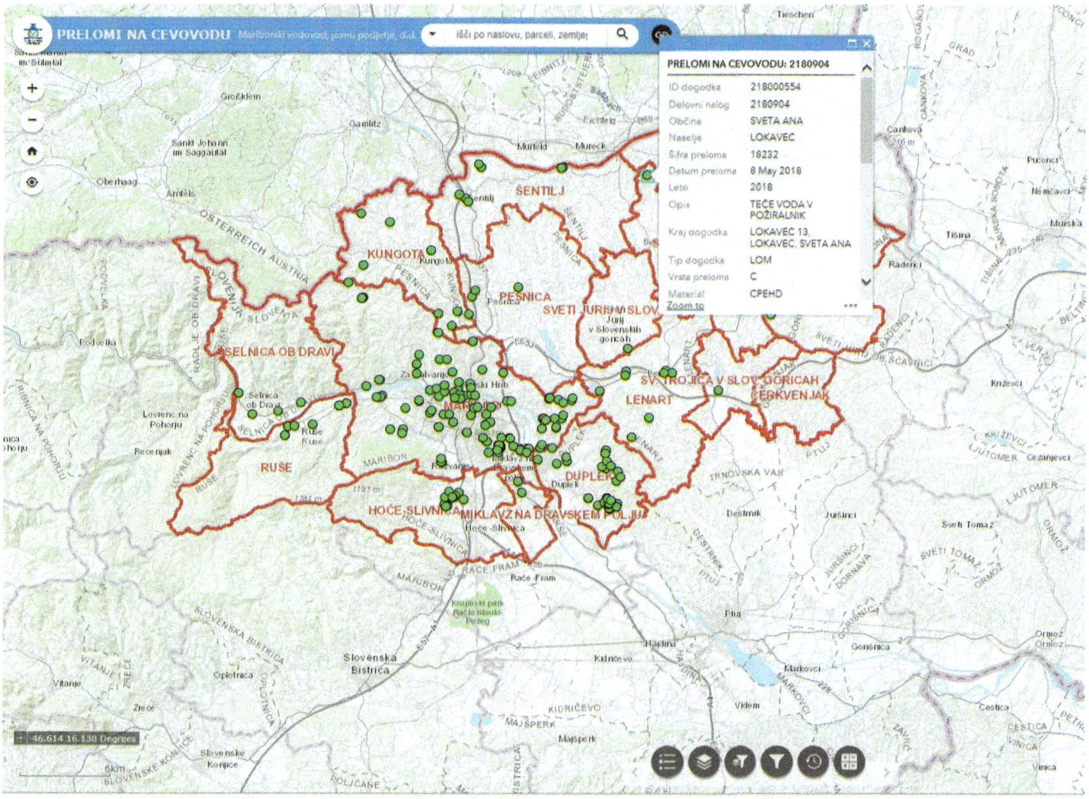

Fig. 12.7 Visualisation of the locations of ruptured pipelines in the specific period and detailed information about certain rupture. (Source: Authors)

Smart metering technology allows collecting of water consumption data in real time and from a distance up to 100 m without entering customer premises. They started a project in 2013 and 2018 95% of customers have smart water meters. Data are collected by employees who are driving prescribed routes which allows that data is transmitted from smart meter to the device in the car (range is 100 m) (Fig. 12.11).

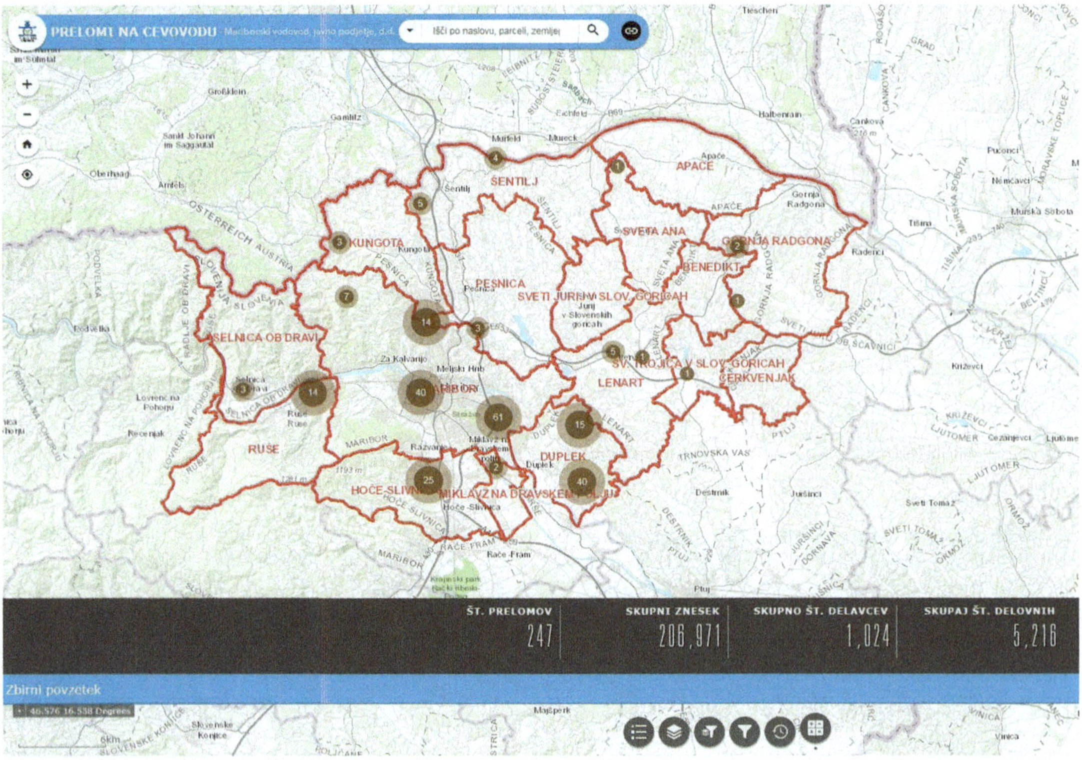

Fig. 12.8 Numerical (statistical) visualisation of pipeline ruptures events in the specific period and estimated repair costs. (Source: Authors)

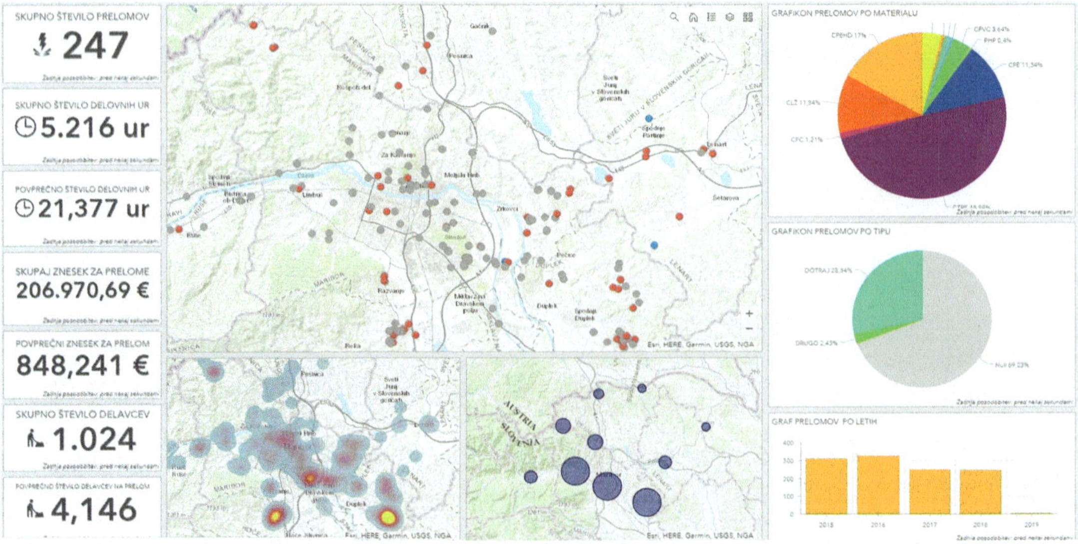

Fig. 12.9 Visualisation of the pipeline ruptures in the network in the form of a dashboard. (Source: Authors)

Fig. 12.10 Functionalities of spatial data enabled management. (Source: Authors)

12.4.3 The Project "Pumping Facilities"

Maintenance of the water supply infrastructure, which includes water reservoirs, pipeline network and associated facilities (pumping stations, stations for water drawing, water storages) is one of the basic activities of the company and essential condition for quality water supply. In order to gain a comprehensive overview of the situation on the water supply infrastructure, for the purposes of maintenance as well as for the municipalities who are owners of the company, they have drawn up a map of the network, which allows overview of all above-ground water supply facilities, display necessary data of a single facility, its maintenance status, photos and photos of the necessary maintenance operations. The map is also available on mobile devices and is an excellent help for maintainers and to employees in the company, who take care of the accounting records of the infrastructure. It is also useful in discussions with the municipalities who are owners on the necessary investments in the reconstruction (Fig. 12.12).

For the effective carrying out of communal services the information system must support spatial data display, mobile business, it must be easy to use, and it must support sharing of information to external users. Having data from the information system at disposal timely and at all times enables the company to carry out all services in a quality manner and thus maintains confidence and reputation to customers. Comprehensive water analytics is needed which allows: insight in usage and loss, information about conservation, indicators regarding operations and maintenance, safety and quality, asset management and efficiency, sourcing efficiency (Fig. 12.13).

12.5 Integration of Advanced Applications and GIS

Mariborski vodovod is conducting intensive digital transformation which was defined as part of their strategy. To achieve these goals they have already implemented some projects of advanced applications while some are still in progress. In several advanced applications, high level of integration with GIS is needed.

12.5.1 Project »Water System Optimisation with Hydraulic Modelling«

Pipeline network which comprises water systems of Mariborski vodovod measures more than 1600 km. The way from the water source to the customer could be in some cases also more than 40 km. For safe and reliable water supply detail insight in the system is necessary. The technology of hydraulic modelling of water systems allows a detail understanding of the system based on the cadastre data and measuring equipment build in the water system. This allows calculation of pressures and through flows which can be compared with measured data. Comparison can be used for different simulations regarding changes in water systems operation. With hydraulic model, it is possible to predict directions of water flows and how old is the water is certain parts of the water system. These allow

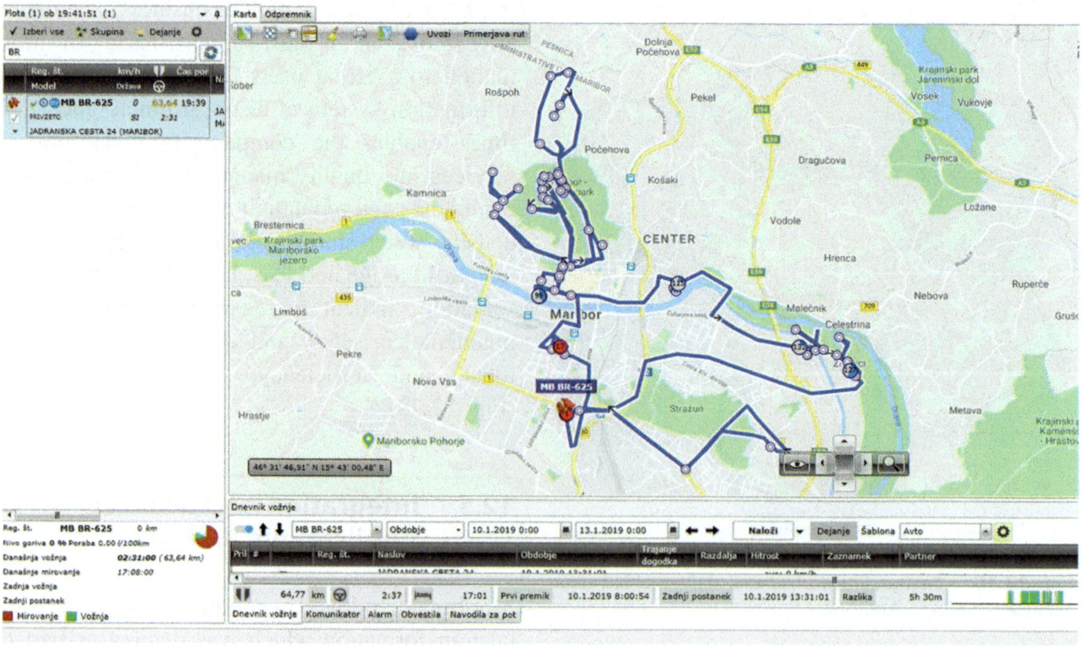

Fig. 12.11 Visualisation of the daily route of the employee collecting data from smart meters driving in his vehicle. (Source: Authors)

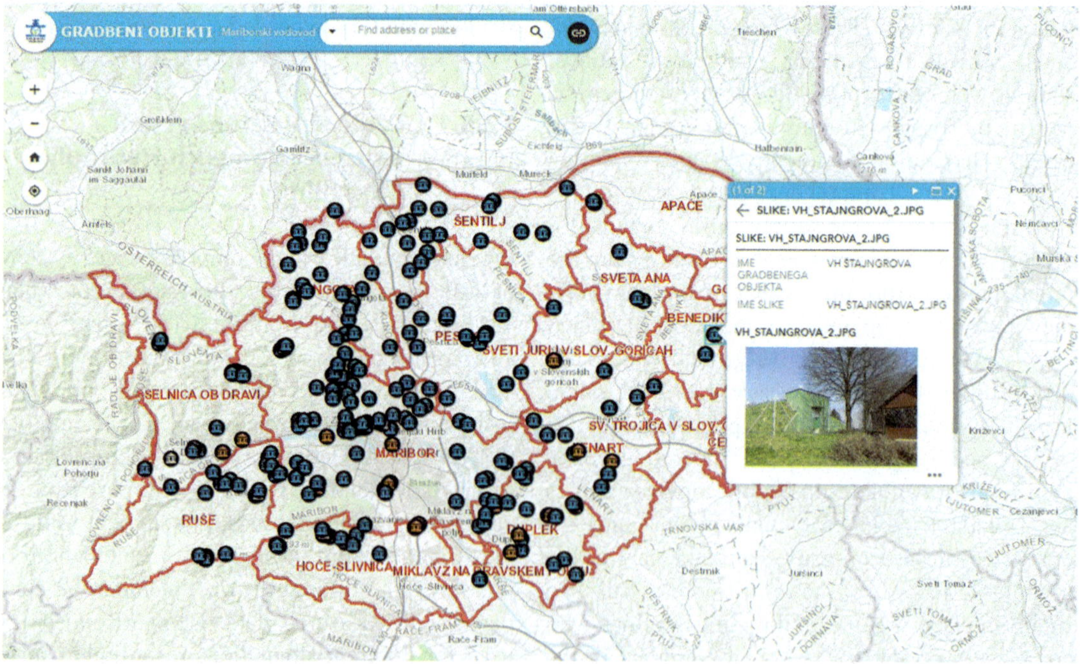

Fig. 12.12 On-line map of pumping facilities allows a comprehensive overview of the maintenance of facilities. (Source: Authors)

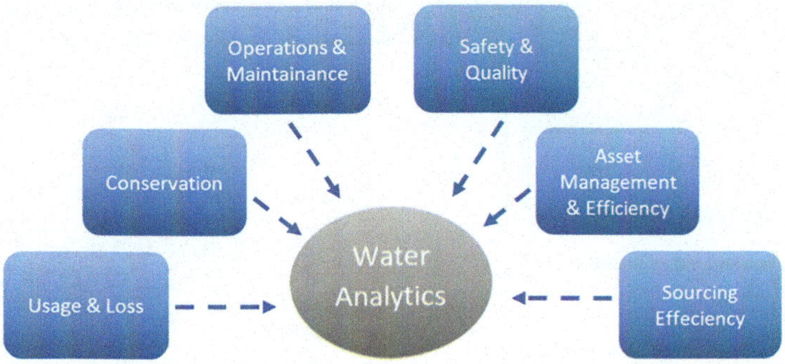

Fig. 12.13 Functionality of water analytics. (Source: Authors)

optimization of the water system and its operation what leads to less water losses and less electricity use. Hydraulic water model is visualized as shown in Fig. 12.14.

12.5.2 Project »Mobile Spatial Data for Employees«

For field staff real-time and up-to-date data is crucial for their work. Mariborski vodovod is using mobile devices and technology to collect data from smart water meters and for maintenance work. ArcGIS ESRI enables the use of cadastre data and maps on the field locations. Employees can upload geo-located photos to document events and conditions related to water system s shown on Fig. 12.15.

12.5.3 Project »Satellite Water Leak Detection«

As mentioned water system of Mariborski vodovod consist of 1600 km of pipelines, Pipelines are of different materials and also different age, so the condition of it is different from location to location. Such complex systems require continuous monitoring and repairs. Spots of extensive leaking can be easily detected while it is seen and therefore can be repaired. More problematic are small leaking which happens underneath and which cannot be seen. Because of pipeline beaks, up to 30% of water is lost.

Pipeline breaks are detected by metering devices which show consumption in different parts of the water system and increase shows that a pipeline break has happened. Field staff is using special detector devices which allow them to trace breaking spots. Because of the water systems complexity, it is impossible to monitor in such a way the whole watering systems by filed force.

To improve monitoring of water system, Mariborski vodovod implemented satellite detecting of water loses. Satellite scans the area using remote sensing and detects humidity up to 3 m below surface. Collected data are then analysed using the maps of water systems and with computer, algorithms showing possible places of pipeline leaking. The potential leaking spot is detected in the circle of 45 m.

Mariborski vodovod got the data based on satellite remote sensing in December 2019 and analysis of collected data to show 114 places with big certainty of pipeline leaking. As shown in Fig. 12.16 the software predicted pealing on a certain street and filed staff find a break in the pipeline there after a few minutes search (Fig. 12.16).

12.6 Conclusion

Use of GIS in organisations and its integration with other information system becomes widespread and is an important issue for organisations. The functionality of GIS is very useful and add value to information systems already used in

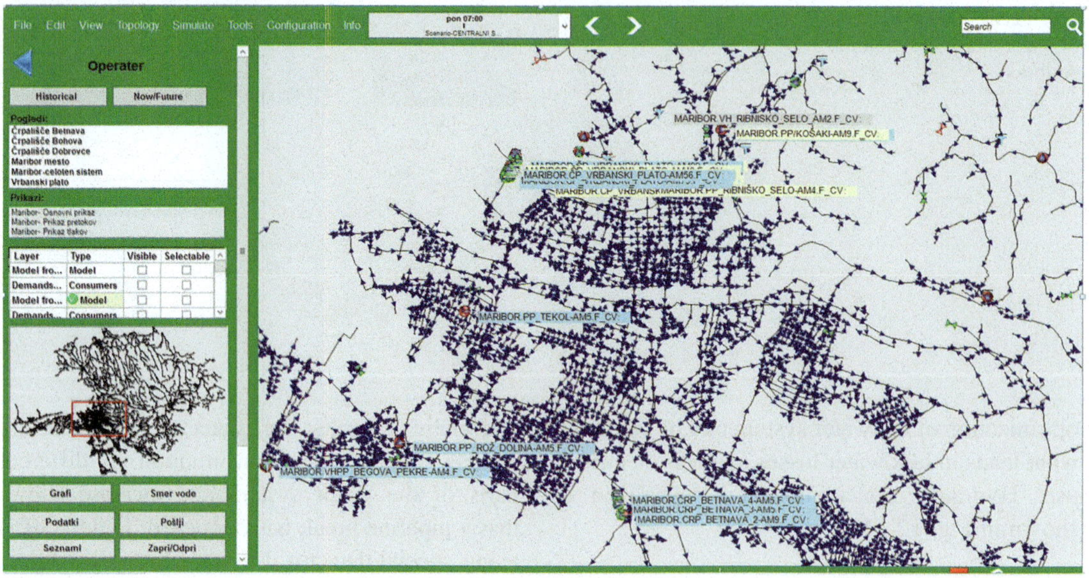

Fig. 12.14 Spatial visualisation of hydraulic water model showing water flows. (Source: Authors)

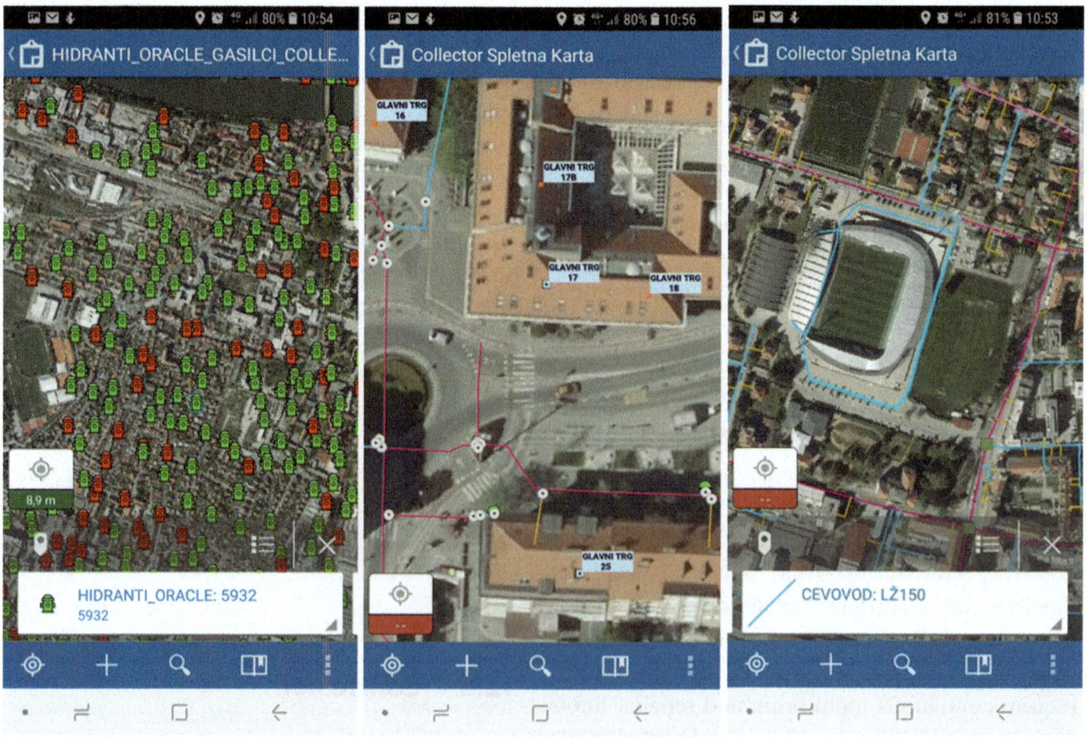

Fig. 12.15 Spatial data related to water system on mobile devices. (Source: Authors)

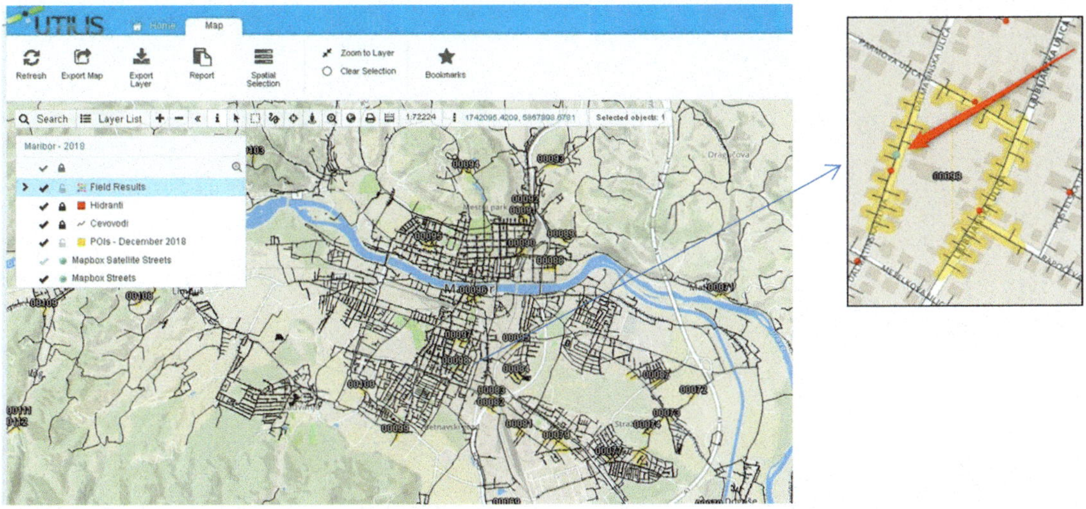

Fig. 12.16 Spatial visualisation of potential pipeline breakpoints based on satellite sensing data and computer algorithms. (Source: Authors)

organisations. Importance of GIS differs from organisation to organization and also from industry to industry. Industries with location based resources and wit location based business events are leaders in GIS implementation and use. There are many opportunities to use GIS in organizations – the company Mariborski vododovod already exploited many of them successfully Examples of GIS use in Mariborski Vodovod show that they are beyond initial GIS use on a basic level and that they use GIS on an advanced level. According to this, we can conclude that they already entered the GIS adoption phase which correspond to stabile and advanced use and that they are approaching maturity phase. Without the leadership of top management – CEO and CIO – this could not be possible.

Application – Site Analysis Furniture Store

13

Nicolai Moos

Abstract

This chapter deals with the virtual situation that a furniture company is searching for a site for a new furniture store. The user is facing several different datasets (e.g. streets, landuse, existing sites, etc.) with the task to process them with a variety of tools and methods. The decisions on when to use which tool with which parameters are subjected to distinct requirements and restrictions given by chosen conditions on the new site.

The workflow illustrates a simple site analysis with data that can be obtained freely from the web and thereby shows the application and combination of few of the most common tools. By revealing a practical approach on how to use geodata to end with a result layer that only contains features that fulfil all requirements as well as they obey all restrictions, this case study is a vivid example of the calculation of both economic and spatial data.

Keywords

Spatial analysis · Georeferencing · Geoprocessing · Logic operators · Table joins

N. Moos (✉)
Geography, Geomatics Group, Ruhr-University Bochum, Bochum, Germany
e-mail: nicolai.moos@rub.de

13.1 Introduction

This chapter is not a classical case study in the narrower sense but gives an applicable and practical example on how the two scientific fields of spatial and economic data acquisition and processing can each be extended and improved by combining them with each other. The simulated workflow is kept basic to focus on the general idea and illustrates a possible routine that can be applied practically to distinct real situations.

It is conceivable that a successful company that produces and sells furniture is looking for a new location for another store in North-Rhine Westphalia (NRW). Several requirements on the new site must be fulfilled, while other parameters need to be excluded. The whole process of combining these inclusions and exclusions is based on spatial data while the thresholds are defined by economical approaches.

For the analysis, the requirements and restrictions for the new site have to be separated from each other, as they have to be processed individually.

Requirements
- maximum distance of 500 m to federal highways
- maximum distance of 1000 m to freeways
- population density of at least 800 inhabitants per square kilometer in respective municipality

V. Pászto et al. (eds.), *Spationomy*, https://doi.org/10.1007/978-3-030-26626-4_13

Restrictions
- must not be in range of 40 km to existing sites
- must not be in settlement areas

The number and structure of requirements and restrictions are only limited by available data. No matter if it is annual income per capita, the prices of certain premises or perfect accessibility to water or electricity – as long as there is the possibility of obtaining data for an issue, it is possible to include the parameters into the analysis.

Given datasets in this exemplary workflow are vector datasets (shapefiles) with municipalities, settlement areas and the street network of NRW, an unreferenced raster map in GIF format and a csv-table with the population data of NRW (for more information on data structure and acquisition, see Chap. 1. Data Sources).

Using this data, the analysis workflow starts with finding the suitable areas that will match the requirements. It then proceeds by finding the unsuitable areas and finally subtracts these areas from the suitable ones to get the result layer that will display all areas matching both, requirements and restrictions.

This analysis can be done in any GIS (ArcGIS, QGIS, MapInfo, etc.). To provide the possibility of a proper replicability, the whole workflow will be processed in QGIS for this case study, as it is open source.

13.2 Project Setup and Data Preparation

At first all datasets that already have a coordinate system need to be uploaded into an empty project with a proper visualization (see Fig. 13.1). The project properties should be checked to make sure that the overall project coordinate system is correct - in this case it is the UTM projection of zone 32 N with the ETRS89 reference frame (EPSG: 25832).

13.2.1 Georeferencing

Once the layer structure is set up the next step is to perform a georeferencing on the raster map that will contain the information about where in NRW the existing sites of the company are located. A proper georeferencing connects points of the image coordinate system with their geographical coordinates of the given reference system. Either the geographical coordinates are picked out of already referenced material (shapefile, raster image, etc.) which is then called point-to-point-referencing or your unreferenced image data is overlaid by a cartographical grid that defines the coordinates of certain points in the map (point-to-coordinate-referencing). As the unreferenced raster map here covers the area of whole NRW as well as the shapefile of the municipalities the georeferencing can be processed via the point-to-point referencing method, using the georeferencer in QGIS. It is necessary to distribute the connection points homogenously over the whole map to make sure that the it is georeferenced equally accurate, as an algorithm is interpolating the coordinates for the rest of the map according to the chosen points. Clustering these points in a certain area will affect results in a negative way. As image coordinates are transferred to geographical coordinates, the order of setting the points is always first point on the unreferenced map, second point on the referenced material (Fig. 13.2).

After at least four well-distributed control points are set and the mean error is in an acceptable range, the software interpolates the coordinates for the rest of the referenced raster image and saves it as a new and overall referenced dataset (for more information on coordinate systems and projections see Part I, Sect. 1.1.2. Spatial Data Models).

13.2.2 Digitizing

After the raster map has geographic coordinates, the locations need to be extracted out of the map by digitizing the coordinates in a point shapefile as until now the locations are only pixels in a continuous dataset (see Part I, Chap. 1. Data Sources). To do this, a new point layer has to be created that has both the corresponding coordinate system and the necessary attribute table field to store the name information of the city where the existing sites are located (Fig. 13.3).

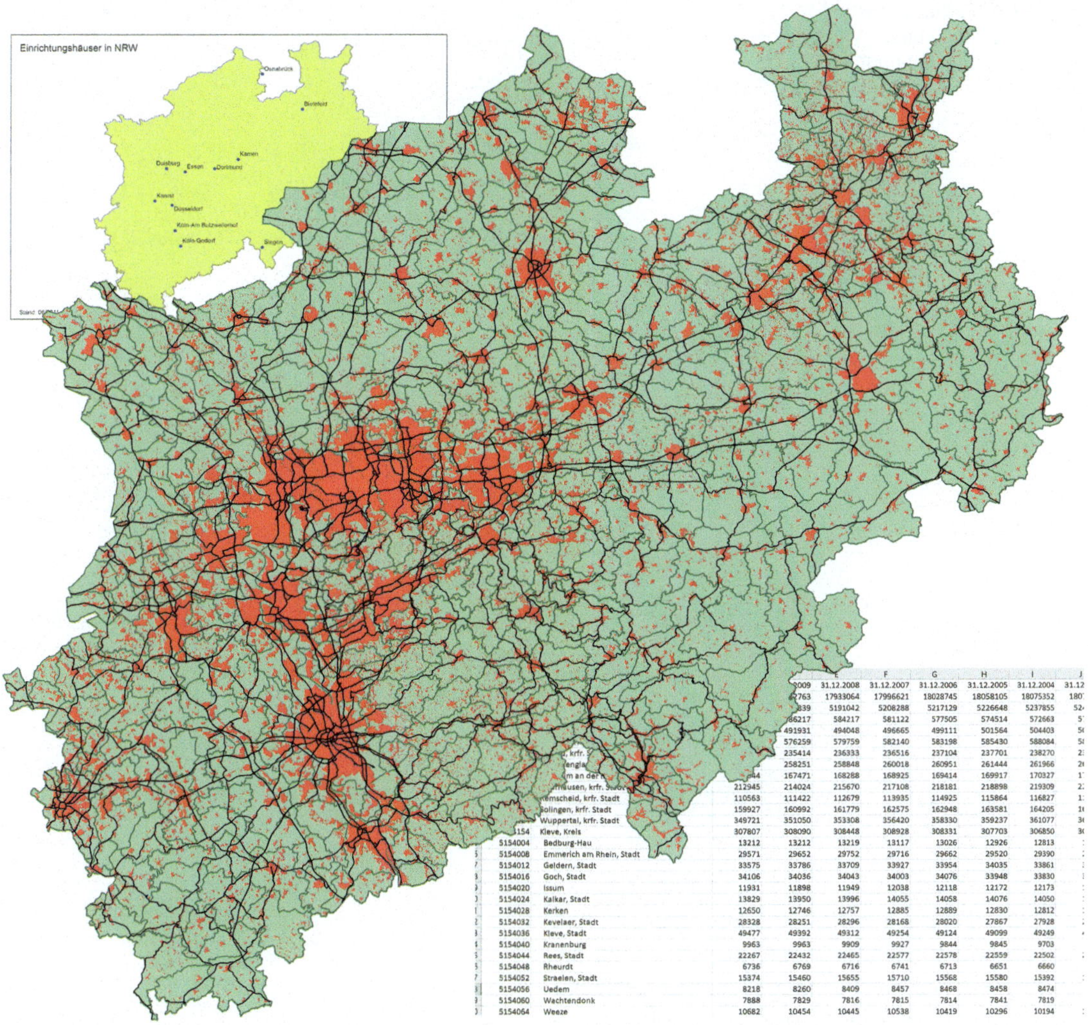

Fig. 13.1 Feature layer municipalities NRW (green), settlement areas (red), streets (black), raster map (top left), csv-table (bottom right)

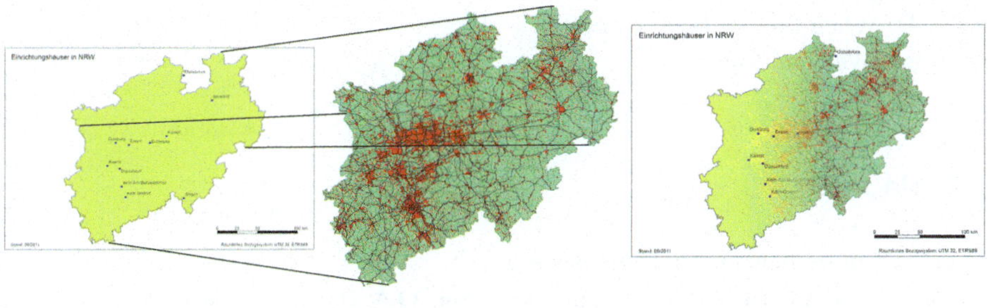

Fig. 13.2 Georeferencing from unreferenced to referenced map and result

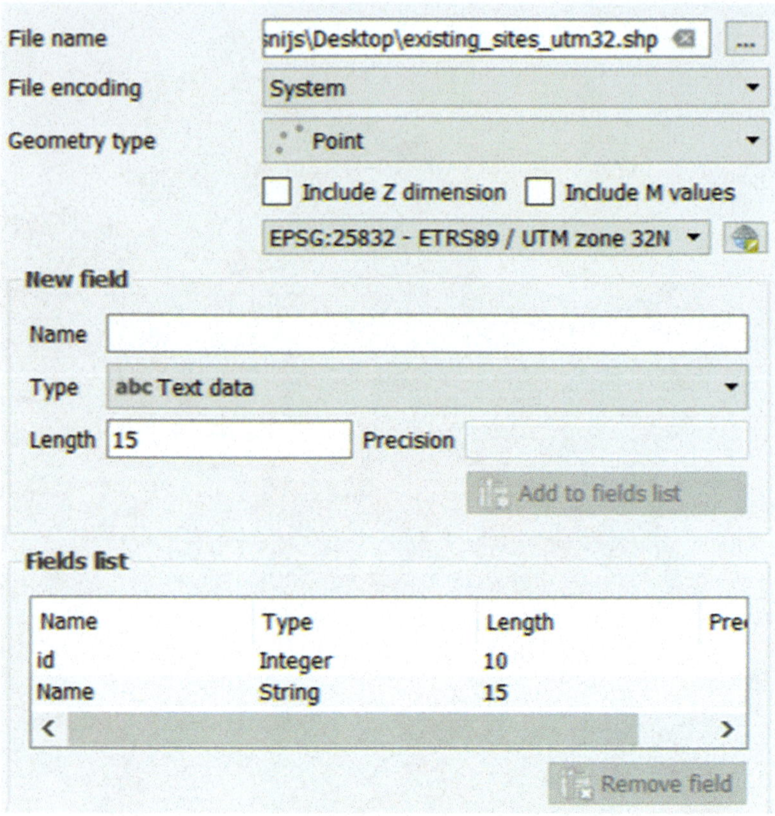

Fig. 13.3 Creating a new point shapefile with a proper attribute field (QGIS screenshot)

Once there is a new and empty shapefile that has the coordinate system and the table field, the edit session needs to be initiated for this particular layer. Only then features and their attributes can be added to it (here: city names). To digitize the features as well as the linked attributes it is necessary to add the point features via the 'add features'-tool and then extend them by the information of the city name in the respective table field (Fig. 13.4). When this is done the edit session needs to be saved and toggled off.

Fig. 13.4 Digitizing existing sites (red crosshairs)

13.2.3 Table Join

One of the given requirements for the analysis is a population density of at least 800 inhabitants per square kilometer in each municipality. As the attribute table of the shapefile that contains the geometries of the municipalities stores neither the information of the relative nor the absolute number of inhabitants per municipality, these values have to be added.

They can be found in the csv-table that contains the (absolute) population data of NRW and needs to be imported into your GIS beforehand so that all the columns and lines are displayed properly.

For the import of the csv-file it is necessary to set several parameters. First there is the delimiter of the different fields (comma, tab, space, etc.), which ensures a proper conversion and puts each value into an individual cell. Second the encoding of the table content has to be set to make certain that all values are displayed correctly. Finally, one has to check if there are coordinates in the table that could be converted into geometries (Fig. 13.5) which isn't the case here. After importing the table, it needs to be saved as either dbf- or spatial lite-format to provide that it's editable.

To then add this imported data to the attribute table of the shapefile, it is necessary to perform a table join. This requires the identification of a key field (unique ID) that connects both tables – attribute and imported table – via identical values of each feature (Fig. 13.6).

Since a table join can only be done successfully if both key fields are of the same data type (text, number [integer, double, etc.], date, etc.), it can be necessary to copy the values of the key field of the imported table into a new field, setting a data type that matches the key field in the attribute table.

In this case, the key field is the field that stores the code of the particular municipality consisting

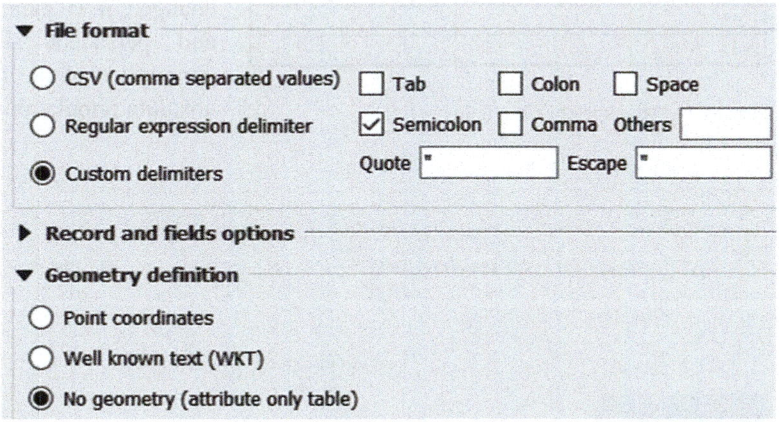

Fig. 13.5 Parametrizing the import of a csv-table (QGIS-screenshot)

area	name	code
250,7	City 1	1
134,8	City 2	2
42,9	City 3	3
1.000,2	City 4	4
782,6	City 5	5

code	avr_income	population
1	1.003,00 €	76.000
2	1.090,70 €	85.000
3	985,70 €	107.000
4	1.307,30 €	40.000
5	790,80 €	95.000

area	name	code	avr_income	population
250,7	City 1	1	1.003,00 €	76.000
134,8	City 2	2	1.090,70 €	85.000
42,9	City 3	3	985,70 €	107.000
1.000,2	City 4	4	1.307,30 €	40.000
782,6	City 5	5	790,80 €	95.000

Fig. 13.6 Example for a table join via a key field

of seven digits specifying the data type as a number (Fig. 13.7). As the data type of the field in the imported table is a text and there are missing several digits in some cells that prohibit an overall join, it is necessary to create a new field and copy the code values of the text field as numbers to this new field. Therefore the table was modified to dbf or spatial lite in the first place to make it editable.

To now add the missing digits to the respective values, all cells that need to be updated are selected by an SQL expression (see Part I, Sect. 3.1. Simple Spatial Analysis). Looking at the values in the cells it can be stated that either they are complete and ready for a join or they are missing three zeros on their right to match the ones in the attribute table of the spatial layer. The task now is to find an expression that filters the table and only selects the specific values that need an update and then multiply them by one thousand which will add three zeros. As all affected values are in between a range of five thousand and six thousand the right expression for the selection here is *"<field name>" > 5000 AND "<field name>" < 6000*. Via the field calculator the selection is multiplied by one thousand as mentioned above (Fig. 13.8).

After that the table join can be done by navigating to the properties of the municipality shapefile and add a join, choosing the key field in the attribute table, then choosing the table whose values should be added and then choosing the key field in that particular table.

To prevent the result dataset from being changed it is then exported into a new single and persistent shapefile that contains the geometries of the municipalities as well as the absolute population data.

categorie	data type	storable range
Numbers	short integer	-32.768 to 32.767
	long integer	-2.147.483.648 to 2.147.483.647
	float	-3,4E38 to 1,2E38
	double	-2,2E308 to 1,8E308
Text		
Date		

Fig. 13.7 Data types in table fields

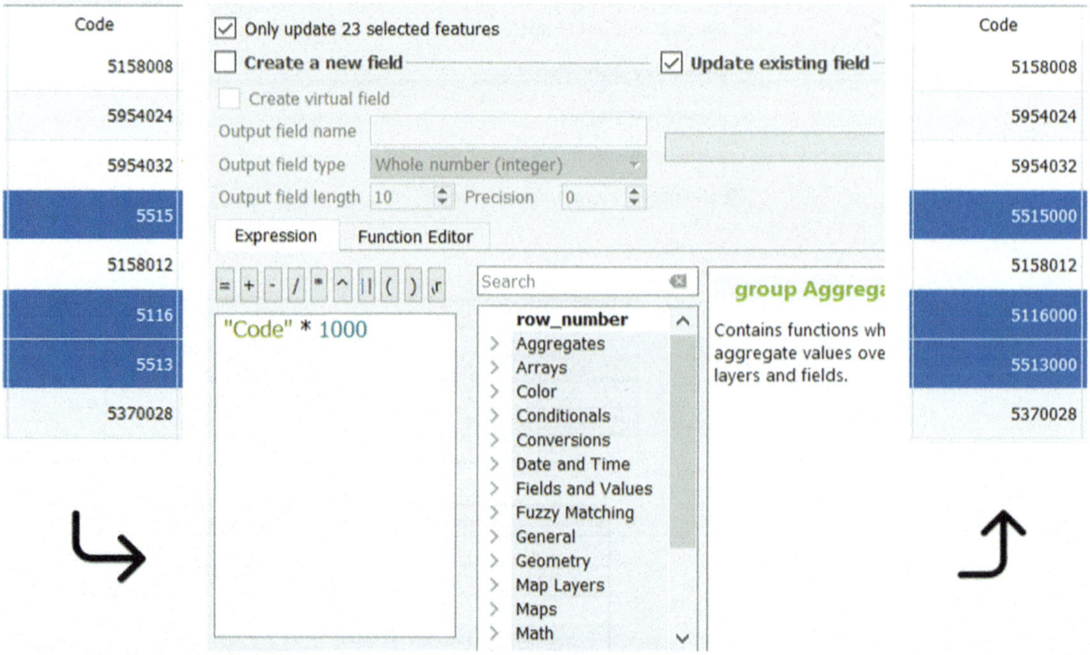

Fig. 13.8 Updating selected features (values in blue) via field calculator

13.3 Suitable Areas

To obtain suitable areas it is necessary to extract particular layers out of the prepared data. For this purpose it is helpful to arrange all the relevant parameters in a table (Fig. 13.9). The suitable areas will be exported into *positive layers* and the unsuitable areas will be exported into *negative layers*.

13.3.1 Freeways

The first positive layer for the analysis contains all areas that are in range of 1000m within freeways. That is a valuable approach to provide an easy access to the new store avoiding small roads as most customers use to go shopping with their own car. To create this layer as a shapefile, all freeway features of the street network layer need to be selected, buffered by 1000 m and then exported to a new shapefile.

The selection is done most accurately and effectively via an expression in structured query language (SQL, see Part I, Sect. 3.1. Simple Spatial Analysis). The relevant information for this selection is stored in a certain field that implies abbreviations of the street classes combined with the individual numeric value of the street. All freeways in NRW are tagged with an A (freeway in German: **A**utobahn) and an individual number. As the expression should manage to select all highways at once, the individual number of the freeways is replaced by '%', changing the '=' to a 'LIKE', as now the value does not equal a certain value but is implying all the different freeways at once, what then leads to the expression *"<field name>" LIKE 'A%'*.

The result layers of all upcoming calculations are saved in two different directories – one called *positive* (for the requirement-layers) and one called *negative* (for the restrictive layers).

To then buffer all selected freeways by the value of 1000 m and save it in the positive-folder as an independent shapefile, one has to open the buffer tool, select the input layer, set the beforehand selected features as the only ones that should be used and define a buffer distance with a value in the unit of the used coordinate system.

If the result features should be accumulated to have one mutual border in case of overlaps, it is necessary to dissolve them (Fig. 13.10). In some versions of the tool (depending on the GIS-software and its version), it is as well necessary to define the number of segments that build the outline of the buffered area (the more segments, the rounder the shape).

13.3.2 Federal Highways

As there are not only freeways but also federal highways that should be in direct range of the new store to assure a good accessibility, the whole process can be repeated in almost the same manner. The only thing that needs an adjustment is the expression for the selection of the particular features. In this case to *"<field name>" LIKE 'B %' OR "<field name>" LIKE 'N%'* since the federal highways are tagged with two different letters. To not select the features that contain both – B *and* N – the logical operator OR connects the

Selection Criteria

Criteria	Dataset	Attribute	Value(s)	Buffer
include:				
population density at least 800 inh./km^2	*municipalities_utm32.shp*	*pop_dens*	*>=800*	-
maximum distance to federal highways 500m	*streets_utm32.shp*	*SHIELDNUM*	*'B%', 'N%'*	*500m*
maximum distance to freeways 1000m	*streets_utm32.shp*	*SHIELDNUM*	*'A%'*	*1000m*
exclude:				
distance to existing sites	*furniture_stores.gif*	-	-	*40km*
outside of settlement areas	*built_up_areas_wgs84.shp*	-	-	-

Fig. 13.9 Table with the selection criteria of included and excluded layers

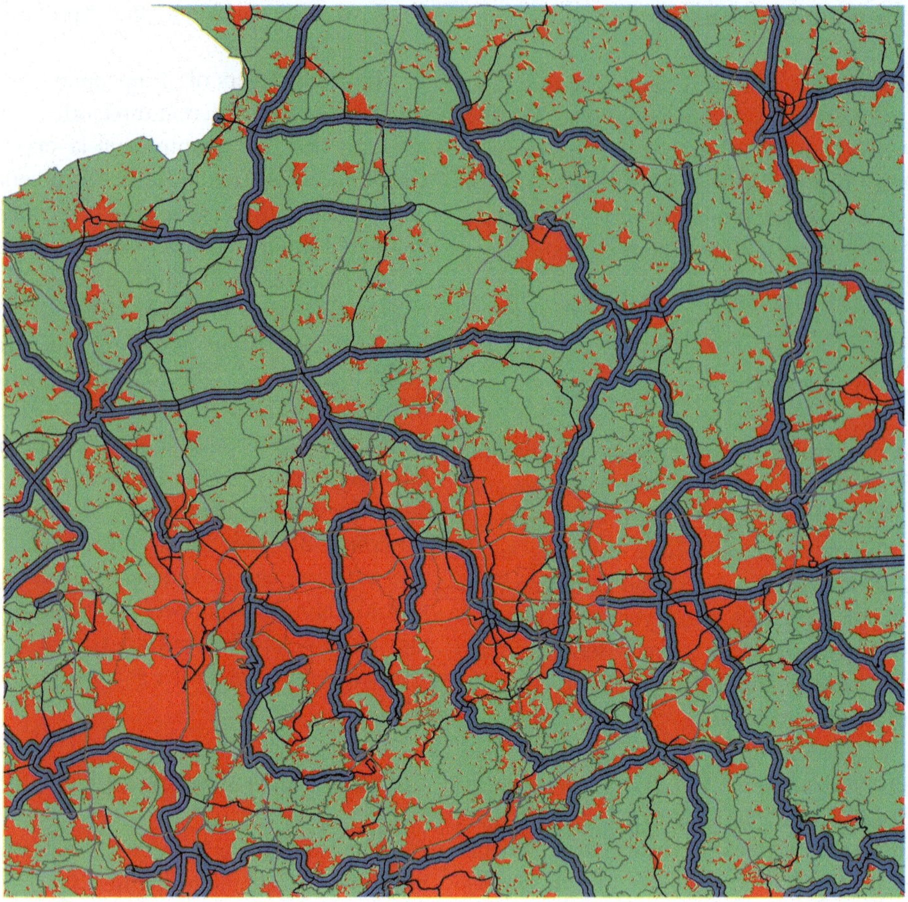

Fig. 13.10 Buffered federal highways (blue)

two expressions, for only one of the two commands has to be true to select the corresponding feature.

Afterwards the selection is buffered by 500 m (Fig. 13.11) and then saved into the positive-folder.

13.3.3 Population Density

For the new site of the store it is desirable to only include regions into the consideration that have a high population density to reach as many customers as possible. To calculate the population density of each municipality in NRW it is necessary to look into the attribute table of the joined layer and detect the table field that contains the absolute number of inhabitants per municipality. In this case, the field with the population data of 2008 is used for the relative population per square kilometer. In order to perform the calculation for all features in one step, again the field calculator is the tool of choice. The new information is stored in a new table field that has to be created with a decimal number field type and a sufficient length. The population density is then calculated dividing the absolute numbers of inhabitants by the area of the certain municipality e.g. *"<field name population>"/area [sqkm].*

The UTM coordinate system is in the metric unit *meter*, which can require another important step (depending on the used GIS) for the calculation of the population density per square kilometer, as the area has to be multiplied by 1.000.000 ($1\ \text{km}^2 = 1\ \text{m}^2 \times \mathit{1.000.000}$) in case the unit of the area can't be defined before the calculation.

Fig. 13.11 Buffered freeways (yellow)

After the population density is calculated properly, all municipalities that are greater or equal than 800 inhabitants per square kilometer need to be selected by another expression.

As the only reasonable operator here is the greater than or equal to operator (>=) the expression for this selection is *"<field name population density>"* > = *800*. The selection then is exported to a separate and persistent layer into the positive-folder (Fig. 13.12).

13.4 Unsuitable Areas

To subtract all restricted areas from the result layer a negative-folder is created to store the two layers which define the areas that need to be avoided for the new location. These are built-up areas, as it is too expensive to build a huge new warehouse in an area that is more valuable for living than for commercial use, and areas that are in between a radius of 40 km of already existing warehouses. This buffer around the existing sites is reasonable as the company rather wants to approach new customers than give already nearby ones a second opportunity where to buy their new piece of furniture. In a more detailed analysis it could be considered to obtain the buffer values not from a simple linear buffer but from a service area that is processed using the street network instead (see Part 1, Sect. 3.3. Network Analysis).

As the built-up areas layer already contains only features that define the built-up areas themselves this layer is used directly and completely as a negative layer for one restriction by copying it into the corresponding directory.

However, the distance to existing sites needs to be processed before the layer is useable for the

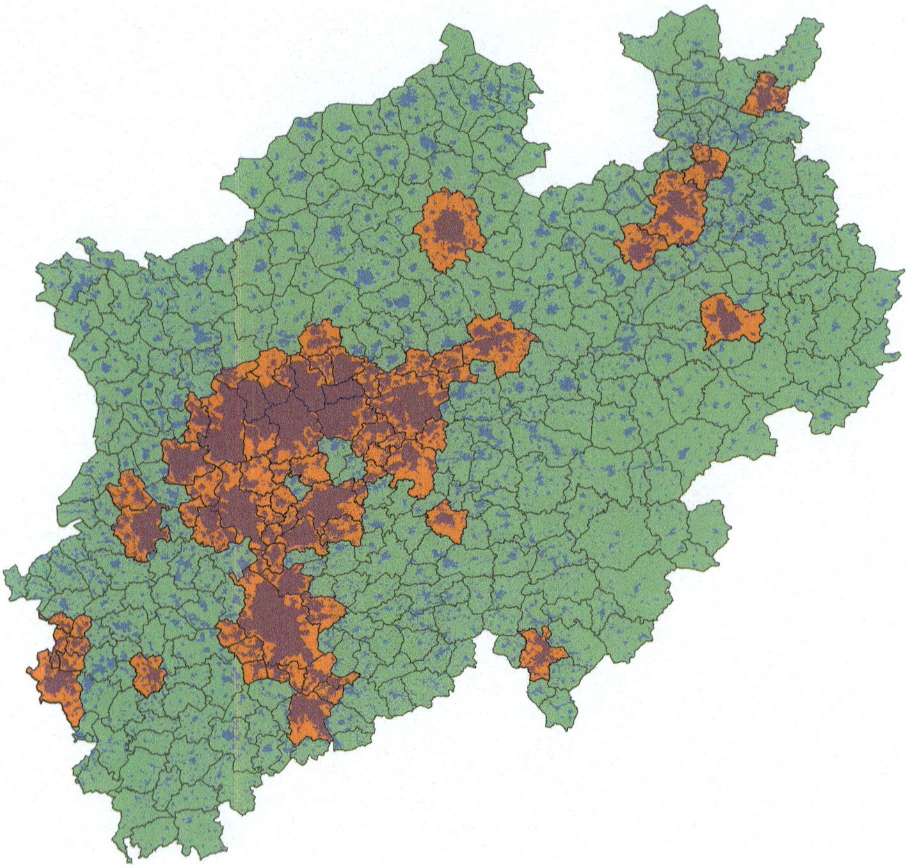

Fig. 13.12 Municipalities with over 800 inhabitants per square kilometer (orange) and built-up sites (light blue) to compare

following calculations and therefore added to the folder containing the two negative layers. The processing here only consists of a single buffering by 40 km that is again done via the buffering tool, setting the beforehand digitized layer of the existing sites as input layer and buffer it by 40.000 m to then save it into the negative-folder (Fig. 13.13).

13.5 Combining Layers

Now that the two folders *positive* and *negative* each contain all the files that are either a requirement or a restriction, all files in each folder need to be combined to one overall file (example of logical operators in Fig. 13.14).

13.5.1 Positive Layer

The combination of all three files from the positive-folder to a single layer is done via an intersection, which is equal to a logical AND ('&'). After simply setting the three layers as input, the output of the intersect-tool only contains areas where all requirements are fulfilled concurrently. Areas where only one or two requirements are accomplished are excluded from the positive result layer (Fig. 13.15).

13.5.2 Negative Layer

For the overall negative layer the single layers have to be combined to an overarching layer via

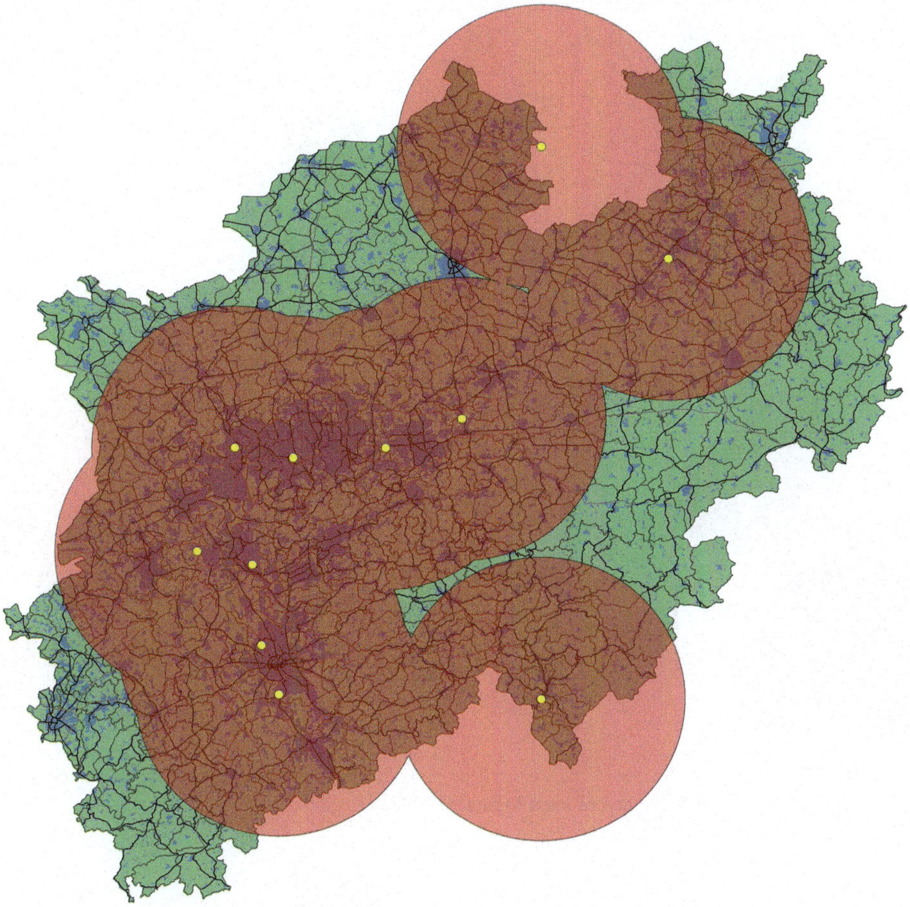

Fig. 13.13 Existing sites (yellow dots) buffered by 40 km (light red)

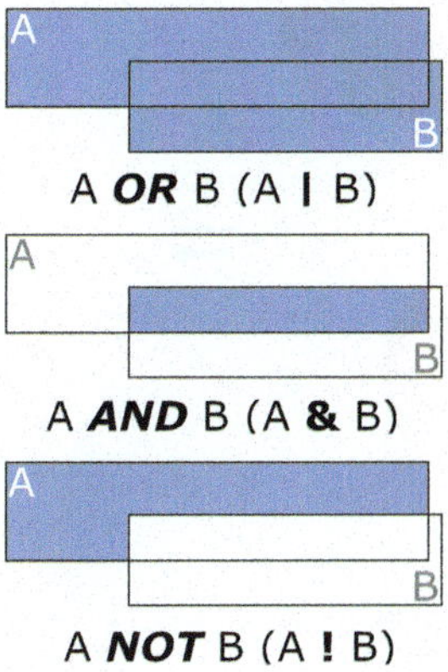

Fig. 13.14 Logical operators

the merge or union tool, which is equal to a logical OR ('|'). No matter if one or all restrictions are regarded – there is no way the new site will be built in an area that is located within a single negative layer (Fig. 13.16).

13.6 Final Result

For the final result the two layers (overall negative and overall positive) need to be combined somehow. The tool of choice here is the difference tool, as it is the same as a logical NOT ('!') and subtracts all unsuitable areas from the suitable ones (Fig. 13.17).

The final result now can be readout in area size and precise location and then put into a map for an overview of what is left and where. Following this, new investigations can be done on the suitable areas, initializing accurately the upcoming search on a smaller scale with further approaches

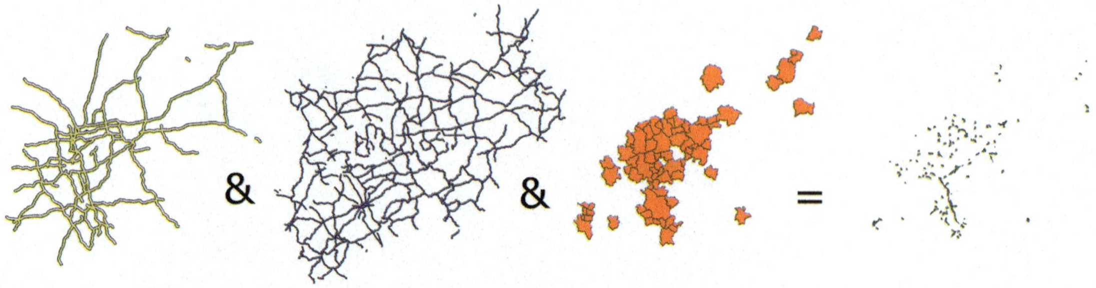

Fig. 13.15 Intersecting all positive layers via a logical AND ('&')

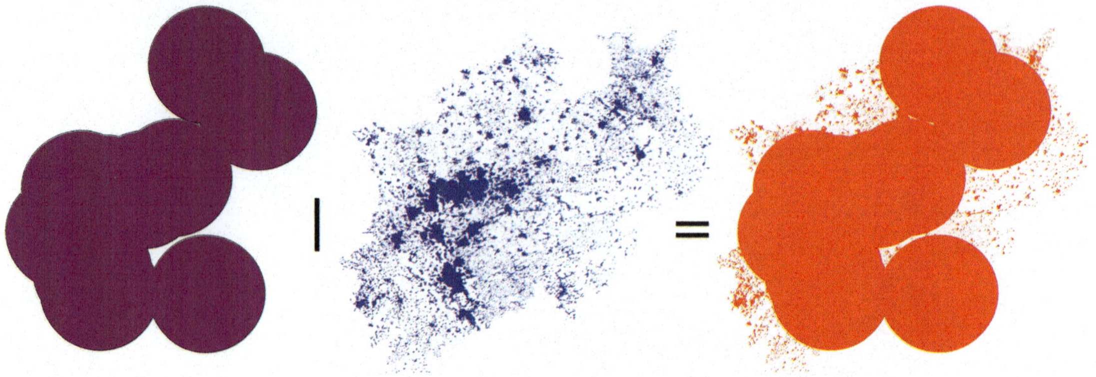

Fig. 13.16 Merging the two negative layers via a logical OR ('|')

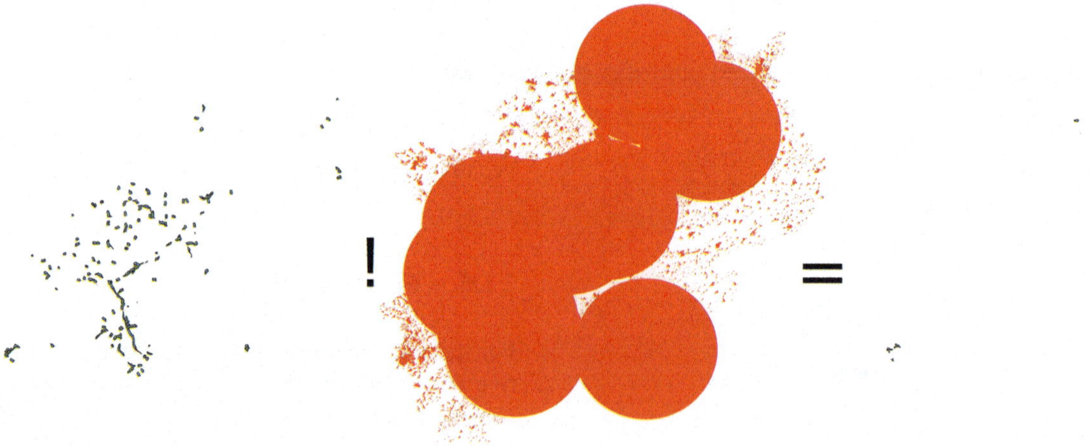

Fig. 13.17 Final result: the difference or logical NOT ('!') of the positive and negative layer

for a property to build a new furniture store (Fig. 13.18).

This short workflow could be one possible scenario for a valuable combination of both spatial and economic approaches. There are certainly several other factors that can be included to the whole process like income per capita, land cost, availability of water and electricity supply etc. To do this, corresponding data needs to be acquired, combined and processed to then set the particular

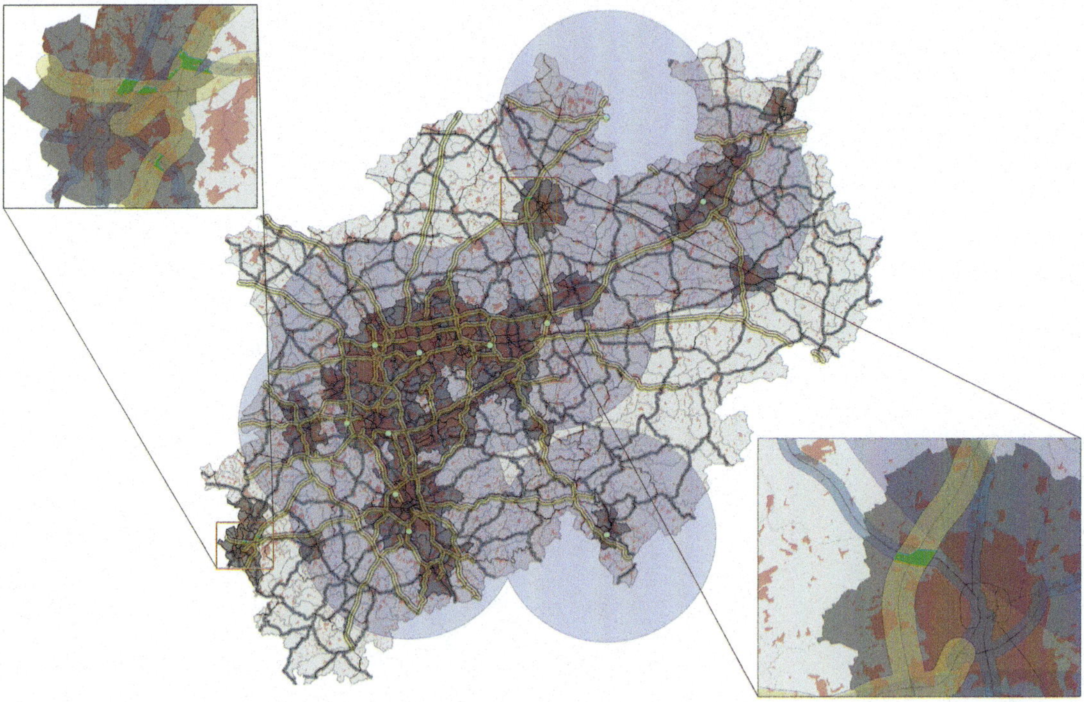

Fig. 13.18 Final result, suitable areas in green

thresholds and extract the layers that should affect the final result.

The two fields of GIS-science and economy require and complement each other in almost every step. This becomes explicit while parametrizing the different calculations and tools, dealing with values that underly business related decision making while the whole processing is done spatially with layers and ends up in a characteristic and solving map.

Demographic Development Planning in Cities

14

Jaroslav Burian, Jarmila Zimmermannová, and Karel Macků

Abstract

Currently, the population ageing is one of the greatest economic, social and environmental challenges facing all EU countries. For this reason, the main goal of this study is analysis of the population ageing and its economic aspects. As the main method we used particular instruments of economic and spatial analysis. As the first step, we evaluated the consumption behavior of seniors. In the next step, a spatial analysis of population ageing in the territory of the Olomouc region was processed, models of demographic projection were used to estimate the number of seniors in future years. This information was connected with social services facilities for seniors – the current capacities, the number of candidates and the ratio of refusals are evaluated. Based on this results, an estimation of possible changes needed in the area of social services in the region was presented. At the same time, the riskiest areas of the region were defined regarding the ratio of the seniors to the social services offered. Finally, the Urban Planner model was used to analyse areas for identifying suitable locations for a new social service facility in accordance with urban planning.

Keywords

Seniors · Ageing · Economic impact · Spatial · Spatial planning

14.1 Introduction

Population ageing is one of the most significant social and economic challenges facing the EU (Burian et al. 2017). At the same time it is a topic that allows to combine economic and spatial aspects of population ageing into one case study.

Regarding actual development of the demographic structure of the population in EU countries, Eurostat published statistics EUROPOP2015 (Eurostat 2017). The median age of the EU-28's population was 42,6 years on 1 January 2016 and it increased by 4.3 years (on average, by 0.3 years per annum) between 2001 and 2016, rising from 38.3 years to 42.6 years. Between 2006 and 2016 the median age increased in all of the EU Member States. Projections foresee a growing number and share of elderly persons (aged 65 and over), with a particularly rapid increase in the number of very old persons (aged 85 and over). These demographic developments are likely to have a considerable impact on a wide range of policy areas with

J. Burian (✉) · K. Macků
Department of Geoinformatics, Palacký University Olomouc, Olomouc, Czech Republic
e-mail: jaroslav.burian@upol.cz; karel.macku@upol.cz

J. Zimmermannová
Department of Economics, Moravian Business College Olomouc, Olomouc, Czech Republic
e-mail: jarmila.zimmermannova@mvso.cz

V. Pászto et al. (eds.), *Spationomy*, https://doi.org/10.1007/978-3-030-26626-4_14

respect to economic and financial issues (Eurostat 2017).

In the Czech Republic, the Ministry of Finance is regularly publishing the Macroeconomic forecast of the Czech Republic, including also the issue of population ageing. According to the Ministry of Finance (2017), it is obvious, that the group of older people over the age of 65 continues to grow and on the contrary, the working-age population declines.

We can look at the problem of population ageing with relative and absolute numbers. On one side, the population began to age relatively – there began to be more elderly people and fewer working-age population and children. On the other side, in connection with the good living conditions of the inhabitants, mortality began to decline, and the average lifespan began to lengthen. The population began to age also absolutely – the number of old people (Arltova and Langhamrova 2010).

Based on the economic view, a very important issue is the development of demand of the group of older people – their consumption patterns. This topic is analysed mainly by marketing scientists, which are focusing on specific markets, for example, the market with tobacco products, medical products etc. Higgs et al. (2009) chart consumption by retired households in Great Britain in two areas; ownership of key consumer goods and key components of household spending. The results demonstrate mainly the growing extent of ownership of key goods in retired households and show the differences in proportional expenditure between retired households and the employed (Higgs et al. 2009). Wolf et al. (2014) describe that people often respond to retirement in conjunction with lifestyle changes; many such changes lead to changes in consumption patterns. Changes in consumption expenditures expose consumers to different offers, leading to changes in consumer preferences (Wolf et al. 2014).

Within the regions and cities, the economic issues are strongly connected with urban policy and urban planning. Population ageing and demographic development tend to increase the demand for health care, social security and the care for the disabled and the elderly. Carbonaro (Carbonaro et al. 2016) examined the links between population ageing and demographic development, local economic development prospects, and the financial implications for urban policy. Demographic change impacts on a variety of components of the urban economy. Some of the key areas are urban labour markets, infrastructure planning and housing. But not only urban areas, but also rural places can be affected by these changes. The border between urban and rural can be difficult to differentiate, especially in connection with suburbanization (Paszto et al. 2015, 2016).

There is a limited number scientific studies focusing in more details on population ageing and its consequences with economic and urban planning issues. Therefore, this research is focused on the economic analysis of consumption behaviour of households of retired, the development of retired in the society (age index), possibilities of retirement homes, sustainable urban planning and economic aspects of this issue (Macků et al. 2018; Burian et al. 2018b).

The main goal of this case study is an evaluation of the demographic development, the consumption behaviour of households of retired and inclusion of expected changes to the urban planning in cities. Within this task, the Urban Planner Model (Burian et al. 2015, 2018a) for selecting suitable locations for households of retired was used. The Olomouc region was selected as a case study since detailed economic and geographic data are available for this region. Besides the main goal, the following objectives are formulated:

- evaluation of current situation and trends in the population ageing in the region;
- identification of the riskiest locations from the population ageing perspective;
- Identification the best places for the future cities development.

14.2 Material and Methods

14.2.1 Data

For the purposes of the analysis of consumption behaviour of pensioners, following data sets were used: Expenditure and Consumption of

Households statistics – Household Budget Survey, precisely Consumption expenditures – annual averages per capita in CZK and detailed development of consumption expenditures of households of pensioners without EA members (Czech Statistical Office 2017). For the prediction of the development of retired in the society, the following data sources were used: Social services establishments, Selected social security data, Five years population structure and Life tables. All these datasets were also provided by the Czech Statistical Office (Czech Statistical Office 2017). A total of more than 120 vector layers covering Urban Planner Model factors (see the section about Urban Planner) were used as the input for land suitability calculations. These layers are collected according to the regulations of Act No. 183/2006 Coll., the Construction Act, by all regional planning offices in the Czech Republic as part of the analytical material for planning (Burian et al. 2016). The source of the data is more than 40 organisations or private companies (e.g. Czech Geological Survey; Czech Hydrometeorological Institute; Czech Statistical Office; Czech Office for Surveying, Mapping, and Cadastre etc.). The spatial accuracy of the data corresponds to the Cadastral Map – scale 1:2000.

14.2.2 Methods

14.2.2.1 Analysis of Consumption Behaviour of Pensioners

Analysis of consumption behaviour of pensioners serves for the estimation of potential areas for additional investments into new social services. For the purposes of this analysis, we use Expenditure and Consumption of Households statistics – Household Budget Survey, precisely Consumption expenditures – annual averages per capita in CZK and detailed development of consumption expenditures of households of pensioners without economic active (EA) members. We provide a comparative analysis of households of employees in total and households of pensioners without economic active) members; the difference in consumption patterns is calculated as both difference in CZK and difference in %.

14.2.2.2 Prediction of the Development of Retired in the Society

Regarding the ageing of the population and the analysis of the consumption behaviour of seniors, the demographic situation in the territory of the Olomouc Region was evaluated in this study (Macků et al. 2018). In the statistical surveys, 65 years is usually used as the age of retirement. This statement also follows the definitions used at international level by organisations such as the UN or Eurostat. Of course, the age limit of 65 years does not always coincide with the actual start of retirement. The demographic projection is used to predict the future development of the population. The most commonly used population prognosis method is a component method, that takes the age structure of the population in 5-year categories as a starting point and shifts population numbers by age group to higher age levels using survival probabilities (O'Neill et al. 2001). These numbers are reduced by the number of deaths and enlarged by the number of births. The unborn population is calculated on the basis of expected development of fertility rates by the age of women. Migration is not considered in this component method.

As a tool for identification of elderly areas, spatial distribution of ageing index is used. The ageing index is calculated as the ratio between the number of people at the age over 65 years per 100 children in age 0–14. It is often used as an indicator of demographic ageing of the population. Values displayed in the map can quickly deliver the information about current state and also development over past decade. In the same way, the spatial distribution of retirement homes with their capacity is evaluated and helps to determine zones within sufficient social services.

14.2.2.3 Urban Planner Suitability Analysis

To identify the suitable (optimal) locations for households of retired within the municipalities, land suitability analysis in the Urban Planner

Model was performed (Burian et al. 2017). The Urban Planner is an analytic extension for Esri ArcGIS for Desktop designed to evaluate the land suitability and to detect the most suitable areas for spatial development. The model was developed at the Department of Geoinformatics, Palacký University in Olomouc (Burian et al. 2015, 2018a). The model uses a multi-criteria analysis, respects the principles of sustainable development, and allows for the creation of several land use and land suitability scenarios. The core of the Urban Planner Model focuses on the evaluation of land suitability according to input data, it's values and weights. Land suitability is analyzed in three levels (pillars, factors, and layers) for the five predefined categories of land use. For the analysis described in this case study, only one category of land-use (specific type housing) was calculated.

The total land suitability is calculated according to the setting of the weights between the three classes (the three pillars): environmental, social, economic. Weights can acquire values from 0 to 100; the sum of the weights of all three pillars must be equal to 100. Each of the three pillars (classes) consists of factors. Factors are divided into three groups – positive, negative and limits, and are assigned to the pillars. As in the case of pillars, the combination of factors is based on weighted overlay method. The following factors were used for land suitability analysis:

- ENVIRONMENTAL PILLAR – Water bodies protection, Nature protection, Mineral resources protection, Soil and forest protection
- SOCIAL PILLAR – Slope aspect, Kindergarten accessibility, Elementary school accessibility, Grocery store accessibility, Public spaces accessibility, Bus stops and public transport accessibility, Trains stops accessibility, Density of forests, Distance of rivers, Density of build-up areas, Air pollution, Noise pollution, Population density, Radon risk, Cultural and social protection
- ECONOMIC PILLAR - Population density, Radon risk, Slope, Electric power supply, Drinking water supply, Gas supply, Wastewater management, Heat supply, Fiber-optic internet connection, Distance from roads, Distance from railroads, Flood hazard, Geological hazard, Specific infrastructure protection (Fig. 14.1)

The weights for input factors and pillars were calculated with the commonly used Saaty's method (Saaty 1983), which makes it possible to define the weights for several criteria as objectively as possible. All selected factors follow the standard layer used in a master plan creation and reflect consumption behaviour of households of retired (e.g. factor grocery store accessibility, public transport accessibility etc.). The weights were tested and calibrated in several regions in the Czech Republic (Olomouc Region, Ostrava Region, Vysocina Region, and Prague Region).

14.3 Results and Discussion

By 31st December 2016, 122,257 people over 65 years lived in the Olomouc Region and there were 148,420 registered recipients of pensions. It is clear that the age of 65 years as the retirement age is currently overestimated, also because many people decide for early retirement (Czech Statistical Office 2015). In connection with increasing of the retirement age, gradual wiping of this difference can be expected. Ageing is proved by a simple trend (Fig. 14.2). The number of persons older than 65 years is increasing approximately by 3500 persons per year. On the other side, amount of pension recipients does not have so clear trend, and in some years, slight decrease can be observed.

The population prediction in the Olomouc region was calculated for each age group at 5-year steps up to the year 2035 (see Table 14.1). The prediction results in both the total decrease in the number of inhabitants in the region and the gradual increase of the age group over 65 years. These two trends clearly indicate the ageing of the population, as also evidenced by ageing index values. Average value of the ageing index in the year 2016 is 128, which is slightly higher than the national average (120). Figure 14.3a) shows its development in last

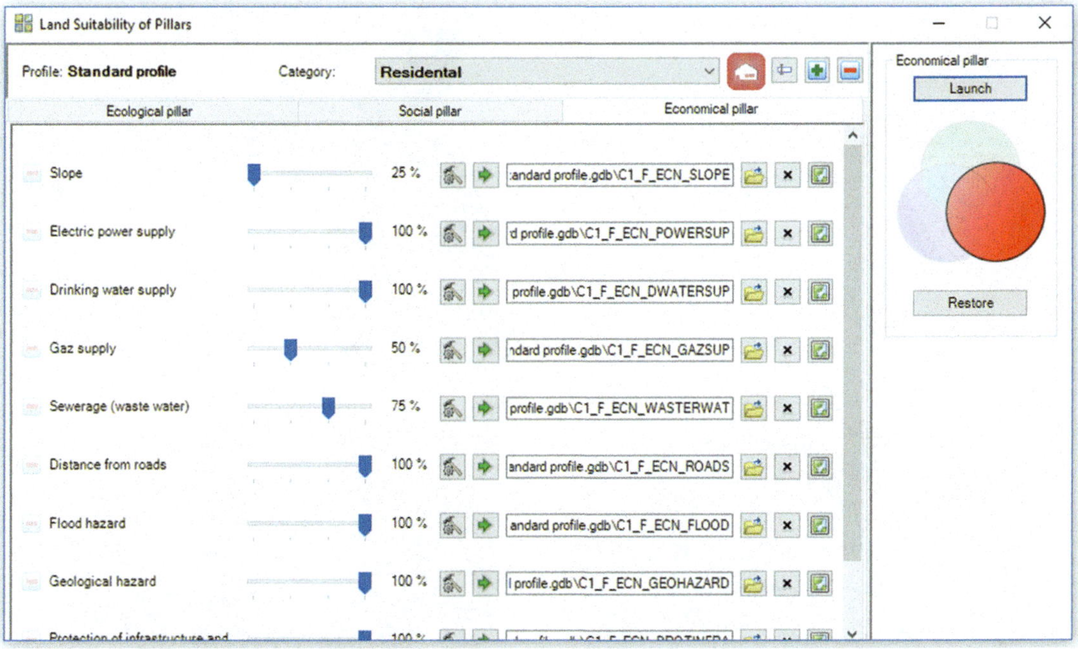

Fig. 14.1 Example of economic factors' weights (Source: Authors)

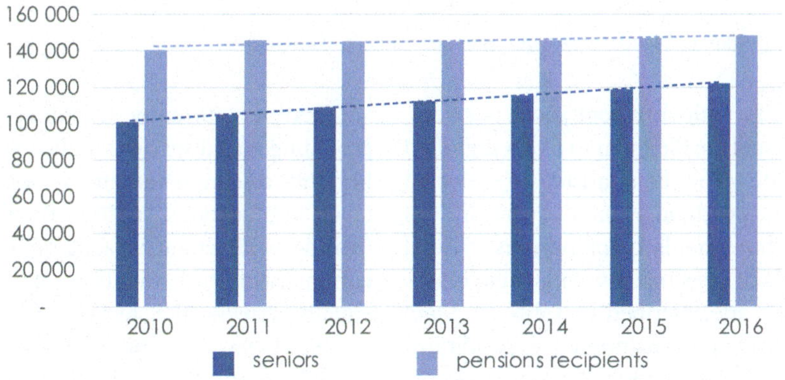

Fig. 14.2 Development of seniors in Olomouc region. (Source: Authors, CZSO 2017)

Table 14.1 Demographic projection

Age group	2015	2020	2025	2030	2035
0–14	95,884	94,781	84,867	75,753	69,509
15–64	420,019	399,827	390,581	378,801	363,866
65+	118,815	132,695	137,858	139,254	137,791

Source: CZSO (2015, Authors)

10 years: only 104 municipalities tend to be younger, the remaining 294 suffer from increase. Average values in these two dichotomous categories are unequal, too – the average decrease is 16%, average increase 38%. Therefore, ageing not only predominates, but it is also more intensive.

With the increasing number of seniors, this research asks if it is possible to provide this

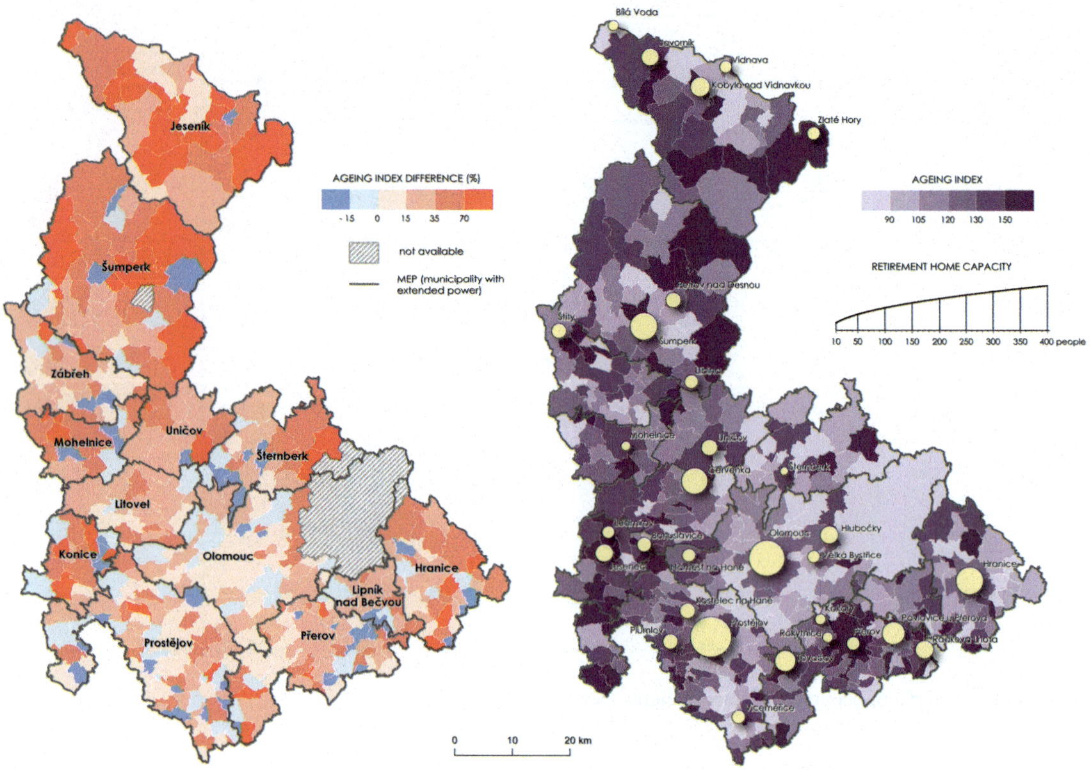

Fig. 14.3 Ageing index and its development. (Source: Authors, CZSO 2017)

population group with sufficient social services, necessary care and enable them to live a dignified and quality life. The comparison of selected consumption expenditures in two groups of households – households of employees in total and households of pensioners without active economic members are described in Table 14.2 and Fig. 14.4. Focusing on consumption expenditures of households of pensioners without economic active members, it is obvious, that there are 3 groups of significantly higher consumption expenditures in comparison with households of employees (food and non-alcoholic beverages; housing, water, electricity, gas and other fuels; health) and also groups of significantly lower consumption expenditures (clothing and footwear; transport; education; restaurants and hotels). This conclusions are used in the Urban Planner Model to set higher weights for these factors to calculate land suitability for housing for seniors.

There are currently several types of facilities providing social services in the Olomouc Region. For this research, retirement houses have been taken into account. In 2016 there were 34 such facilities with a total capacity of 2674 people. The current spatial distribution of social facilities equally covers the territory of the Olomouc Region. Less accessibility is only in the northern part of the region between MEP (the municipality with extended power) Jeseník and Šumperk, as well as MEP Mohelnice and Zábřeh do not have high capacity (see Fig. 14.3b). The facilities of social services are (as expected) located in municipalities with a higher number of inhabitants or a higher proportion of people aged 65+, but the dependence between the number of inhabitants and the total capacity of the facility is not significant.

Evaluation of demographic data indicates long-term ageing in many of Olomouc region municipalities. The highest increase in the ageing

Table 14.2 Comparison of selected consumption expenditures in 2016 in the Czech Republic

Consumption expenditure	Households		Differences in consumption patterns	
	Employees in total	Pensioners without EA members	Difference in CZK	Difference in %
Food and non-alcoholic beverages	23,854	29,244	5390	122.60
Alcoholic beverages, tobacco	3572	3757	185	105.18
Clothing and footwear	7444	3836	−3608	51.53
Housing, water, electricity, gas and other fuels	24,175	34,063	9888	140.90
Furnishings, household equipment etc.	8059	7526	−533	93.39
Health	2912	4565	1653	156.77
Transport	16,108	8459	−7649	52.51
Communication	5310	4562	−748	85.91
Recreation and culture	12,768	10,553	−2215	82.65
Education	916	34	−882	3.71
Restaurants and hotels	8573	4209	−4364	49.10
Miscellaneous goods and services	16,468	11,191	−5277	67.96

Source: CZSO (2017, Authors)

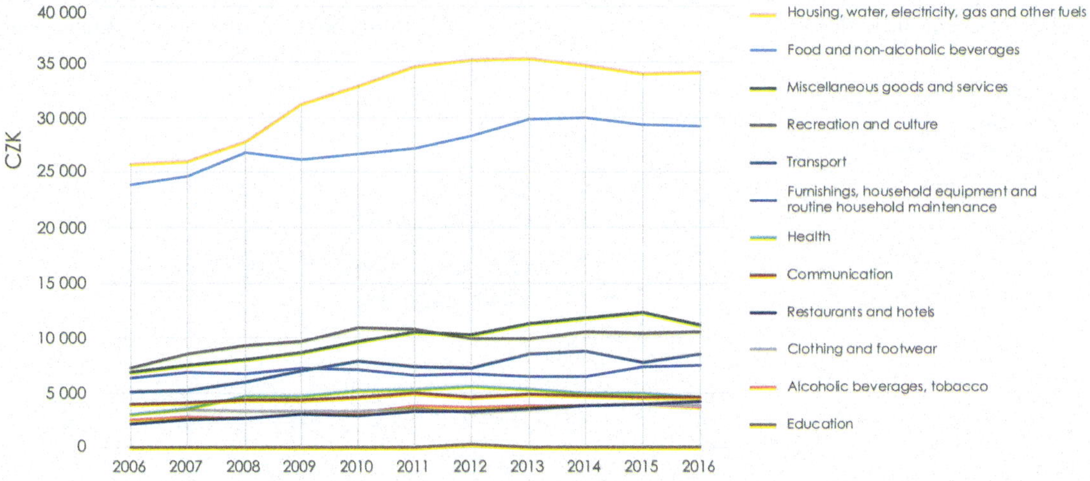

Fig. 14.4 Consumption expenditures of households of retired, annual averages per capita in CZK. (Source: Authors, CZSO 2017)

index is in the northern part of the region and at east-north borders (municipalities Norberčany, Rejchartice and Mírov). On the other hand, biggest decrease is typical for small municipalities (less than 500 citizens – e.g. Šlégov, Hačky and Provodovice). These villages are located close to bigger cities, so the changes can be attributed to suburbanization processes.

Regarding this demographic development, new investments in social facilities should be expected. Considering the simple dependence between the current number of seniors and the capacities of retirement houses, a very simplified estimate can be made to illustrate how many new facilities with sufficient capacity will need to be built in the future.

Expecting approximately the same proportion of seniors using these services, it will be necessary to increase the capacity of the facility by approximately 415 places, keeping the average capacity of 70 people per facility, six new social service facilities will be needed. Unfortunately, retirement homes are not able to meet the needs of all applicants, much remains to be refused. The Ministry of Labor and Social Affairs records the rejected applications mostly in the category of retirement homes (65,764 in the Czech Republic, 2016). There are no indicators comparing rejected applications to all applications, rejected applications can be relativized e.g. to 1000 seniors. In this case, the worst situation is in Jihomoravský region (56 rejected per 1000 seniors), Olomouc region registered 29 rejected applications per 1000 seniors, which is slightly worse than the national average (25 per 1000).

Focusing on economic aspects of population ageing and consequent planning of facilities for seniors, Fig. 14.5 shows us increasing expenditures (in thousand CZK) on homes for seniors, homes with a special regime and homes for people with disabilities.

Since the highest yield of expenditures is represented by homes for seniors, it is interesting to focus on more details on the structure of ownership of these homes. As you can see in Fig. 14.6, the number of state, regional and municipal houses for seniors is decreasing, on the other hand the number of non-state houses for seniors is increasing.

Based on the development of ownership of houses for seniors, the question of financing of planning and building of new social services facilities represents key issue for responsible local authorities. Dealing with increasing number of seniors and simultaneously higher demand for new facilities and houses for seniors, we can expect significant financial burden on both public and private budgets. On the other hand, based on the survey, almost 80% of seniors would prefer to finish one's days at home. This statement opens new discussion about enhance of investments into

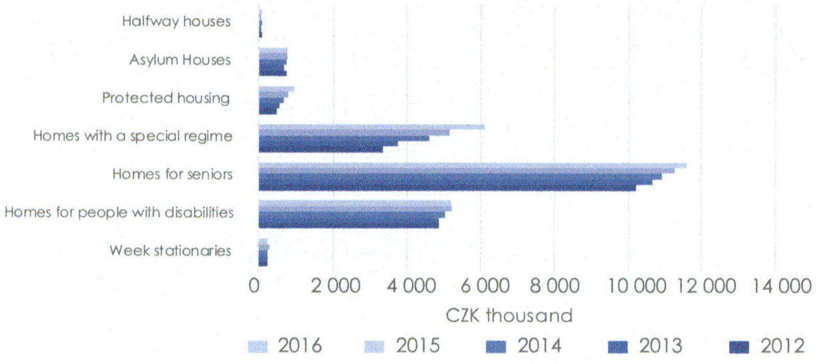

Fig. 14.5 Social services expenditures in thousands CZK. (Source: Authors, CZSO 2017)

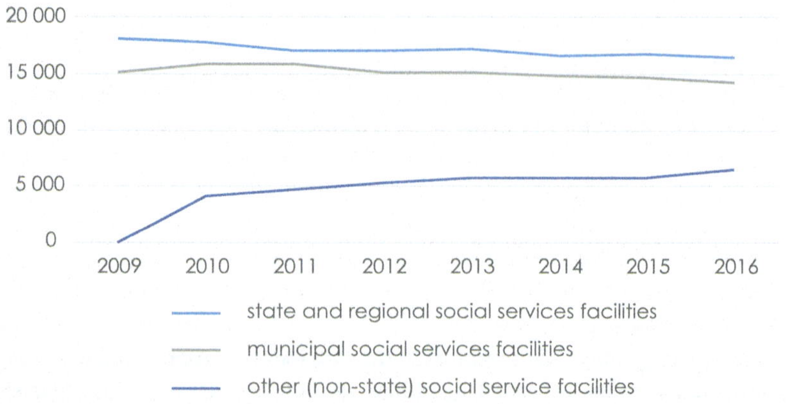

Fig. 14.6 Retirement homes – capacity. (Source: Authors, CZSO 2017)

terrain social services, that would provide only the necessary help to the old people.

Comparing our results with other scientific studies, there is a lack of studies focusing simultaneously on economic aspects of population ageing and urban planning. We can find mainly studies focusing on development and testing of different models of healthcare planning based on more variables, including demographic variables, technological innovations, epidemiological changes and socioeconomic factors (Chernichovsky and Markowitz 2004; Birch et al. 2013; Mason et al. 2015; Basu and Pak 2016; Birch et al. 2017). Based on these scientific studies and results of our research, it is obvious, that we should plan social and healthcare services facilities not only with respect to demographic development; therefore the model should be more complex, including different variables which can be part of particular decision-making.

As the last step within this case study, Urban Planner Model was used to calculate land suitability for housing. Information about consumption expenditures was used to define input factor and set their weights. The final results of the land suitability calculations are raster layers, that can be visualised in the map. Because of the raster resolution (10 m/pixel) and the study area (whole Olomouc Region), the simple web map application in Esri ArcGIS Online was created (Fig. 14.7). The application allows to zoom in to see detailed situation within the area of each municipality in the region (Fig. 14.8). For better orientation, in larger scale, boundaries of existing and proposed build up are being used. The values of land suitability range from 0 (the lowest suitability) to 100 (the highest suitability). Areas with no values represent places disqualified for suitability due to limits (e.g., flooding areas or protection zones). The map can be used for the detection of the most suitable places for new households. Similar results from Urban Planner Model (raster layers of land suitability) has been used in several regions in the Czech Republic (e.g. Olomouc Region, Ostrava Region, Vysocina Region) as supporting material for planning decisions at urban planning offices.

14.4 Conclusion

The main goal of this study was to analyse the population ageing issue with detailed focus on Olomouc region and to discuss its economic and spatial aspects. For the purpose of the main goal achievement, we used economic and spatial analysis. Regarding the evaluation of current situation and trends in the population ageing in the region, we can say that our analyses prove a trend of population ageing in the region. In comparison

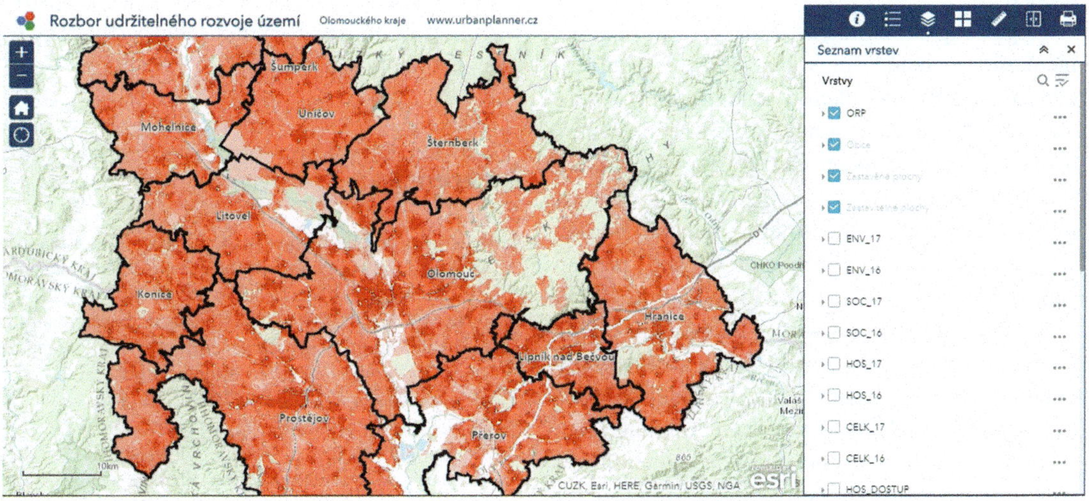

Fig. 14.7 Web Map Application with land suitability for housing. (Source: Authors)

Fig. 14.8 Web Map Application with land suitability for housing. (Source: Authors)

with national numbers, the situation in the Olomouc region is a little worse than the national average (ageing index, retirement homes rejected applications etc.).

Focusing on identification of the riskiest locations from the population ageing perspective, thanks to spatial analysis and visualization, the most problematic areas can be discovered. At the same time, the spatial distribution of retirement homes is assessed – they are appropriately placed across the region. In connection with home capacity and the current trend of rejection of applications, many new facilities will be probably needed.

Regarding economic aspects, we can observe increasing demand for social service facilities, connected with population ageing. Since there are not only state, regional or municipal owners of these facilities, moreover the share of private facilities is increasing, particularly in case of houses for seniors, the future financial burden will be divided between public and private stakeholders.

Finally, the land suitability assessment of the Olomouc region was analysed. All results were obtained by the Urban Planner Model using a multi-criteria analysis as the main computational method. The highest consumption expenditures were used to refine default input factor and their weights to reflect demands of the retired group. The final output from the model was the raster layer of land suitability, that was visualised in the web map application. The application can be used to detect the most suitable places for the new household in each municipality in the Olomouc region.

The future research can be focused on deeper spatial analysis, e.g. network analysis describing exact coverage of region by particular retirement home. At the same time, sophisticated models for urban planning can be used for detailed searching of suitable locations for new social facilities (e.g. Urban Planner). This kind of approach allows to include another economic data together with spatial information and then perform better complex analysis.

References

Arltova, M., & Langhamrova, J. (2010). Migration and ageing of the population of the Czech Republic and the EU countries. *Prague Economic Papers, 19*(1), 54–73.

Basu, K., & Pak, M. (2016). Will the needs-based planning of health human resources currently undertaken in several countries Lead to excess supply and inefficiency? *Health Economics, 25*(1), 101–110.

Birch, S., Mason, T., Sutton, M., & Whittaker, W. (2013). Not enough doctors or not enough needs? Refocusing health workforce planning from providers and services to populations and needs. *Journal of Health Services Research and Policy, 18*, 107–113.

Birch, S., Tomblin, M. G., MacKenzie, A., Whittaker, W., & Mason, T. (2017). Will the need-based planning of health human resources currently undertaken in several countries Lead to excess supply and inefficiency? A comment on Basu and Pak. *Health Economics, 26*(6), 818–821.

Burian, J., Stastny, S., Brus, J., Pechanec, V., & Vozenilek, V. (2015). Urban planner: Model for optimal land use scenario modelling. *GEOGRAFIE, 120* (3), 330–353.

Burian, J., Brychtova, A., Vavra, A., & Hladisova, B. (2016). Analytical material for planning in Olomouc, Czech Republic. *Journal of Maps, 12*(4), 649–654.

Burian, J., Zimmermannová, J., & Macků, K. (2017). Urban planner as a tool for demographic development planning in the cities. In Proceedings from the 8th annual international scientific conference "Region in the development of society 2017." Mendel University in Brno.

Burian, J., Stachova, M., & Vondrakova, A. (2018a). Land suitability assessment of the Olomouc region: An application of an urban. *Planner Model Journal of Maps, 14*(1), 73–80. https://doi.org/10.1080/17445647.2018.1493407.

Burian, J., Macků, K., Zimmermannová, J., & Kočvarová, B. (2018b). Spatio-temporal changes and dependencies of land prices: A case study of the city of Olomouc. *Sustainability, 10*(12), 4831. https://doi.org/10.3390/su10124831.

Carbonaro, G., Leanza, E., McCann, P., & Medda, F. (2016). Demographic decline, population aging, and modern financial approaches to urban policy. *International Regional Science Review, 41*(2), 210–232.

Chernichovsky, D., & Markowitz, S. (2004). Aging and aggregate costs of medical care: Conceptual and policy issues. *Health Economics, 13*(6), 543–562.

CZSO (Czech Statistical Office). (2015). *Seniors in Olomouc region*. https://www.czso.cz/

CZSO (Czech Statistical Office). (2017). *Household budget survey*. https://www.czso.cz/

EUROSTAT. (2017). *Population structure and ageing*. http://ec.europa.eu/eurostat/statistics-explained/index.php/Population_structure_and_ageing

Higgs, P. F., Hyde, M., Gilleard, C. J., Victor, C. R., Wiggins, R. D., & Jones, I. R. (2009). From passive to active consumers? Later life consumption in the UK from 1968–2005. *The Sociological Review, 57*(1), 102–124.

Macků, K., Zimmermannová, J., & Burian, J. (2018). Economic and spatial analysis of population ageing. In P. Dvořáková & P. Baisa (Eds.), *Proceedings of the 22nd international conference current trends in public sector research*. Brno: Masaryk University.

Mason, T., Sutton, M., Whittaker, W., & Birch, S. (2015). Exploring the limitations of age-based models for health care planning. *Social Science & Medicine, 132*, 11–19.

MFCR (Ministry of Finance of the Czech Republic). (2017) *Macroeconomic forecast*. http://www.mfcr.cz/en/statistics/macroeconomic-forecast/

O'Neill, B. C., Balk, D., Brickman, M., & Erza, M. (2001). A guide to global population projections. *Demographic Research, 4*, 203–288. https://doi.org/10.4054/DemRes.2001.4.8. ISSN:1435-9871.

Paszto, V., Brychtova, A., Tucek, P., Marek, L., & Burian, J. (2015). Using a fuzzy inference system to delimit rural and urban municipalities in the Czech republic in 2010. *Journal of Maps, 11*(2), 231–239. https://doi.org/10.1080/17445647.2014.944942.

Paszto, V., Burian, J., Marek, L., Vozenilek, V., & Tucek, P. (2016). Membership of Czech municipalities to rural and urban areas: A fuzzy-based approach. *Geografie-Sbornik CGS, 121*, 1.

Saaty, T. (1983). Priority setting in complex problems. *IEEE Transactions on Enigneering Management, 3*, 140–155.

Wolf, F., Sandner, P., & Welpe, I. M. (2014). Why do responses to age-based marketing stimuli differ? The influence of retirees' group identification and changing consumption patterns. *Psychology & Marketing, 31* (10), 914–931.

15 Selected Economic and Environmental Indicators in EU28 Countries Connected with Climate Protection

Jarmila Zimmermannová and Vít Pászto

Abstract

The main goal of this chapter is to present the development of selected economic and environmental indicators in EU28 countries connected with climate protection in the period from 2005 to 2015. European Union Emission Trading Scheme (EU ETS) was introduced in 2005. Currently, the EU ETS is in operation for more than a decade; moreover, in 2018, the European Commission adopts rules for the next 4th trading period. Is there visible any improvement in CO_2 emissions development? Can it be connected to the changes in the macroeconomic indicators? The methodological part presents the data sources and methodological background of the research. Possible geographical patterns in the development of selected indicators within the EU28 are indicated. Detailed analysis of economic and environmental data of EU28 countries is provided, with the use of (geo)visual analysis of the data. The results of the (geo) visual analysis show that CO_2 emissions within selected EU countries were decreasing in the chosen period 2005–2015, with some exceptions (e.g. Iceland and Latvia). As the development of CO_2 emissions in all selected EU countries is not similar, the other economic and environmental indicators were included (e.g. GDP, Investments) into the analysis to reveal a typical (geographical) pattern and explain the current situation.

Keywords

Carbon dioxide · Emissions · Spatial pattern · (Geo)visual analysis · Maps

15.1 Relationships Between Economic and Environmental Indicators

The development of economic and environmental indicators in particular countries can be influenced by different factors. The researchers and analytics can observe the development of specific indicators in the same direction, increase in both kinds of indicators (positive development) or decrease in both types of indicators (negative development). Regarding the reduction of the environmental burden of the economy, the best possibility is positive development of economic indicators and negative development of environmental indicators; referred to as "decoupling".

Decoupling environmental pressures from economic growth are one of the main objectives

J. Zimmermannová
Department of Economics, Moravian Business College Olomouc, Olomouc, Czech Republic
e-mail: jarmila.zimmermannova@mvso.cz

V. Pászto (✉)
Department of Informatics and Applied Mathematics, Moravian Business College Olomouc, Olomouc, Czech Republic

Department of Geoinformatics, Palacký University Olomouc, Olomouc, Czech Republic
e-mail: vit.paszto@gmail.com

V. Pászto et al. (eds.), *Spationomy*, https://doi.org/10.1007/978-3-030-26626-4_15

of the OECD Environmental Strategy for the First Decade of the twenty-first century, adopted by OECD Environment Ministers in 2001. Decoupling occurs when the growth rate of an environmental pressure is less than that of its economic driving force (e.g. GDP) over a given period. Decoupling can be either absolute or relative. Absolute decoupling is said to occur when the environmentally relevant variable is stable or decreasing while the economic driving force is growing. Decoupling is said to be relative when the growth rate of the environmentally relevant variable is positive but less than the growth rate of the economic variable (OECD 2002).

Except for some pressures, decoupling is usual in OECD countries, and further progress seems possible. The evidence presented in the OECD Report "Indicators to Measure Decoupling of Environmental Pressure from Economic Growth" shows that relative decoupling is widespread in OECD Member countries. Absolute decoupling is also quite common, but for some environmental pressures, little decoupling is occurring. The evidence also suggests that further decoupling is possible since absolute decoupling was recorded in at least one OECD country for all but two of the decoupling indicators examined at the national level.

The OECD report explores a set of 31 decoupling indicators is covering a broad spectrum of environmental issues. Sixteen indicators related to the decoupling of environmental pressures from total economic activity under the headings of climate change, air pollution, water quality, waste disposal, material use and natural resources. The remaining 15 indicators focus on production and use in four specific sectors: energy, transport, agriculture and manufacturing. Some indicators have also been decomposed to highlight the extent to which various factors (e.g. technological factors, structural changes) have contributed to reducing or adding to environmental pressures in recent years.

Regarding our analysis, we will deal with selected economic and environmental indicators in EU countries, with a focus on climate protection and CO_2 emissions development.

15.2 The EU Emissions Trading System Background

The European Union established a scheme for emission allowances trading, the EU Emissions Trading System, also called as the EU ETS. Currently, the EU ETS is in operation for more than a decade; moreover, in 2018, the European Commission adopts rules for the next 4th trading period. The initial EU Emissions Trading System was based on Directive 2003/87/EC, which established a fundamentally decentralised system for the pilot phase of emissions trading (2005–2007) and the Kyoto Protocol commitment phase (2008–2012). The key instrument was the preparation of National Allocation Plans (NAPs) (Wettestad et al. 2012).

Currently, the EU ETS is the most significant emissions market in the world. Based on Directive 2009/29/EC, the EU ETS is in Phase III (2013–2020), the post-Kyoto commitment period.

The regulatory framework of the EU ETS was mostly unchanged for the first two trading periods of its operation, the beginning of the third trading period in 2013 brings changes in standard rules (based on Directive 2009/29/EC), which should strengthen the system – from the year 2013 the most important yield of the emission allowances is auctioned. Sectorial differentiation was introduced, with (initially) far more auctioning of allowances for energy producers than energy-intensive industries. Also, free allocations were further harmonised, to be based on joint state-of-the-art technology benchmarks (Wettestad et al. 2012, p. 73). Policymakers give firms an incentive to move towards production that is less fossil-fuel intensive (Aatola et al. 2013).

In the last years, CO_2 became a significant member of the European commodity trading market. However, there is a fundamental difference between trading in CO_2 and more traditional commodities. Sellers are expected to produce fewer emissions than they are allowed to, so they may sell the unused allowances to someone who emits more than the allocated amount. Therefore, the emissions become either an asset or a liability for the obligation to deliver allowances to cover those emissions (Benz and Trück 2009).

Generally, the market price of the allowances is determined by supply and demand. However, there can also be other so-called "price drivers". Both in the first and the second trading period, the EU emission allowances were traded mostly on the BlueNext trading exchange (BlueNext 2012). In the third trading period, there has only been one significant exchange which can be used for emission rights trading – European Energy Exchange – EEX (EEX 2018).

EEX has offered to trade of emission allowances on the base of the EU ETS since 2005. EEX currently runs a secondary market for continuous trading on a Spot and Derivatives basis for EU ETS allowances (European Emissions Allowances – EUA, European Aviation Allowances – EUAA) and Kyoto credits (CER, ERU). In addition to the secondary market, EEX conducts large-scale primary auctions of emissions allowances on behalf of the EU Member States as well as for Germany and Poland, held 4 days per week. In the framework of these auctions, emission allowances are issued to the market participants for the first time (EEX 2018).

The EU ETS covers more than 11,000 power stations and manufacturing plants in the 28 EU member states as well as Iceland, Liechtenstein and Norway. Aviation operators flying within and between most of these countries are also covered. In total, around 45% of total EU emissions are limited by the EU ETS (European Commission 2013). The EU ETS includes both European Emissions Allowances – EUAs (since 2005) and European Aviation Allowances – EUAAs (since 2012). The market price of the allowances is determined by supply and demand at the exchange.

Generally, the first period (2005–2007) of the EU ETS was a three-year pilot period for the preparation for the second, Kyoto based, period (2008–2012). Emission allowances were allocated for free (grandfathering), based on the National allocation plans and historical emissions. The first period aimed to establish a carbon market, determine the market price of carbon and build the necessary infrastructure for monitoring, reporting and verifying actual emissions. The data generated from the first period subsequently filled the information gap and helped to set national emission limits (caps) for the second phase. The EUA spot price fluctuated between 25 EUR/t CO_2 at the beginning of the period and the nearly zero level at the end of the period.

The second period (2008–2012) corresponds with the targets set under the Kyoto Protocol. The European Union committed itself to achieve an overall 8% reduction in CO_2 emissions in the period 2008–2012 compared to 1990 levels. Based on the verified emissions reported in the first period, the volume of emission allowances allocated in the second period was reduced by 6.5% compared to the level of Y2005. The EUA spot price fluctuated in the range 6–25 EUR/t CO_2.

The development of EUA price in the first and the second period of EU ETS (2005–2012) is presented in Fig. 15.1.

In the third, post-Kyoto period (2013–2020), the conditions for the functioning of the EU ETS have changed in connection with so-called Climate and Energy Package, based on the amendment of Directive 2003/87/EC by Directive 2009/29/EC. Moreover, the new directive on CO_2 geological storage was adopted, and the European Commission presented the EU's energy and climate change targets for 2020 (known as the 20–20–20 targets). One of these objectives was also to reduce EU greenhouse gas emissions by 20% compared to 1990 levels. Since the EU emission allowances were previously grandfathered, from the year 2013, the significant yield of the emission allowances is auctioned. Grandfathering was widely criticised, mostly because it introduced significant distortions to the EU ETS (Falbo et al. 2013). Auctioning is the most transparent method of allocating allowances and puts into practice the polluter pays principle (European Commission 2013). Sectorial differentiation was also introduced, with (initially) far more auctioning of allowances for energy producers than energy-intensive industries. The development of EUA price in the

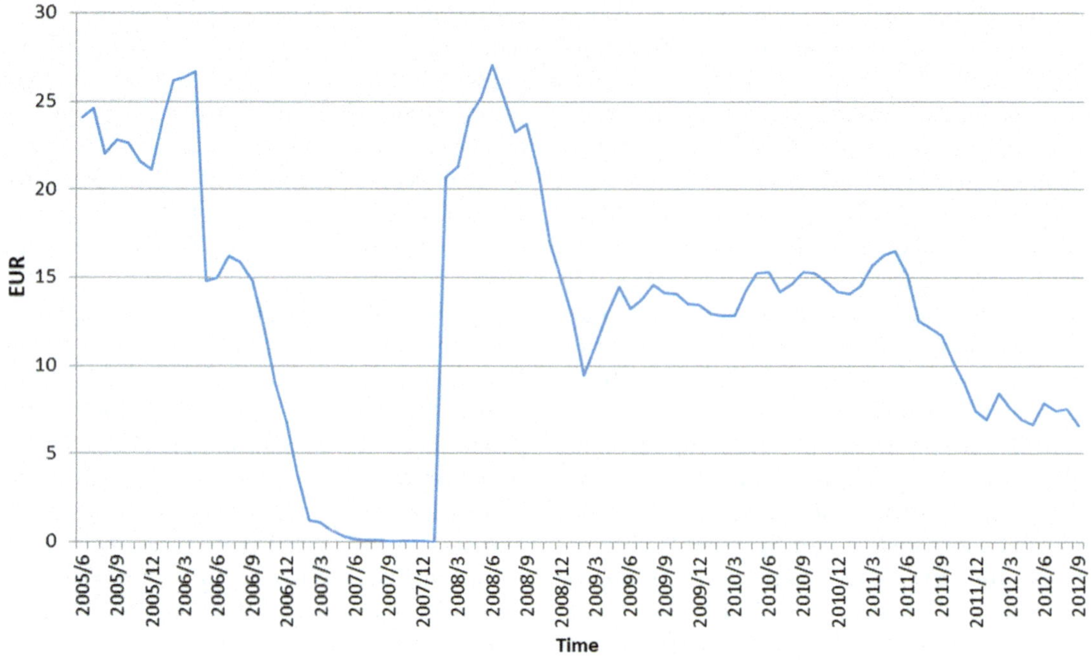

Fig. 15.1 EUA price development 2005–2012. (Source: BlueNext 2012; EEX 2018)

third period of EU ETS (2013–2018) is presented in the following Fig. 15.2.

Currently, the fourth phase of the EU ETS (2021–2028) is prepared, known as the "Post-2020 Reform of the EU Emissions Trading System". At the beginning of Y2018, the fourth phase of the EU ETS has been approved by both the European Parliament and the EU Council. On 19 March 2018, the final text of Directive 2018/410/EU amending Directive 2003/87/EC to enhance cost-effective emission reductions and low-carbon investments was published in Official Journal of the European Union.

The key questions are – (1) Was the amount of CO_2 emissions increasing or decreasing during the previous periods of EU ETS, and (2) Is it possible to observe geographical similarities within the EU countries connected with their economic and environmental indicators?

This chapter focuses on the evaluation of the development of CO_2 emissions and selected economic indicators of EU28 countries in the period from 2005 to 2015 – the year 2005 is the first year of the EU ETS introduction and year 2015 represents the significant year with CO_2 emissions available data. As a next task, the chapter will examine and evaluate possible geographical pattern in the development of selected indicators within the EU. Analysis of a geographic pattern and spatial distribution of countries emitting pollution is essential due to a common geopolitical context of such countries. The chapter will provide a detailed spatial analysis of economic and environmental data of EU28 countries, with the use of (geo)visual analysis of spatial data and spatial statistics (grouping analysis). Obtained results will be presented using analytical maps.

15.3 Methods and Data

For the analysis, the Eurostat database contains greenhouse gas emissions and corresponding macroeconomic data that were used to cover the years 2005 and 2015 (Eurostat, 2018). Namely, all sectors' indirect CO_2 emissions in total, fuel combustion in energy industries, Gross domestic product at market prices, and Gross capital formation. Geographically, all indicators were

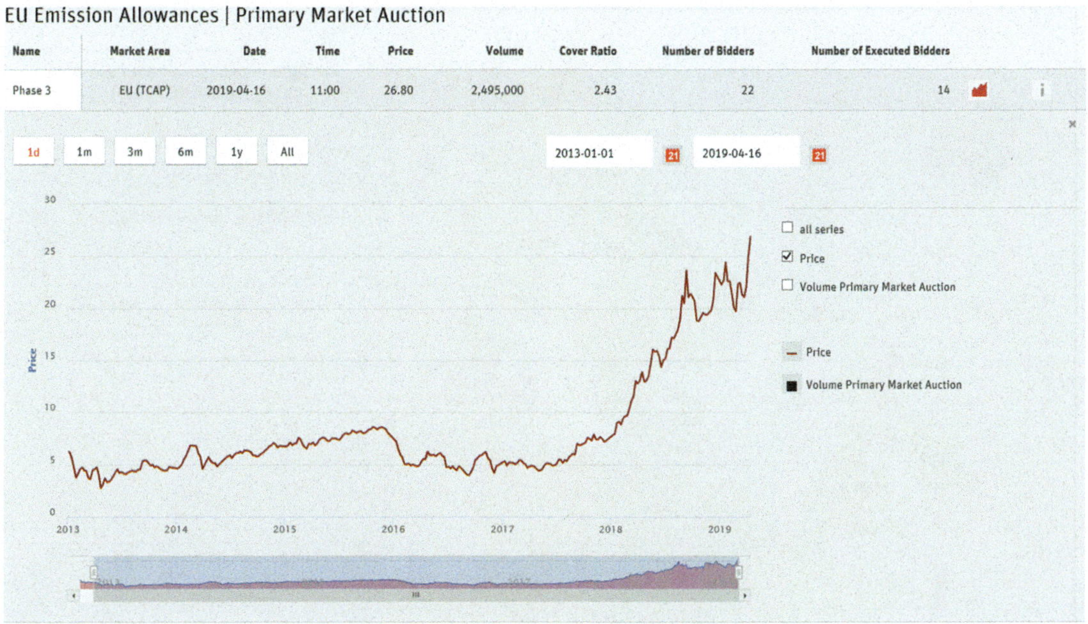

Fig. 15.2 EUA price development 2013–2018 (auction). (Source: EEX 2018)

available on the country level, while some indicators were not available for all EU28+ countries (e.g. GDP for Liechtenstein).

Reference spatial data covering study area of EU28+ countries were obtained from Eurostat as well, specifically from its subordinate unit for geographical data management – GISCO (Geographic Information System of the COmmission). These data represent the last officially valid release from 2014.

The absolute data (instead of relative) were used for the analysis because the initial emission target (the emission cup) was set as % decrease of the total amount of the greenhouse gas emissions. Emission target was set for the EU as a whole – the EU ETS follows a "cap-and-trade" approach: the EU sets a cap on how much greenhouse gas pollution can be emitted each year, and companies need to hold European Emission Allowance (EUA) for every ton of CO_2 they emit within one calendar year.

Geovisual analytics was used to evaluate the development of greenhouse gas emissions and complementary macroeconomic indicators spatially. For this purpose, data from 2005 (EU ETS system came into a force) and 2015 were visualised. Geovisual analytics is described as the science of analytical reasoning and decision-making with geographic information, facilitated by interactive visual interfaces, computational methods, and knowledge construction, representation and management strategies (Andrienko et al. 2007). Geovisual analytics was performed with the use of two cartographical approaches – (1) categories (colours assigned to each qualitative information, or group of information sharing common attribute), and (2) proportional symbol technique (symbol size varies according to the attribute–quantitative measure).

For the first case (categories), Figs. 15.3, 15.4, 15.5 and 15.6, colours were complemented with the number expressing a percentage difference between 2005 and 2015, i.e. yellow colour stands for a decrease, and violet colour stands for the increase. Colours are chosen to stimulate reading the map by highlighting the countries values. As for the proportional symbol technique, for figures in Appendix, intervals were set with the use of Jenks method (natural breaks), which maximise differences between intervals, and at the same time minimise differences inside intervals (Jenks, 1967). Target five intervals were adjusted

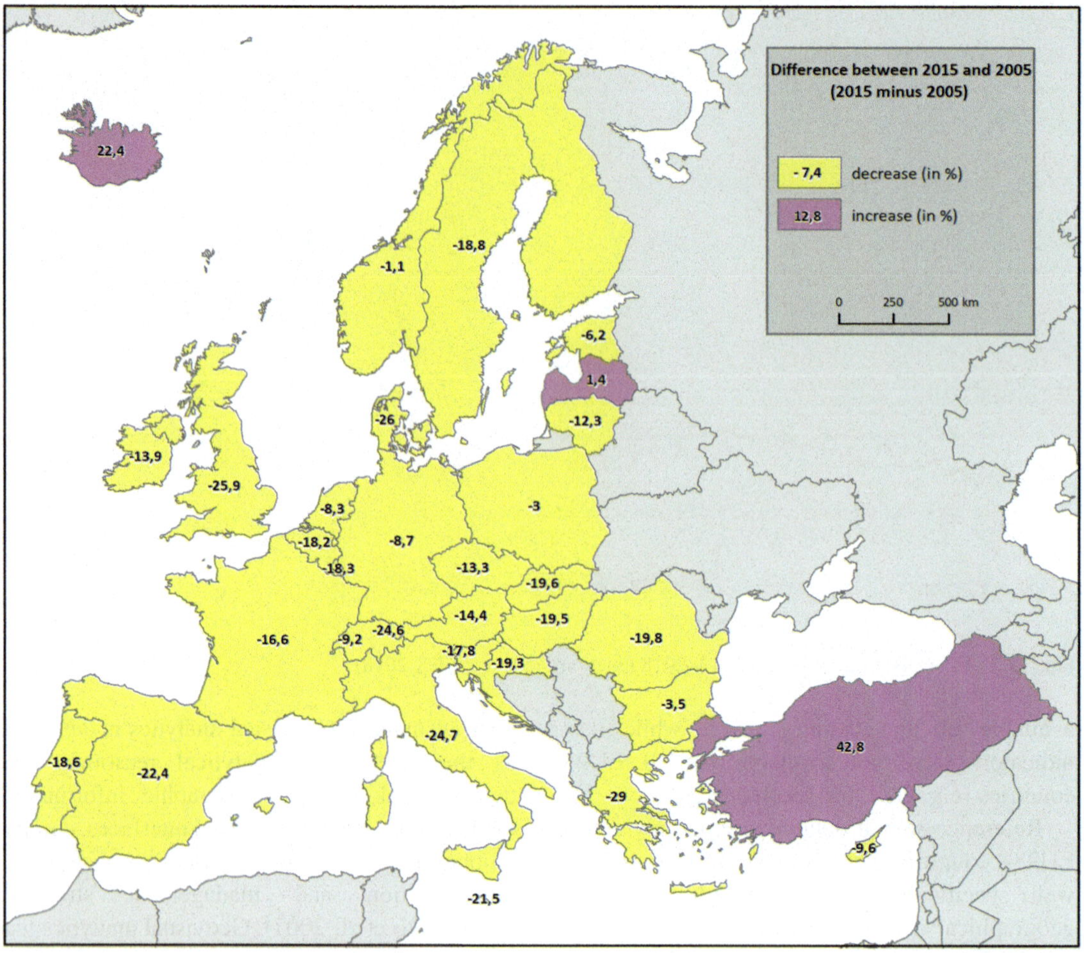

Fig. 15.3 Total CO_2 emissions (difference 2005, 2015). (Source: Eurostat 2018; Authors)

according to cartographical rules for interval border-values (Voženílek et al. 2011). Abovementioned basic methods of thematic cartography allow to display, analyse, and understand source data more efficiently due to the geographical context inherent in data.

15.4 Results

Geovisualisation of greenhouse gases emissions in two reference years – the year 2005 (EU ETS introduction) and year 2015 (after 10 years of EU ETS) is presented in the Appendix, specifically in Figs. 15.7a, 15.7b, 15.8a and 15.8b. These maps show the total amount of CO_2 emissions and the number of emissions in the energy sector. As a relative increase and decrease in the values of monitored indicators is the main subject to evaluation, and do not refer about total sums (values), it is appropriate to take into account also absolute values. This applies mainly in the case of small countries (geographically or economically). Therefore, the Appendix contains maps with the absolute values of the monitored indicators using a mentioned proportional symbols method. It is interesting to confront the (geo)visual analysis presented in the next paragraph with the corresponding maps in Appendix.

Figures 15.3 and 15.4 show the differences in the amount of greenhouse gas emissions in the individual countries involved in the EU ETS

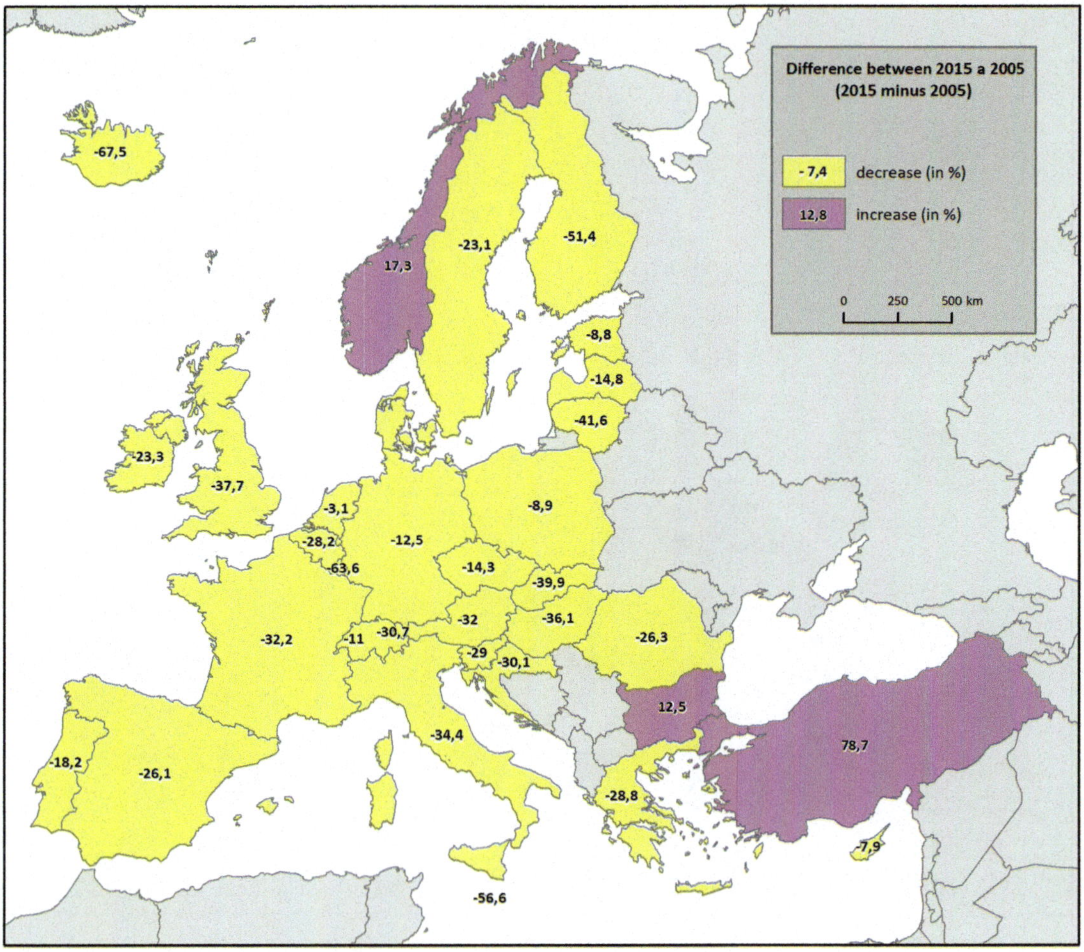

Fig. 15.4 CO_2 emissions in the energy sector (difference 2005, 2015). (Source: Eurostat 2018; Authors)

between 2005 (the year of the EU ETS introduction) and 2015. Firstly (Fig. 15.3) total CO_2 emissions are displayed and in Fig. 15.4, the difference in the amount of greenhouse gas emissions in the energy sector only is depicted. Regarding the development of CO_2 emissions in the analysed period, it is clear that the amount of greenhouse gas emissions were reduced in all countries except Turkey, Iceland and Latvia. Overall, the decrease ranges from 29 (Greece) to 1.1% (Norway). The Czech Republic with a total CO_2 emissions reduction of 13.3% belongs to the group of countries with a smaller decrease. However, neighbouring Poland and Germany showed even lower cuts (3.0, and 8.7 respectively).

On the other hand, Turkey significantly (by 42.8%) increased overall CO_2 production over the 10 years. Interestingly, Iceland, which is characterised by environmental friendliness, showed a 22.4% increase in total CO_2 emissions over the period under review. What's more surprising is that in the production of CO_2 in the energy sector, Iceland is the first in terms of its reduction – a drop of 67.5%. It must be noted that in the case of Iceland, the absolute emission values are low (compared to other countries). That is why it is important also to confront these findings with the absolute numbers displayed in maps in Appendix. The amount of CO_2 decline, as regards only the energy sector (Fig. 15.4), varied from 67.5 (Iceland) to 3.1% (the

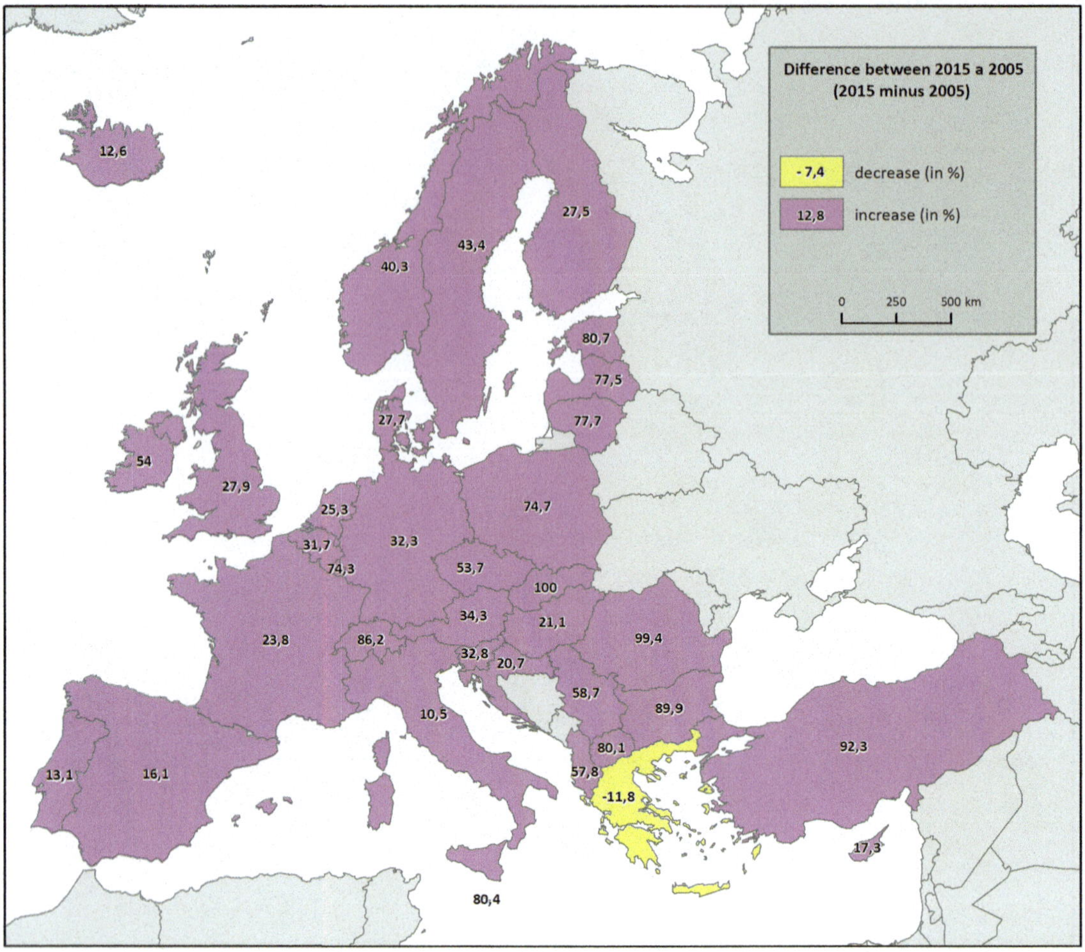

Fig. 15.5 GDP difference 2005 and 2015. (Source: Eurostat 2018; Authors)

Netherlands). The Czech Republic is part of the group of countries with a low decline. On the contrary, emissions in Turkey rose again, up 78.7%, which is alarming. Increase of emissions can also be observed in Bulgaria (12.5%) and Norway (17.3%) but again – taking absolute values into account, it is not that dramatic.

Within the framework of the (geo)visual analysis, selected macroeconomic indicators were also examined. Figure 15.5 shows the difference in GDP as a whole and Fig. 15.6 shows the difference in investments (gross capital formation), again in the period 2005 and 2015 in the European countries (Fig. 15.9a and 15.9b, respectively Fig. 15.10a and 15.10b in Appendix). Gross domestic product (GDP) grew in all states, except for Greece, where a decrease of 11.8% within the observed period is recorded. It should be noted, however, that the growth in the southern European countries (Portugal, Spain, Italy, Cyprus) was the lowest among all other countries (from approximately from 10% to 17%). These low ratios are significant, especially in comparison with the younger EU member states and Turkey – all these economies increased their GDP in tens of per cent (often exceeding 50%). Moreover, Slovakia and Bulgaria have even doubled their GDP in the monitored period. The low performance of southern European countries can be linked with the financial crisis affecting most of the "western" economies and which took place in 2008 (roughly in the middle of the observed

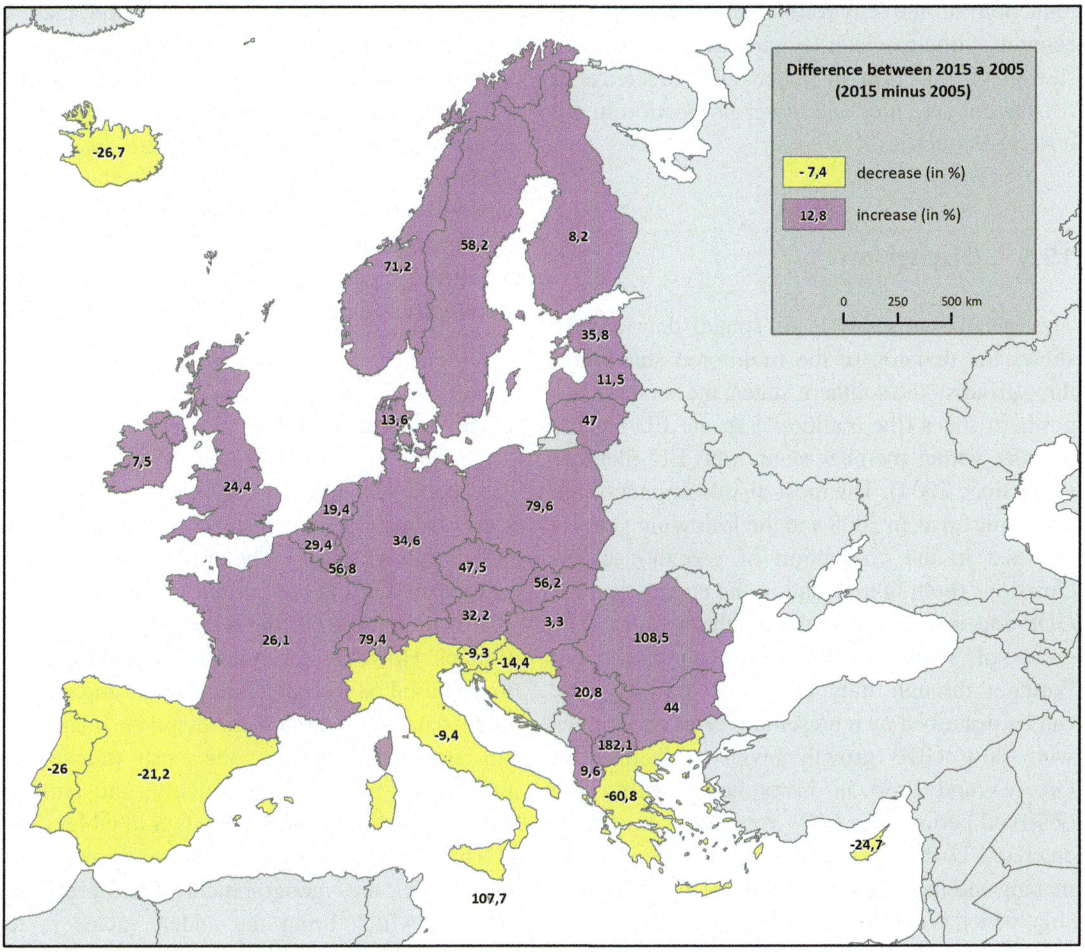

Fig. 15.6 Investments difference 2005 and 2015. (Source: Eurostat 2018; Authors)

period). In traditionally strong economies, the growth was lower (around 20% to 30%) as those countries have a smaller potential for growth in comparison with new member states. However, economies in Switzerland, Luxembourg and Ireland showed an increase above 50 per cent.

As for the investments (Fig. 15.6), the situation is slightly different, but it follows the previous one since this indicator (investments) are connected to the GDP. In many countries, there is a decline in the number of investments, especially in southern European countries - Portugal, Spain, Italy, Greece, and Cyprus. These countries have simultaneous lowest GDP growth (or a decline in the case of Greece). Other countries with declining investments are Iceland, Slovenia and Croatia. However, the cause of the fall in investments in these countries may be different or not so closely linked to GDP growth/decline. Generally, in the traditionally strong economies (e.g. the United Kingdom, France, Germany, Austria, and Benelux countries), the increase in investments is around 19 to 35 per cent. Significant increase in investments (more than 70%) in the 10 years can be observed in Norway, Switzerland, Poland, Romania and North Macedonia. The reasons leading to this increase differs from country to country and should be studied in more detailed together with other socio-economic data.

The key question is, whether the observed decline in CO_2 emissions is related to the reduction of the environmental impact of the economy, due to clean technologies and energy savings introduction, or rather due to decrease in investment activities and lower production in the observed countries.

15.5 Discussion

The (geo)visual analysis of spatial data clearly shows the division of the monitored states into three groups - the southern states, the western and northern states (the traditional strong EU states), and the young member states (the EU member states since 2004). The most visible impact of the economic crisis in 2008 and the following years is reflected in the production of greenhouse gas emissions (both in total and in the energy sector), GDP and investment in the southern states. The group of southern states, i.e. countries from Portugal through Italy and Greece to Cyprus, can be described as representatives of economies with slow GDP growth (even a decline for Greece) and drop in investments within the observed period. The EU's younger states (those entered EU in 2004 and onwards) seem to be making the most of the benefits of EU membership, which is confirmed by both GDP and investment growth (and may be linked to the reduction of CO_2 emissions through the possible promotion of eco-technologies). Traditional EU countries have not such growth potential as the younger states, but the decrease in emissions, as well as the growth of GDP and investment, is also evident in these countries. However, this trend is not as dynamic as in the group of young EU countries.

In the light of the spatial character of this economic instrument, further research on tradable emission allowances should be a more comprehensive spatial analysis focusing on the possible different effects of EU ETS in the individual Member States and a difference or similarity in the cost of avoiding pollution. Moreover, another set of indicators can be added into the analysis (e.g. sectoral employment rates, measures on the quality of life, entrepreneurial data etc.) which would require the application of some multivariate statistical methods. Especially the cluster analysis would be of perfect use since it could be applied both non-spatially and spatially. Following comparison of results might reveal some geographical dependencies among countries and shed light on the whole EU ETS system.

15.6 Conclusions

The chapter describes a very contemporary topic from various aspects – historical, legal, economic and also geographical. The primary methodological approach lies in the "simple" (geo)visual analysis of the indicators' representation in the form of maps. However, the "simplicity" of interpretation is based on a proper cartographic depiction of given data. If geographical displays (maps) are cartographically correct, they can transfer the information more easily, quickly, and comprehensively than presenting the data in tables.

This contribution combines purely economic data with the geographical (cartographical) methods which bring the added values in the analysis of the (spatial) pattern of environmental pollution and related economic issues. I have been clearly shown which regions of EU and associated countries are responsible for high CO_2 emissions, what is their progress in this manner throughout the studied period, and how these emissions are inter-connected with the major (and basic) economic indicators. This study demonstrates benefits of the "spationomy", i.e. the fusion of GIScience (geomatics, geography, cartography and other disciplines) and economy (and its data sources), which could be very effectively employed in any research in the field of spatial exploration of economic data.

Appendix

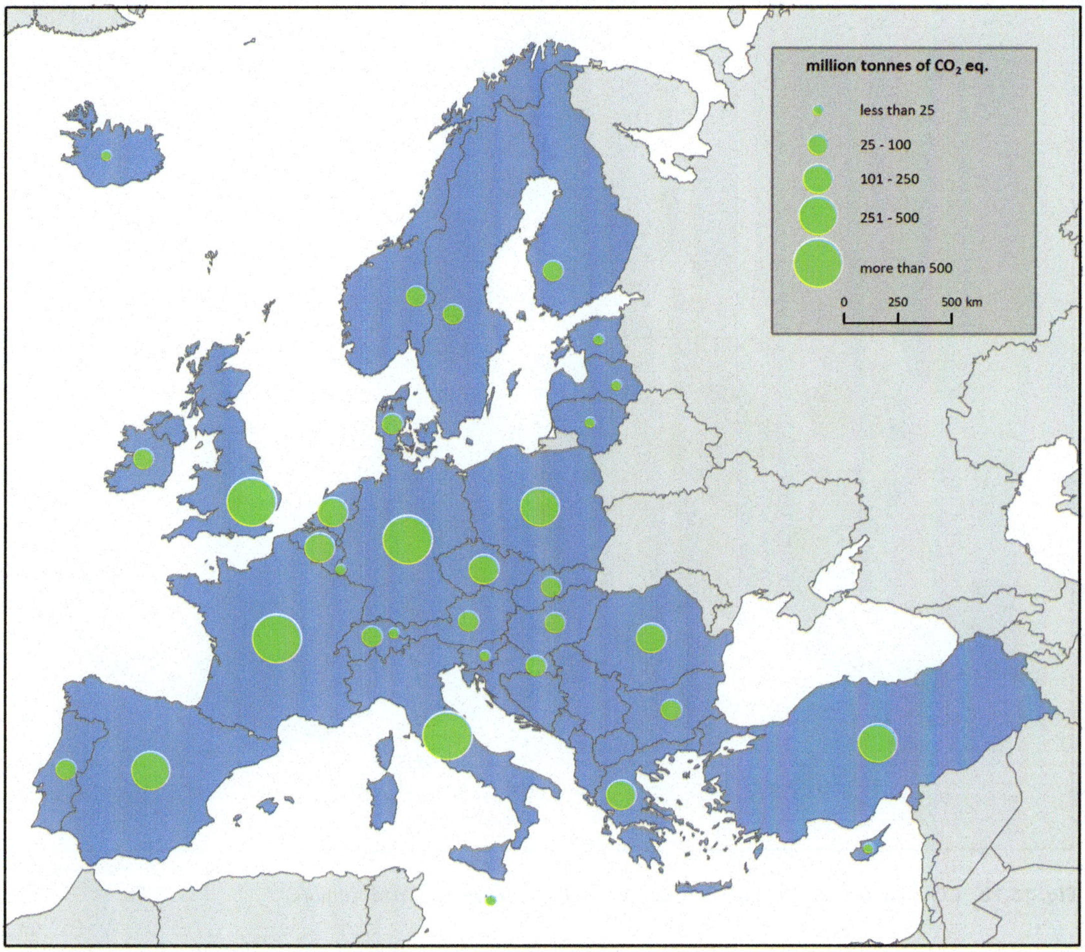

Fig. 15.7a CO_2 emissions in EU countries in 2005. (Source: Eurostat 2018; Authors)

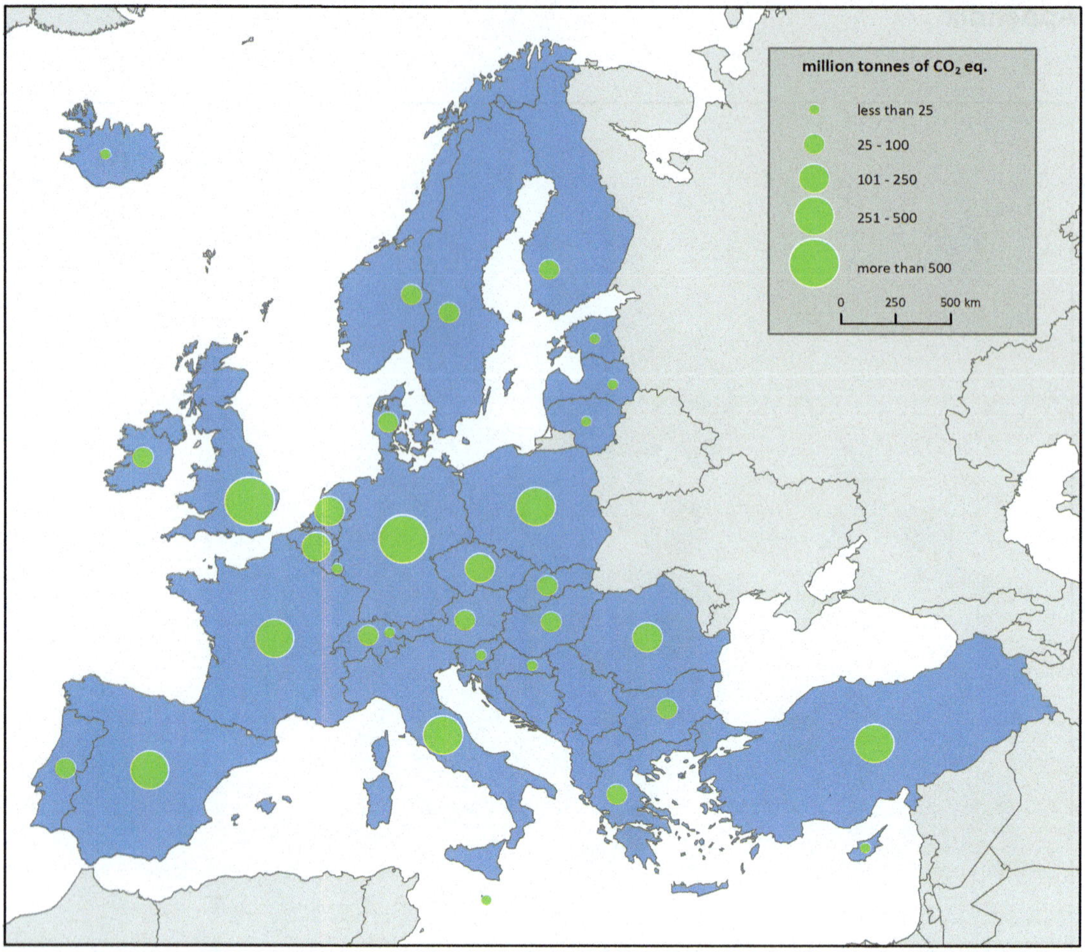

Fig. 15.7b CO_2 emissions in EU countries in 2015. (Source: Eurostat 2018; Authors)

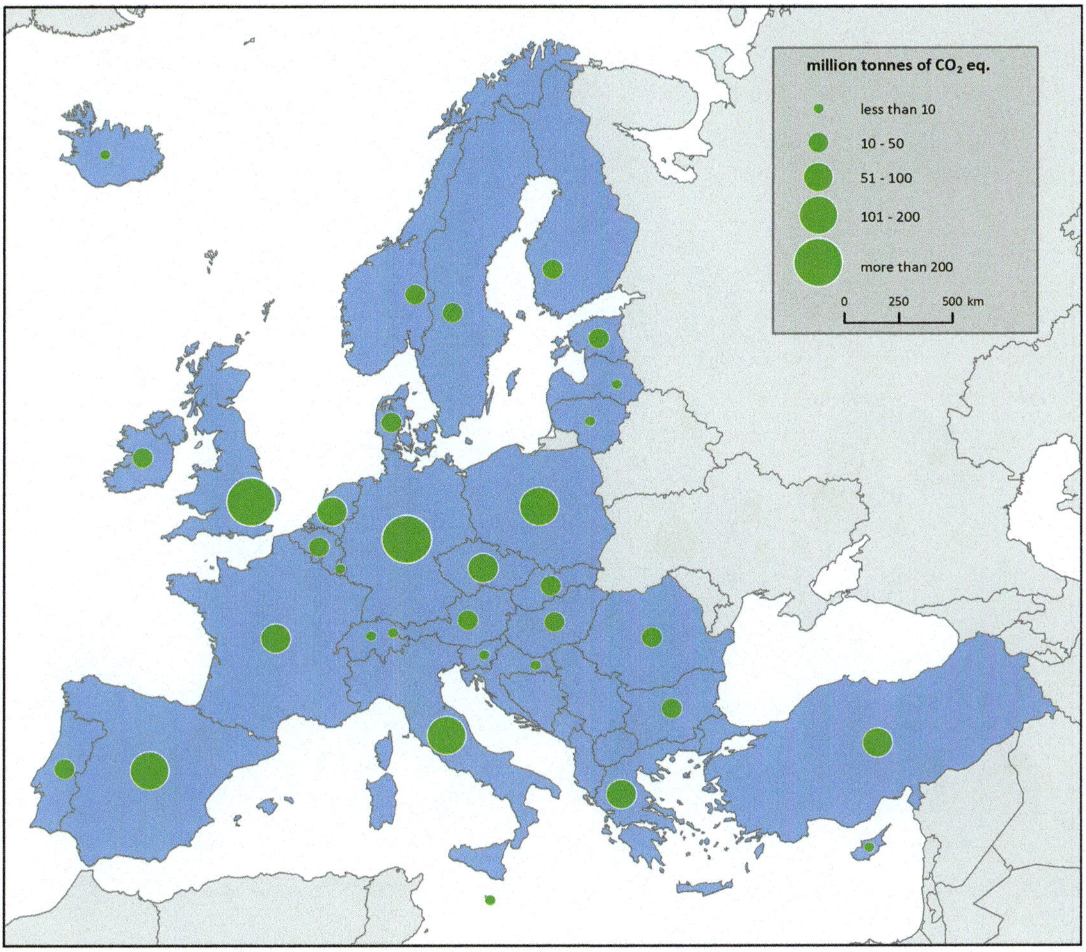

Fig. 15.8a CO_2 emissions in energy sector in EU countries in 2005. (Source: Eurostat 2018; Authors)

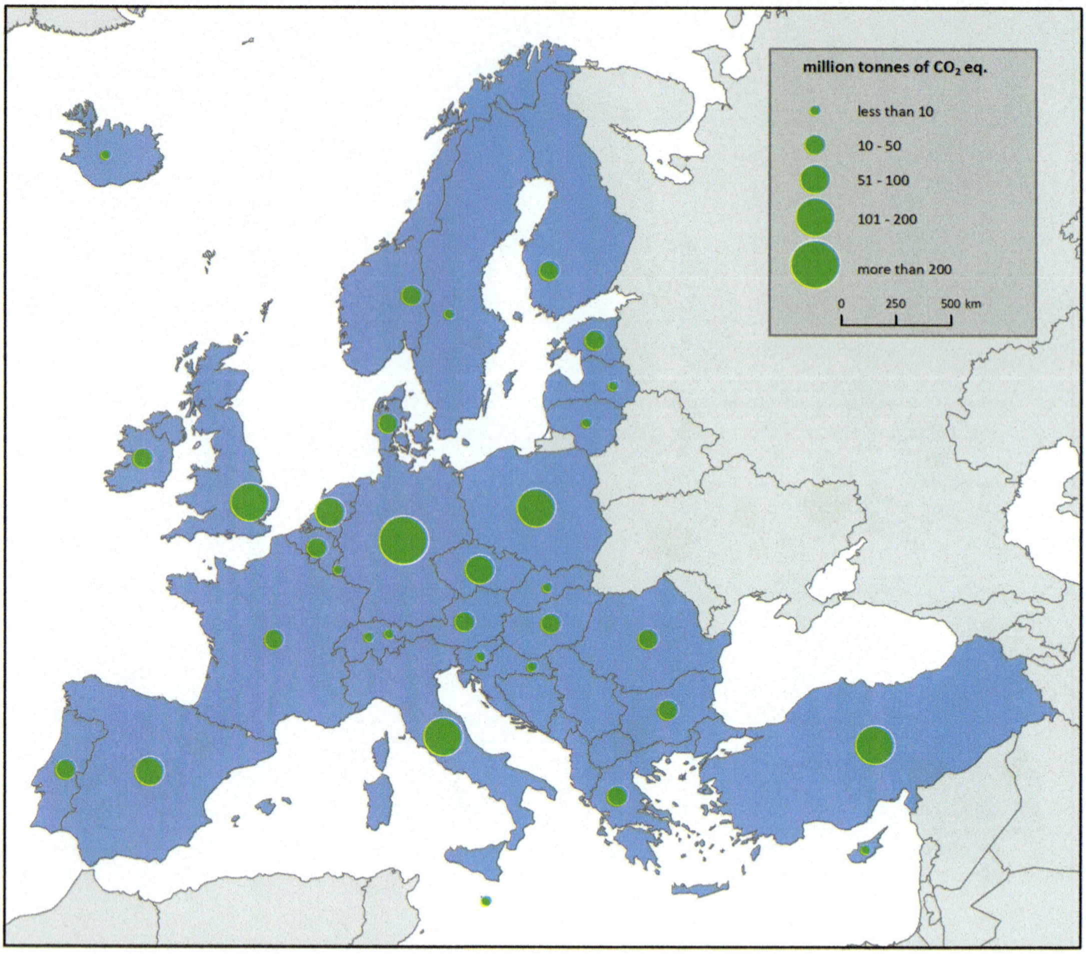

Fig. 15.8b CO_2 emissions in energy sector in EU countries in 2015. (Source: Eurostat 2018; Authors)

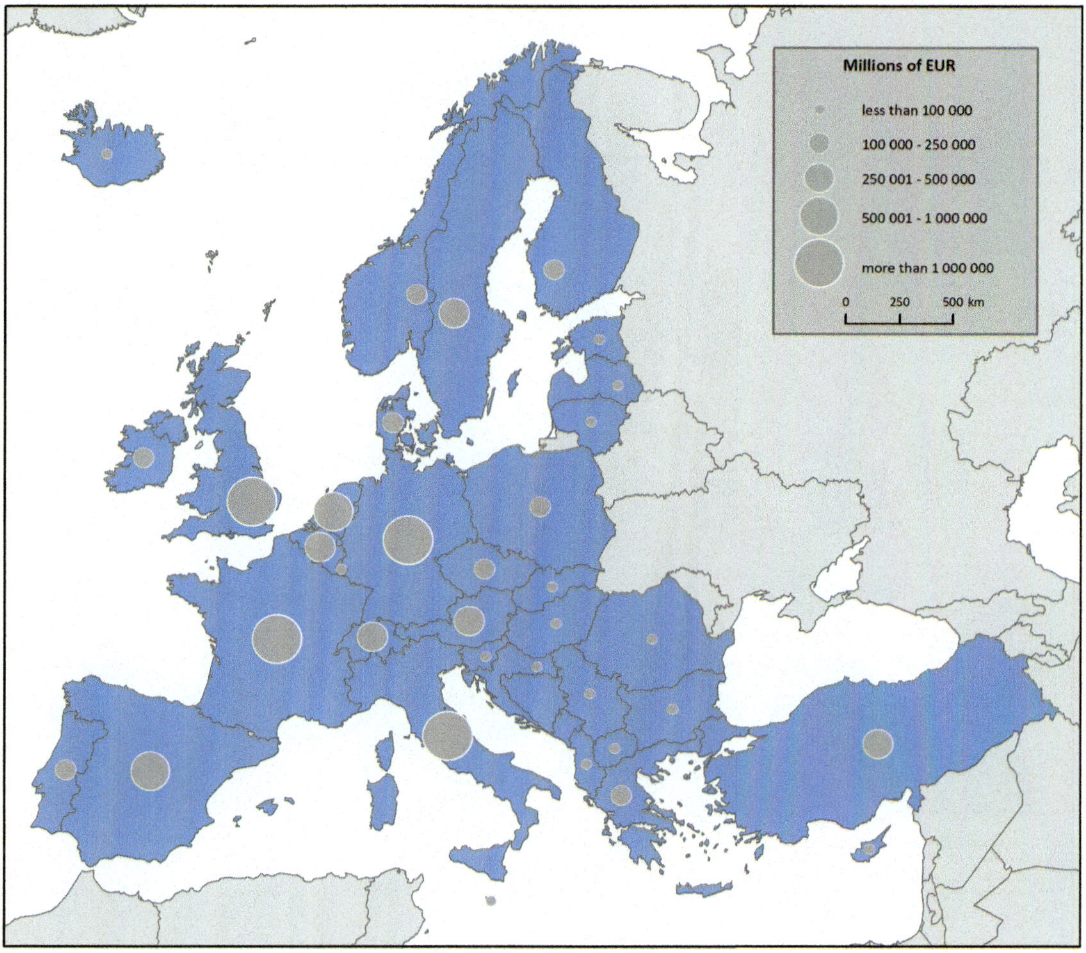

Fig. 15.9a GDP in EU countries in 2005. (Source: Eurostat 2018; Authors)

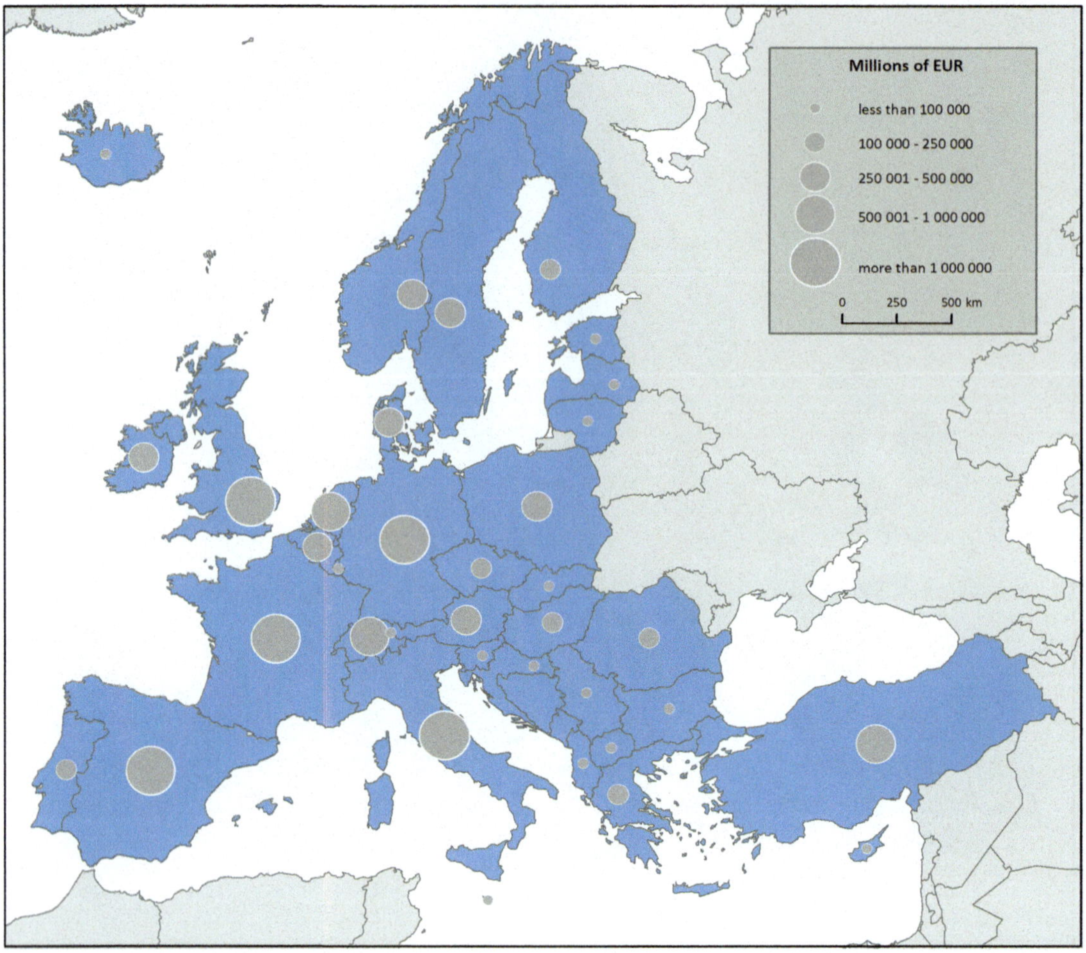

Fig. 15.9b GDP in EU countries in 2015. (Source: Eurostat 2018; Authors)

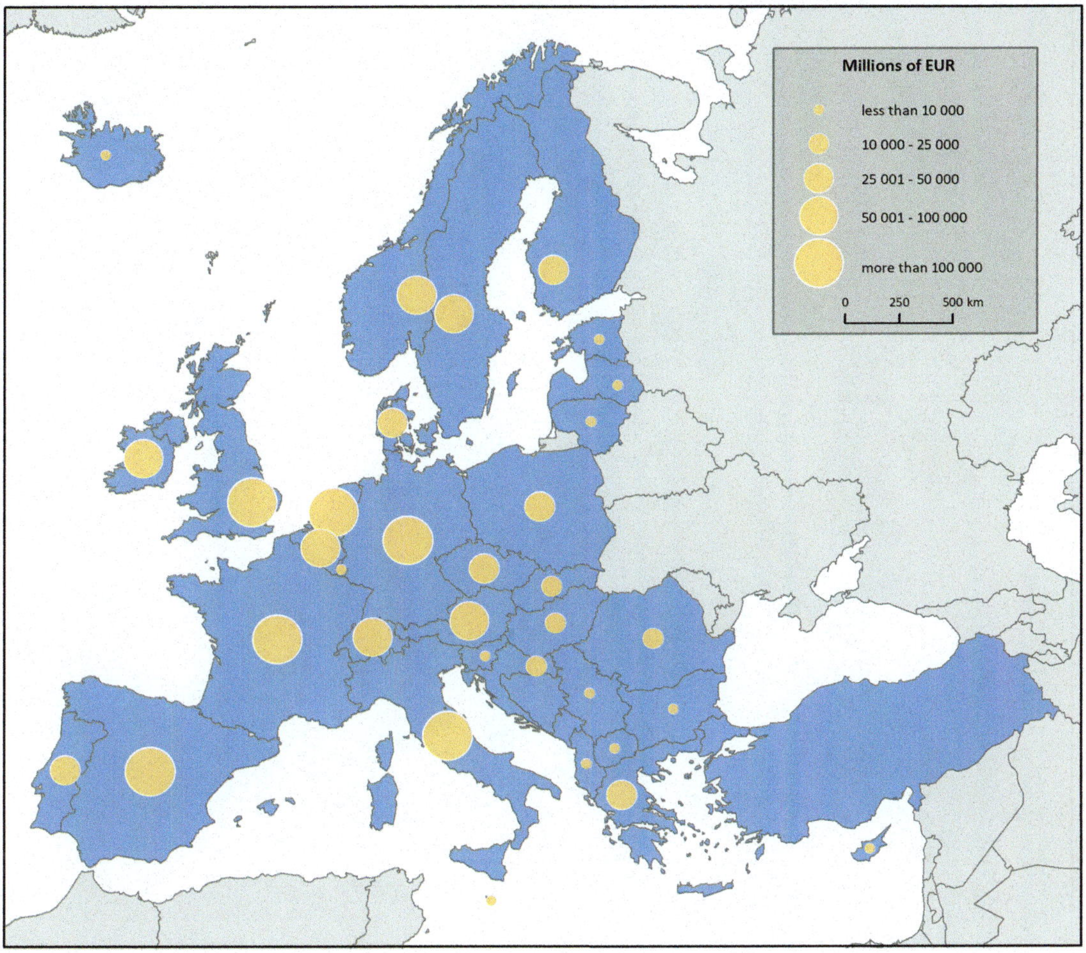

Fig. 15.10a Investments in EU countries in 2005. (Source: Eurostat 2018; Authors)

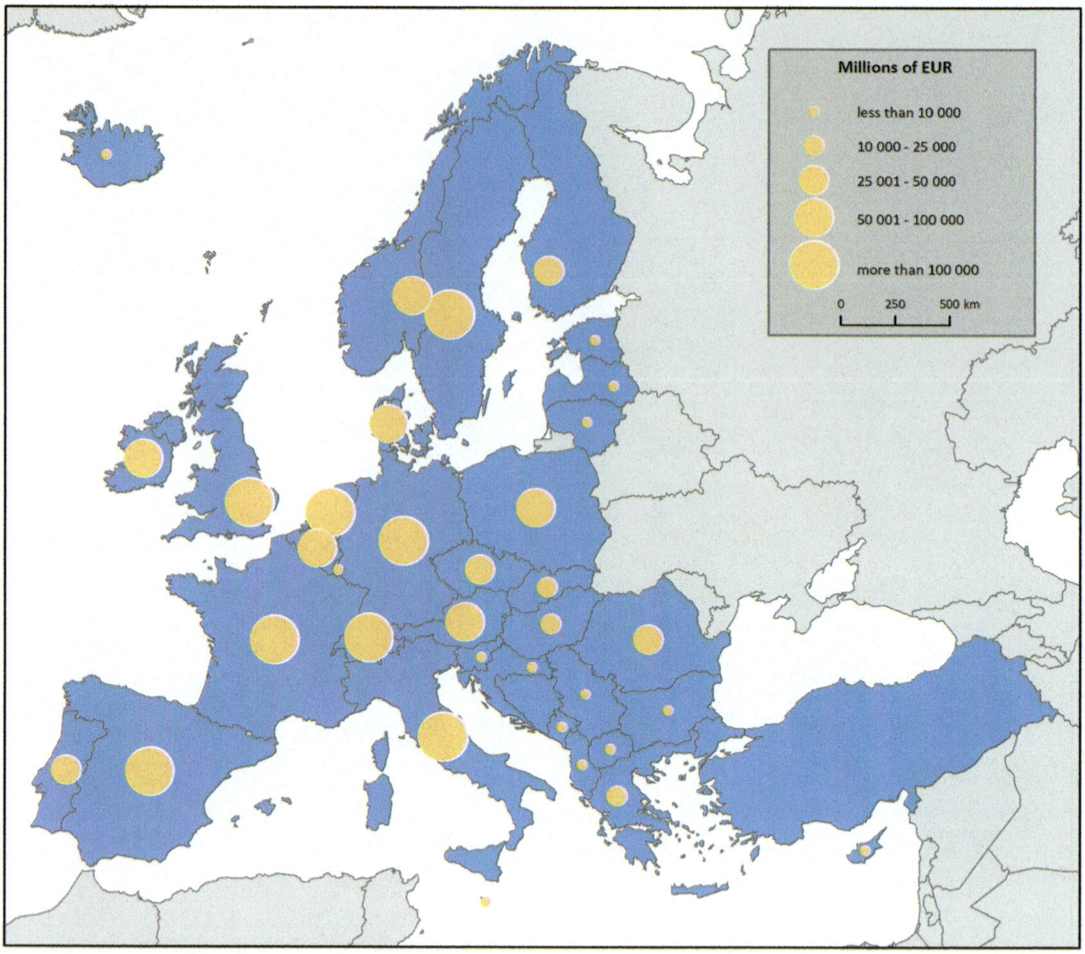

Fig. 15.10b Investments in EU countries in 2015. (Source: Eurostat 2018; Authors)

References

Aatola, P., Ollikainen, M., & Toppinen, A. (2013). Price determination in the EU ETS market: Theory and econometric analysis with market fundamentals. *Energy Economics, 36*, 380–395.

Andrienko, G., Andrienko, N., Jankowski, P., Keim, D., Kraak, M. J., MacEachren, A., & Wrobel, S. (2007). Geovisual analytics for spatial decision support: Setting the research agenda. *International Journal of Geographical Information Science, 21*(8), 839–857.

Benz, E., & Trück, S. (2009). Modeling the price dynamics of CO_2 emission allowances. *Energy Economics, 31*(1), 44–15.

BLUENEXT. (2012). *EU emission allowances databases*. Available at: http://www.bluenext.eu/

EEX. (2018). *EU emission allowances – Prices and trading volumes*. Available at: https://www.eex.com.

European Commission. (2013). *The EU emissions trading system (EU ETS)*. European Union. *October, 2013*. Available at: http://ec.europa.eu/clima/publications/docs/factsheet_ets_en.pdf.

Eurostat. (2018). *Greenhouse gas emissions. GDP and its components*. Available at: http://ec.europa.eu/eurostat/

Falbo, P., Felletti, D., & Stefani, S. (2013). Free EUAs and fuel switching. *Energy Economics, 2013*(35), 14–21.

Jenks, G. F. (1967). The data model concept in statistical mapping. *International Yearbook of Cartography, 7*, 186–190.

OECD. (2002). *Indicators to measure decoupling of environmental pressure from economic growth* (SG/SD (2002)/1/FINAL). Available at: http://www.oecd.org/officialdocuments/publicdisplaydocumentpdf/?doclanguage=en&cote=sg/sd(2002)1/final

Voženílek, V., Kaňok, J., et al. (2011). *Metody tematické kartografie-Vizualizace prostorových jevů.* Univerzita Palackého v Olomouci. 216 p.

Wettestad, J., Eikeland, P. O., & Nilsson, M. (2012). EU climate and energy policy: A hesitant supranational turn? *Global Environmental Politics, 12*(2), 65–84.

Part IV

Playing the Spationomy Simulation Game

Spationomy Simulation Game 16

Vít Pászto and Jiří Pánek

Abstract

The last part of the book is dedicated to the simulation game framework. What is meant by the simulation game within the Spationomy project will be described in the following Sect. 16.1. Conceptual thoughts, as well as the motivation and goals of the Spationomy simulation game, will be briefly introduced as well. Also, the overall theoretical framing about educational aspects of simulation games (or gamification in general) with a focus on geography or economy will be provided with the first section. After this introductory chapter, each of the Spationomy simulation game rounds will be described in detail and also original instructions assigned to the students will be presented (Sect. 16.2). The reason for this chapter is to give a chance to the reader to use, extend and employ these materials in his/her teaching activity. The last two chapters are devoted to feedback and evaluation of the simulation game. Section 16.3 is about teacher/project staff experiences during the creation of simulation game, playing the game itself, modifying it during its evolution and also about "what worked and what did not". In the same section (Sect. 16.3), we will provide the reader with students' feedback and evaluation. Students' authentic commentaries and suggestion are essential for further improvements of the game, and their insights also prove that the whole idea of making spatial economy playful is highly appreciated and fruitful. The main goal of Part IV of the book is to allow anyone to adopt and replicate the simulation game framework in the learning and teaching process.

Keywords

Playful learning · Gamification · Spationomy · Geography · Economy

16.1 Spationomy Simulation Game Concept

One of the Spationomy goals was to assess the potential of playful, experiential and simulation game-based learning in the context of interdisciplinary learning. By doing so, it was needed to establish a simulation game scenario to model real-world problems and their solutions. It turned out that this has been the greatest achievement of the Spationomy project so far and proves the fruitfulness of the project. Simulation game-based learning appears to be much more playful

V. Pászto (✉)
Department of Informatics and Applied Mathematics, Moravian Business College Olomouc, Olomouc, Czech Republic

Department of Geoinformatics, Palacký University Olomouc, Olomouc, Czech Republic
e-mail: vit.paszto@gmail.com

J. Pánek
Department of Development and Environmental Studies, Palacký University Olomouc, Olomouc, Czech Republic
e-mail: jiri.panek@upol.cz

V. Pászto et al. (eds.), *Spationomy*, https://doi.org/10.1007/978-3-030-26626-4_16

and experiential compared to traditional teaching. Hence, participating students and also staff members have an opportunity to test out new methods and create outputs that are necessary to implement those methods. Within the simulation game, students learn and adopt joint methodologies, techniques, and tools. Students serve as actors in an economic/business analytics game with important spatial (geographical) aspects. The game is deployed to structure group-based and student-led investigations of advanced economic data analyses. Students are active agents in stimulating economic/business analytics issues from the real world via a series of game rounds. Unconventional learning in the simulation-based game entails more attractive and relevant pedagogy than lecture or seminar based approaches. Following sections allow readers to inhale basic principles of the simulation game in an educational context and to uncover key features and conceptual settings of the Spationomy simulation game.

16.1.1 Theoretical Background

The idea of the gamification of a learning process is not new in (higher) education. It was famous Czech philosopher and educator Jan Ámos Komenský (John Amos Comenius, 1592–1670), and his well-known work "Schola Ludus" (School by Play) in which he promoted an entertaining way of learning. Indeed, we all know that learning is easier if it contains playful features. It is then more pleasant to acquire new knowledge and to master our skills. If we dare to jump over several centuries into the twentieth century, we can find works and discussions of how simulations (in general) are used or could be used in (geographical) education. For instance, McCormick (1972) discusses the definitions of simulation and explains basic steps in the construction and application of simulation in education. He also mentions the broad definition of simulation formed by Twelker (1970) – simulation is a means for letting learners experience things that otherwise might remain beyond their imagination, a means to practise skills safely and without embarrassment, and perhaps even discover insights into actual problems. An essential aspect in this definition is the fact that simulation is a method by which the learner or participant can be involved at the centre of the problem under investigation (McCormick 1972). McCormick (1972) also notes very interestingly, and indeed a true aspect of simulation – the role of a teacher/supervisor is blurred since there is no central element in simulations, and the main workload is then spread to learners (students). Moreover, participating students are much more "mentally" pulled into the simulation process, so they somehow tend to ignore (or mishear, better to say) orders from a teacher. As the Spationomy playful activity - the simulation game - contains both words "simulation" and "game", it is worth to mention the main difference between "simulation" and "game" (based on McCormick 1972; Abt 1969; Nesbitt 1971; Strum 1969). In general, when playing games, there is usually a set of firm rules based on which a winner is determined. In games, the key agents are competitors (individuals, teams, environments, computers), who strive to reach an objective (e.g. complete a game round task) with actions framed by pre-set rules. Moreover, a game often includes a feature to make it enjoyable for a player, while still serious enough to keep the attention and focus. As indicated above, there could be different types of competitors. Besides an individual or a human team player, we can play against ourselves, environment, or computer (artificial intelligence). Whereas the simulations do not necessarily produce a winner (winners). Simulation is more about the investigation of possible ways to solve a real-world problem. Both games and simulations include a very key element, which is decision making. Allowing a learner to make decisions is the most valuable attribute of playful teaching, and it encourages the participant to think and to act upon his/her conclusions (McCormick 1972). Playful methods, whether or not it is gaming or simulations, can be framed in a concept of "learning by doing", from which a learner can carry off more into his/her reality outside the school environment. Spationomy simulation game represents such a methodological approach. Moreover, it contains all four features of "games" as identified by Crawford (1982): Representation, Interaction, Conflict, and Safety. According to Egenfeldt-Nielsen et al. (2008), Representation means that games model external situations but they are not part of these situations. As Crawford (1982) states, a game subjectively represents a subset of reality. Interaction deals with a level of player engagement, and Conflict (direct or indirect) is the idea that a game has a goal that is blocked by obstacles (Egenfeldt-Nielsen et al. 2008), which are permanently present to stimulate the challenging nature of games. Finally, Safety feature in the games is about the fact that the actions taken during the game result into (game) situations that do not affect a player's real-life (i.e. destroying a car in a virtual race). Egenfeldt-Nielsen et al. (2008) refer to two accepted and short definitions of games. Salen et al. (2004) suggest that a game is a system in which players engage in an artificial conflict, defined by rules, that results in a quantifiable outcome, while Juul (2003) defines a game as a rule-based formal system with a variable and measurable outcome. More about gaming concept can be found in Egenfeldt-Nielsen et al. (2008), thus will not be further discussed here.

Games and simulations as teaching techniques were noticed in the late 1960s (e.g. the pioneering work by Gould 1963) and 1970s, and were treated as an innovative approach in the learning process (e.g. McCormick 1972). In geography, this was followed by descriptions of how games and simulations could be used in (tertiary) geography teaching. For example, Conolly (1981) describes how four different games were modified and integrated into geography classes at that time. Walford (1981) refers to how geographic games and simulations started replacing traditional teaching methods, which appeared to be dissatisfactory. Building on traditions of simulations (e.g. in management and business studies) and using experiential learning (see, e.g. Healey and Jenkins 2000) as a trend that days, Walford (1981) advocates how simulations and games are valuable in the learning process. He points out that games and simulations (1) improve student motivation, (2) create a better teaching atmosphere, (3) help to fulfil more rigorous and ambitious educational goals, and (4) encourage the more effective learning of subject-matter.

On the other hand, Walford (1981) critically argues the disadvantages of games and simulations, mainly its preparatory phase, de-briefing part, and overall fit into a study programme. In the same year as Conolly and Walford published their work, another author is touching the topic of games and experiential learning; King (1981) discusses "to play or not to play" in geography teaching. Interestingly, taking an example from Ancient Chinese proverb, King (1981) emphasise an act of "doing" as a way how to understand a problem under the investigation. In today's words, the "learning-by-doing" approach seems to be applicable with no regards to time (leading us back to the Czech philosopher Comenius and his "school by play"). Nowadays, there are a vast number of research papers and books is dealing with games and simulations connected with (geography in) higher education, and it is not the aim of this section to provide throughout list and review of them. However, it can be observed that modern technologies play an important role in the gamification of lectures. Typical ground-breaking product has been Google Earth and its set of satellite and aerial imagery, although not being a game nor simulation, rather "just" a tool. One of the most known examples of proper simulation games useful for geography teaching could be SimCity (e.g. Kim and Shin 2016) or Transport Tycoon (e.g. Minović et al. 2011, or Raghothama and Meijer, 2013). The latter connects simplified geographical environment, specifically transportation, with the business/economy and local geopolitics. It is also worth to mention a phenomenon developed in 2009 – Minecraft (e.g. List and Bryant 2014; Nebel et al. 2016; Scarlett 2015). This platform can very effectively serve as a learning tool, and Microsoft released Minecraft's educational version in 2016. A detailed study on how Minecraft is used in education and research is provided by Nebel et al. (2016). In the last 5 years, significant proliferation of augmented and virtual reality applications into education has taken place; and it seems that such tools and applications will be a flagship in modern playful teaching methodologies.

As regards (pure) business simulation games, a comprehensive overview of such games is given in an article by Faria et al. (2009). The authors explored papers published in the Simulation & Gaming journal, where they mention the most important advancements in business gaming in the history of the last 40 years before 2008. Authors touch the topics covering technology of business games, how business games were administered, and about current (note in 2008) nature of business simulation games. Interestingly, Faria et al. (2009) note that other researches traced a history of business games back nearly 5000 years when the first board and war games were developed. The first modern business simulation game was developed by Mary Birshtein (Russia) in 1932. The "game" simulated the assembly process at a typewriter factory to train managers on how to handle production problems (Gagnon 1987). On the other side of the Globe, in Northern America, the first business simulation game was launched in 1955, when RAND Corporation (prestigious American research institution) came with simulation application focusing on U.S. Air Force logistics system. The exercise goal was to train participants as inventory managers in a simulation of the Air Force supply system (today, it is a typical example of the role of business managers). Since then, a vast number of business simulation games were developed (including more than 40 versions of Mary Birshtein's game). In 1961, it was estimated that more than 100 business games were in existence in the United States; as for 1969 there were more than 190 business simulation games, and in 1980 around 230 business simulation games in use in the United States (Faria et al. 2009). Faria et al. (2009) also report that in Eastern Europe in 1980, more than 30 business simulations were used, and approximately 200 business games were in use in German-speaking countries in 1985. Table 16.1 summarises the most significant in the history of business simulation games as mentioned by Faria et al. (2009).

As mentioned by Raphael Heath (Head of Geography, Royal High School Bath), in his presentation for Geographical Association Conference workshop 2017 (Heath 2017), the most important reasons to use games in education (in this case of geography education, but it also applies for other fields) are (1) Active and fun lesson activities, (2) Creating a memorable experience for students, (3) Develop a range of skills – discussions, numeracy, teamwork, negotiation, problem solving etc., (4) Promotes thinking skills, (5) Opportunities for gaining knowledge through discovery and experience, (6) Engaging for students and simulating realistic experiences,

Table 16.1 Major business simulation games in the twentieth century

Name	Developer	Year
Birshtein's simulation game	Mary Birshtein (Bureau for the Scientific Organization of Work)	1932
Monopologs	RAND Corporation	1955
Top Management Decision Simulation	American Management Association	1956
Business Management Game	Greene and Andlinger (McKinsey & Company)	1957
Top Management Decision Simulation	University of Washington	1957
New York University Business Game	New York University	The 1970s
Imaginit	Barton	1974
Compete	Faria, Johnstone, & Nulsen	1974
Capstone	Capsim Management Simulations, Inc.	1985/1998 (online)
Marketplace	Innovative Learning Solutions, Inc.	The 1990s (computer version)

Source: Faria et al. (2009)

and (7) Developing increased empathy with issues.

Heath (2017) also identifies some drawbacks connected to the gamification of lectures. It is often (1) time consuming to create a new own gaming lecture, (2) it takes time and resources to produce sets of games, (3) it consumes lots of lesson time and sometimes needs longer than lesson periods, (4) organising the classroom space and design for a game or finding an ample space to conduct it might be difficult, and (5) learners' behaviour management could also be challenging.

There exist plenty of games with educational and geographical context, both online and offline, for example, as board games or in-class-played games. These games strive to encourage learners to adopt terminologies, geographic names, or to improve spatial thinking. Some of the activities are not even games in their true meaning, such as GeoGuessr, Kahoot, Qiuzlet or other quiz-based games. However, we mention three most interesting simulation games, two of them are online, and one is the analogue board game. By accident, the first two of them deal with disaster management, where geography plays a dominant role though; and the third game is about energy transition (from fossil to renewable sources of energy, and energy savings) and decision making in spatial planning. Firstly, the Playgen company, settled in the United Kingdom, created and runs a simulation application called FloodSim. This game focuses on the implementation of various measures (e.g. building on-site flood barriers, rising citizen awareness about floods, applying governmental policies etc.) to prevent and minimise flood damage (Fig. 16.1).

Secondly, the Stopdisasters is an initiative and online game from the United Nations Office for Disaster Risk Reduction (UNISDR). Five game modes/scenarios can be played – hurricane, floods, earthquake, wildfire, and tsunami. Moreover, it is possible to adjust scenarios geographically (e.g. Europe, Asia, Australia) which gives a better context for learners and helps them to understand the risks in their local settings (see floods in the European context in Fig. 16.2). Lastly, there has been a paper recently published by Ampatzidou and Gugerell (2019) about a serious game called Energy Safari to support learning

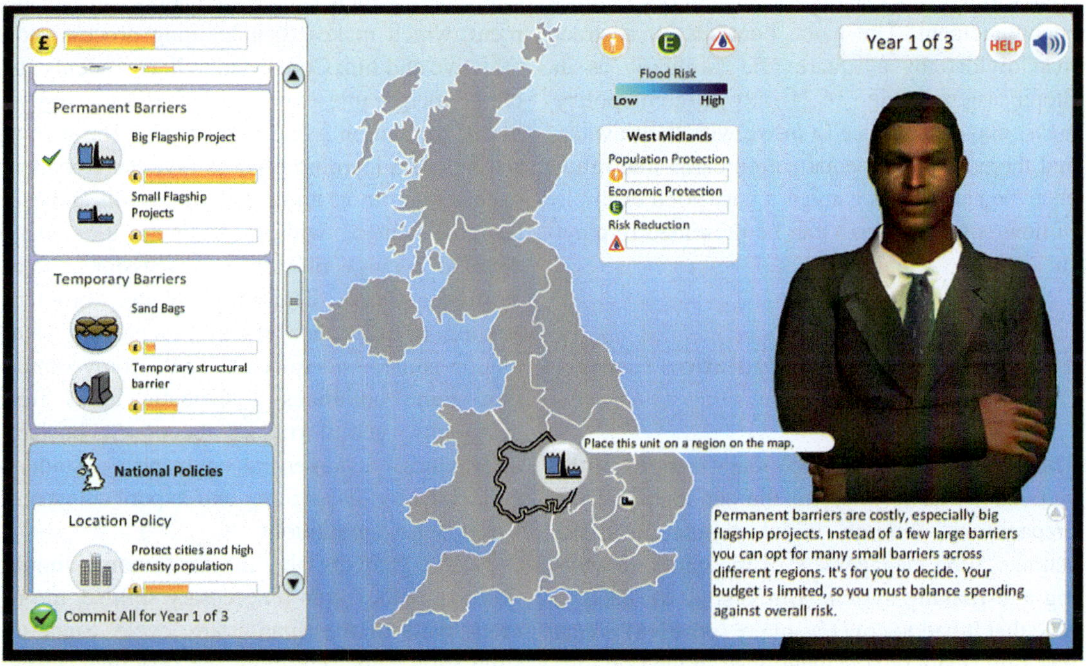

Fig. 16.1 The user interface of the first round in the FloodSim simulation game. (Source: playgen.com/play/floodsim)

Fig. 16.2 Gaming environment in the Stopdisasters game in the topic of floods in Central European settings. (Source: stopdisastersgame.org)

processes in urban and spatial planning when it comes to a renewable energy and energy savings encouragement. The game board is an abstract map divided by a square grid (similarly as in Stopdisasters – Fig. 16.2) into different policy areas, players roll dice to move with their avatar, and there are also special action cards that influence the game. This game is a very good example of how complex topics can be presented in playful settings.

16.1.2 Spationomy Simulation Game Key Features

This section summarises the main features of the simulation game - from the students' teams preparations, through initial simulation game settings, time management of the game, to scoring and ranking system. But firstly, we need to note that it had to cover as many (positive) points made by Heath (2017) mentioned in the previous section. Apart from traditional teacher-focused classes, students were involved in an activity that requires their full participation and engagement, which makes their learning process more effective and fun. Connected to that, students also create memorable experiences by being infused into a simulation game momentum. The gaming settings of a learning process give the students a chance to promote and enrich their skills and team thinking. Often, without knowing it, students develop a range of skills – discussions, brainstorming, team decision making, problem-solving, or time management – by playing a simulation game. And since the Spationomy simulation game combines a thematic focus from geography, geoinformatics, economy, business informatics, management and others, students gain skills by discovering and experiencing real-world issues simulations.

Before students play the Spationomy simulation game, they gain a necessary skill set in previous parts of the Spationomy cycle. Since the simulation game represents the very final activity of the one Spationomy cycle (preceded by an intensive lecture week, semestral project elaboration, and workshops during a summer school), students are stimulated enough to be able to play the game. Students' knowledge and skills gradually increase as the project activities evolve, which allows them to be fully equipped for the game itself. Students are also instructed what is required from them and how they should perform to maximise their learning outcomes. The simulation game represents a final and most comprehensive activity where students apply everything practically what they previously learnt.

From the very beginning of the joint gathering during the preparatory intensive lecture week (a drill part), students are divided to form an international team (a four-member team consists of one student per each partner university). This international setting also enables to work in an interdisciplinary environment, as each student possesses a different specialisation. Moreover, students are also mixed from the perspective of being in various stages/levels of their study. Some of the students just begin their studies, while others nearly finish and graduate, which leads to an exciting interaction within the team when it comes to individual team member roles. On top of it, less experienced students can learn from those with more experiences, and vice versa – younger students often come with less "distorted" ideas by their ongoing studies, albeit sometimes a bit naïve, however somehow inspiring their older colleagues.

The thematic of the simulation game rounds is designed in a way that keeps a balance between "geographical" and "economic" focus. That is ensured by sharing responsibilities within the staff team, so every project partner designs an independent and versatile simulation game round. All simulation game rounds (except the initial setting-up round) are created in a way that allows them to be played at any stage of the game – the rounds are not dependent on previous (or next) round which gives them desired versatility.

From the technical point of view, computers are a necessary precondition of the simulation game. Either more geographically or economically oriented round requires specific software tools. Thus, any means of an electronic device with such tools is inevitable. From the geographical perspective, students mostly work within Geographical Information System environment (such as ArcGIS for Desktop, or QGIS). For the other tasks from the economic and management part, students use table editors (such as Google Sheets, MS Excel, Open Office), statistical/computational software (e.g. R-Project, or SPSS), and for design, a variety of desktop and offline tools are employed. For most of the rounds, there is no limitation on a specific software tool usage and students can decide on their own what tools they prefer.

As regards the simulation game timing, every round lasts around 90 minutes in which students must deliver the required output. The only exception is the initial round that lasted a half-day (i.e. 180 min of lecture time) where students need to solve more tasks, specifically to find an optimal location for their business factory, to decide on a number of employees, or to design their corporate identity. This time settings lead to a two-day gaming session. As indicated above, after each round, students have to report their progress by sharing the required output with game masters (staff members). Depending on a type of the round, students provide a map project, pictures, presentations, results of the deal with another team, or "just" numbers (based on calculations in given rounds) and so on. All these outputs are double-checked by the game masters and creators of the given game round with the use of a simple and single table to keep the scoring transparent and clear (see example in Fig. 16.3).

It must be emphasised that the students' teams compete with each other to win the game. The winning team was ensured by defining objective measures (scores) which serves as a proxy for overall rankings. However, sometimes it is not possible to obtain numeric scores for certain game rounds. Therefore, game masters (staff members) collectively and expertly rank the

Team	Heritage (START)	Size m2	EMP min	EMP max	Empl	Salaries	Operational	Production	Price	Sales	Pollution	Investments	Cash balance
1	3 000 000 €	350	35	105	100	1 119 600 €	951 660 €	1200	2 100 €	3 024 000 €	1200	182 000 €	3 770 740 €
2	3 000 000 €	680	68	204	79	884 484 €	751 811 €	948	2 100 €	2 388 960 €	948	2 126 000 €	1 626 665 €
3	3 000 000 €	700	70	210	200	2 239 200 €	1 903 320 €	2400	2 100 €	6 048 000 €	2400	770 000 €	4 135 480 €
4	3 000 000 €	342	34	102	100	1 119 600 €	951 660 €	1200	2 100 €	2 898 000 €	1200	547 200 €	3 279 540 €
5	3 000 000 €	51	5	15	10	111 960 €	95 166 €	120	2 300 €	331 200 €	120	74 460 €	3 049 614 €
6	3 000 000 €	180	18	54	21	235 116 €	199 849 €	252	2 200 €	637 560 €	252	547 200 €	2 655 395 €
7	3 000 000 €	5574	557	1671	184	2 060 064 €	1 751 054 €	2208	2 100 €	5 332 320 €	2208	1 449 240 €	3 071 962 €

Assets	%	Operational Cash	%	Employees	%	Pollution	%	Location bonuses	%	Cash balance	%	Total		RANK
182 000 €	9%	952 740 €	50%	100	50%	1200	10%	0,2	100%	3 770 740 €		91,2%	310%	4
2 126 000 €	100%	752 665 €	40%	79	40%	948	13%	0,2	100%	1 626 665 €		39,3%	331%	3
770 000 €	36%	1 905 480 €	100%	200	100%	2400	5%	0,2	100%	4 135 480 €		100,0%	441%	1
547 200 €	26%	826 740 €	43%	100	50%	1200	10%	0,15	75%	3 279 540 €		79,3%	283%	6
74 460 €	4%	124 074 €	7%	10	5%	120	100%	0,2	100%	3 049 614 €		73,7%	289%	5
547 200 €	26%	202 595 €	11%	21	11%	252	48%	0,15	75%	2 655 395 €		64,2%	234%	7
1 449 240 €	68%	1 521 202 €	80%	184	92%	2208	5%	0,15	75%	3 071 962 €		74,3%	395%	2

Fig. 16.3 Illustration of the scoring system complexity managed in a simple template using tabular software. (Source: Authors)

students' teams in those specific rounds. In the end, the best team is announced and wins the simulation game.

16.2 Simulation Game Rounds

The initial story that students work with is that they virtually inherit a certain amount of money from their ancestors. They have various tasks (mandatory and voluntary) to invest money from the inheritance. They are obliged to buy retail properties with a specific function. Students use spatial analyses (location and allocation analyses) to choose the best place within the area of interest (city or large town). Simulation game goes on with "accidental" events happening (organised in an individual game round) – such as natural hazards (floods), drops of economic parameters (economic crisis) and so on. They have to apply spatial and economic knowledge, skills, and tools to cope with the newly emerged situation. The simulation game-based learning is focused on the practical application of geospatial tools as well as quantitative economy skills and knowledge to solve given tasks. Each project year/cycle, the game contained several rounds.

In this chapter, we will shortly describe the setting of each round that was played during the first two years/cycles of the Spationomy project. In the next paragraphs, we will present what information was given to the students and what deliverables we expected. During the first year, students had a chance to play four rounds of the simulation game. In the second year of the project, the simulation game was extended and played in six rounds in total. Some of the rounds from year one were also used in the second year, while we added some brand new rounds to enrich the simulation game. The whole simulation game was geographically situated in Olomouc (hometown of the two Czech Spationomy partners) – see overview map for the initial round (Sect. 16.2.1) in Fig. 16.4. The rounds here are presented exactly in the form they were given to students.

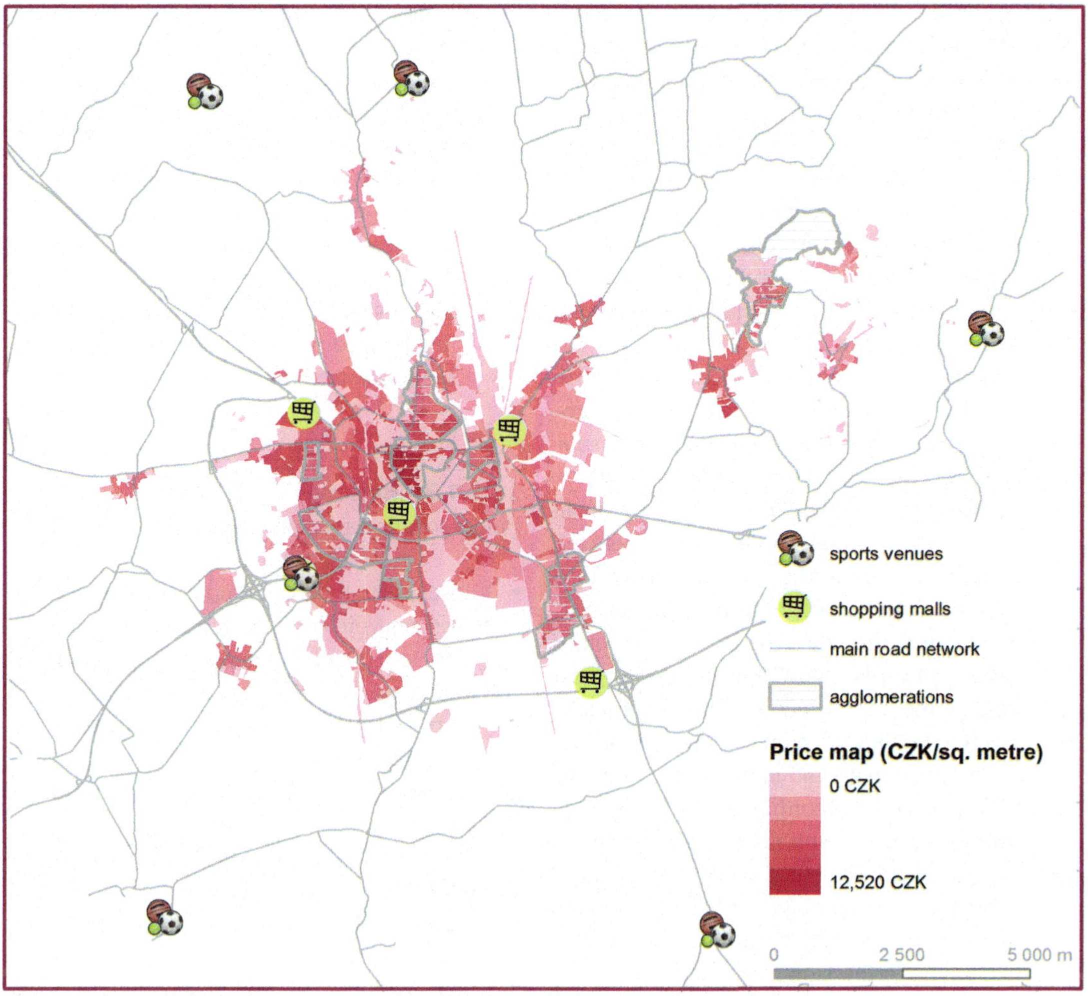

Fig. 16.4 Geospatial setting an overview for the first round. (Source: Authors)

16.2.1 Establishing a Bicycle Company

Each year, this was the starting round where students were allocated a budget, and their task was to find a suitable location for building new bicycle factory, as well as choosing the size/production capacity of the factory. Their initial allocation decision affected the rest of the game and their success in the simulation game. Students were not aware of what are the next rounds/tasks within the game, in order not to suggest any locations within the city.

16.2.1.1 Abstract

You inherited 3,000,000 EUR from your grandmother, who was a successful bicycle racer. Her last wish was that you would establish a bicycle enterprise producing brand new and modern bicycles. The round number 1 of the simulation game is a „set-up“ round in which you will decide on the location, size, production, name and brand (marketing) of your bicycle enterprise. Round 1 is about strategy, (spatial) planning, and teamwork, aiming at the decision-making process within your team. Following rules/tasks are obligatory, and you need to decide which way to go. Your decisions made in this round will have an impact on your business throughout the simulation game. So better think twice.

16.2.1.2 Set-Up Rules

- To build one square meter of your factory costs you an amount according to zones of a price map of Olomouc (see Fig. 16.4)
- Build the enterprise manufacture hall using the zones of price map:
  - you are not allowed to build the enterprise on the roads/railways/rivers/lakes/no-go-zone and other places generally not suitable/accepted (use your common sense)
  - if you plan to build the enterprise in an "open space" area (field, non-urbanized area etc.), you need to pay the price for a square meter according to the price map
  - you can build the enterprise in a place where the other building(s) already exist (according to the aerial image), but you need to buy up the property first (according to the price map) and then pay the price for a square meter according to the price map. In other words – you need to pay twice
  - if your enterprise is located in two or more land prices parcels, you need to calculate the cost by respective parts for each piece according to the land price map
  - if there is no data about the land prices, the median price (1,100 EUR) is paid
- Only one storey manufacturing halls are allowed and should be a rectangular shape.
- Every 10 m^2 need at least one employee, but a maximum of 3 employees per 10 m^2 are allowed (three-shift work).
- You will pay your employees an average monthly wage in the Olomouc region which is 933 EUR.
- One employee can produce one bicycle per month.
- One bicycle is sold for:
  - 2,300 EUR if you are micro-sized enterprise
  - 2,200 EUR if you are small-sized enterprise
  - 2,100 EUR if you are medium-sized enterprise
- Operational expenditures (other than salaries) are 85% of wages per year.
- Factory is producing pollution. The total amount of pollution is: no. of bicycles made per year $*$ 1 ton.
- At the end of the second round, you can change the number of employees.
- Your enterprise will be categorised into three classes:
  - Micro-sized (no. of employees up to 10)
  - Small-sized (no. of employees up to 50)
  - Medium-sized (no. of employees up to 250)

16.2.1.3 Bonuses

Find a location for your enterprise.

- Closeness to the main road network (according to the centroid of the manufacture):
  - up to 100 m – annual income from sales increase by 10%
  - up to 150 m – annual income from sales increase by 5%
  - more than 150 m – no annual income increase.
- Closeness to market (agglomerations, sports venues and shopping malls):
  - up to 100 m – annual income from sales increase by 10%
  - up to 250 m – annual income from sales increase by 5%
  - more than 250 m – no annual income increase

16.2.1.4 Marketing

- Create the name of your company.
- Create a motto for your company.
- Design a logo of your company.
- Create a dashboard for your company.

16.2.1.5 Data

Spatial data are available on the Spationomy website. Economic data are to be derived from your enterprise settings.

16.2.1.6 Deliverables

At the end of this round, you will provide us:

- size and location of the manufacture (ideally a map).
- economic indicators (preferably in Excel):
- number of employees (and sum of their salaries)
- cash balance
- pollution amount (in tons)
- investments (money invested in building your manufacture)

At the beginning of the second round (the other day morning), you will provide us:

- marketing part (name, motto, logo, dashboard)

16.2.2 Floods

Historically, Olomouc was hit by several severe floods in the past 30 years. Hence one of the rounds was the simulation of a natural disaster of this kind. We took the flood zones from 1997 (the largest floods in the modern history of the city) and slightly expanded the flooded areas. This task included two sub-tasks:

- geographical – finding the optimal evacuation route
- economic – calculating the cost of floods on the factory and production

This round combined economical as well as the geographical background of our students, the same way as the whole Spationomy project is designed.

16.2.2.1 Abstract

The floods have come. This disaster was caused by the combination of extreme spring rainfall and saturated soils from previous snow melting. The hazard was magnified by the incomplete anti-flooding system in the region. You need to calculate how much do floods affect your company according to the rules below. Moreover, you must evacuate the employees of your company to the evacuation centre with minimal costs.

16.2.2.2 Rules for Floods Damage

Floods are affecting your enterprise in this way:

- If your company is located in Zone 1 (the most severe damage):
  - your annual production decreased by 50%
  - you need to repair your manufacture with the cost of 20% of your initial investment (i.e. you pay 20% of the manufacture price)
  - your pollution amount decreased by 50%
- If your company is located in Zone 2:
  - your annual production decreased by 40%
  - you need to repair your manufacture with the cost of 10% of your initial investment (i.e. you pay 20% of the manufacture price)
  - your pollution amount decreased by 40%
- If your company is located in Zone 3:
  - your annual production decreased by 30%
  - you need to repair your manufacture with the cost of 0% of your initial investment (i.e. you do not have to pay anything)
  - your pollution amount decreased by 30%
- If your company is located in Zone 4 (no floods at all):
  - your annual production decreased by 10%
  - you need to repair your manufacture with the cost of 0% of your initial investment (i.e. you do not have to pay anything)
  - your pollution amount decreased by 10%

16.2.2.3 Rules for Evacuation

You need to find the cheapest way to any of the evacuation centres:

- If you are in Zone 1, 2 and 3 you must evacuate your employees from the manufacture to the evacuation centre at minimal costs.
- If you are in the Zone 4, you do not have to evacuate any of your employees, but you need to bring them to your manufacture from the evacuation centre at minimal costs, which is then halved.

One employee to evacuate costs you

- 0.5 EUR per kilometre
- 4 Cents per second
- The formula for calculation is:
  - number of employees * (no. of kilometres * 0,5 EUR + no. of seconds * 4 Cents)

Floods are affecting the road network in this way:

- All roads located in Zone 1 (the most severe damage):
  - transportation speed decreased by 90% of the maximum allowed speed
- All roads located in Zone 2:
  - transportation speed decreased by 75% of the maximum allowed speed
- All roads located in Zone 3:
  - transportation speed decreased by 50% of the maximum allowed speed
- All roads located in Zone 4 (no floods at all):
  - transportation speed stays the same as maximum allowed speed

16.2.2.4 End of the Round

- You can change the number of employees
- You need to recalculate all of the indicators

16.2.3 Pollution Allowances

Another round of our simulation game was designed the way so students can experience some action and bargaining about the pollution allowances. The whole round simulated the real events regarding the allowances distribution. According to the evaluation (see below in Sect. 16.4 in this part) the pollution allowances round was one of the most playful and successful.

16.2.3.1 Abstract

Your production is also a source of emissions of pollutants – greenhouse gases (CO_2). Government of your country decided to introduce new environmental legislation. Currently, you are involved in the Olomouc Emission Trading System, and you must decide how you will balance your environmental and economic indicators. The emission limit for the whole area is 70%. There is no time for environmental investments to reduce emissions (e.g. ecological improvements).

You must buy emission allowances in auctions from the state (environmental exchange) to cover all of your emission by allowances. You are obliged to cover all of your pollutions by emission allowances, or you will be penalised.

You are producing the amount of pollution according to your production of bicycle (1 bicycle = 1 ton). Your production of pollution is the same as before floods (your company has fully recovered from the floods).

16.2.3.2 Sub-round 1

Auction (15 min)

- You will join the exchange market and try to buy emission allowances in the auction. The officer will sell only 70% of pollution emission allowances. Your production of pollution is the same as before floods (your company has fully recovered from the floods).
- Firstly, in your team, you should discuss the price you will offer in the auction. The minimum bid is the current price at the real exchange market that is 5 EUR per ton. You should also discuss the number of emission allowances you would like to buy (1 emission allowance can cover one ton of pollution).
- You will offer your bid to the exchange market officer. You do not know the offers of other groups.
- After all, bids are made, the auction will be closed, and the officer will sell the allowances.
- Not all groups (companies) will obtain the emission allowances they need to cover their pollution production – the officer will sell the emission allowances to the bidders with the highest bids (price per ton). If there are two (or more) equal bids, the time/order of the bid is decisive.

- You can also speculate – to buy more allowances than you need and try to sell them to other groups in the next round.
- As soon as you are ready to bid, come to the exchange market officer with your offer.

16.2.3.3 Sub-round 2

Spot Market (30 min)

- As a reminder – you are obliged to cover all of your emissions by emission allowances. In case you have not enough allowances for your emissions, you will pay sanctions to a government regulator. You can try to buy emission allowances from other groups.
- If you have a deal with the other group(s), you have to come to the exchange market officers to register you deal/trade.

16.2.3.4 Sanctions

Olomouc government decided to introduce sanctions per 1 ton of CO_2 for those companies who did not cover all of the pollutions with their allowances. The sanction is 200 EUR per ton. The sanctions will be calculated at the end of this round.

16.2.4 Market Share

The topic focuses on the application of gravity modelling, namely the Huff gravity model within a GIS environment. The tool enables to calculate "service" areas of given event points (bicycle factory locations in this case) and also to include potential new location of another company branch to recalculate the new services areas (market share). The idea of this round was that every team could extend its market share (service area) by moving or splitting their current factory location. However, teams should also take into account the other teams' possible moves, so this round also supports strategic decision making and anticipation of other's behaviour.

16.2.4.1 Abstract

Your company is in the process of expansion, and you want to increase your market share. Your task is to calculate a service area of all the companies using gravity modelling to see what is your market share, i.e. how big is your service area at the moment.

You can temporarily change the location of your company into up to three sub-branches respecting the total "attractiveness" of your company in terms of annual bicycle production. Your market share will be evaluated based on the new customers you gain by changing/not changing the location of your company.

16.2.4.2 Example

The company produces 100 bicycles, which is the attractiveness measure. I want to split it into two sub-branches with new "attractiveness" of 40 and 60 bicycles. And I choose the location for both sub-branches.

Huff Model Settings (for the GIS Plugin)

- Store attractiveness field: production
- Distance calculation: leave blank
- Huff model options
  - distance friction coefficient: 2
  - Generate Market Areas: BOTH
  - Generate probability surfaces: yes
- Origin Locations and Sales Potential – Districts_Olomouc + Pop density

16.2.4.3 Deliverables

After you decide on your strategy, you will provide us with the spatial data (point layer) about the location(s) of your sub-branches with given attractiveness (production number) in the attribute table.

16.2.5 Location for Reseller Shop

One of the most common geospatial tasks is to find an optimal location for placement of a new shop, cash machine, new house etc. Location analysis provides tools to select the optimal location for such purpose based on given restrictions/limits (e.g. 100 m from the road, on a slope less than 10% and so on). Geographical information systems allow performing such analysis very effectively and precisely. This round was mainly geographically-focused, therefore more demanding for "geo" team members of the student teams.

16.2.5.1 Abstract

The sale of your bikes is running not too bad, but since your competitors are not asleep and the different brands are flooding the stores, you had the idea to make the experience of shopping your bikes more special and unique and decided to start selling your bikes in your very own reseller shops.

To find a suitable location for your first shop, there are several requirements that need to be fulfilled to make sure that you reach as many customers as possible. As you cannot be sure that your rival companies have the same idea and the number of suitable sites is limited, there is no time to outsource the site analysis what forces you to do the calculations on your own. And you better do it fast. Sold stores are off the market, and you can only get what is left.

The rush on properties is at this moment declared to be open – hurry up!

16.2.5.2 Requirements and Rules for Finding a Location for your First Shop

- A new shop has to be located in a district with more than 3000 inhabitants per square kilometre
- A new shop has to be located within a distance of 250 m to roads of network class 1, 2 or 3
- Size of the floor area of the new shop has to be between 400 and 1000 m^2
- Permitted types of use for buildings:
  - commercial
  - garage
  - industry
  - other
- Locations will influence each other because the choice of a site by another group can reduce the catchment area of the site chosen by you.
- The customer potential for your site will be calculated in a number of inhabitants within Voronoi polygons processed using the locations of all groups and existing shops.
- The price for the new site depends on the size of the floor area multiplied by the price identified in the price map (round 1).
- The income will be calculated depending on the total number of inhabitants within the catchment areas.
- Sold stores will be displayed immediately on the projector.

16.2.5.3 Deliverables

Once you decide on the location of the new shop, come to the game masters to report your new location (providing spatial data in shapefile or geodatabase).

16.2.6 Investment into Renewable Energy Sources

This round focuses on a decision making the process of the team based on quantitative methods used for investments and their revenues. As this round is focused mostly on economic issues, the "economic" members of the team will take actions. However, decision making must be done collectively within the team.

16.2.6.1 Abstract

Green energy comes from natural sources such as sunlight, wind, rain, tides, plants, algae and geothermal heat. These energy resources are renewable, meaning they are naturally replenished.

As investment in green energy is more accessible than ever before, your task is to invest in selected green energy. Below is data about investments into solar plant and wind power plant. Your company has received two competitive offers for solar power plant and one offer for wind power plant. Accounting department had already prepared investment plans. Cash flows are based on solar plant/wind power plant efficiency (see geo-part for determination of effectiveness depending on the geographical position of your company). As manager of a company, you need to select the best provider (supplier). Expected rate of return (discount rate) is 6%. Take into account that your company can sell the solar/wind equipment as scrap material at the end of investment maturity, for 10% of starting investment amount.

16.2.6.2 Data about Solar Power Plant Investment Options (Alternatives)

Provider A

Your company has received an offer for a construction power plant that will cost 100,000 €. As provider A is known for its quality maturity of an investment is 10 years, meaning that after 10 years the solar plant will stop working. Accounting department had also prepared a plan of cash flows for entire investment maturity (see Table 16.2).

Provider B:

Your company had also received another competitive offer from provider B. Provider B is cheaper, but the quality of their solar plants is also a bit lower, so investment maturity, in this case, is only 7 years. Accounting department had prepared a plan of cash flows for entire investment maturity (see Table 16.3).

Provider C

One of your employees has suggested that there is also a possibility to get in contact with a provider that can construct a wind power plant. You have contacted this provider. Based on the received offer wind power plant will cost you 90,000 €. Cash flows are presented in Table 16.4.

Table 16.2 Investment alternative – provider A (solar plant)

Efficiency based on location	Investment amount	Cash flows	Year									
			1	2	3	4	5	6	7	8	9	10
1/3	−100,000	Revenues (savings)	17,000									
		Maintenance cost	0	0	0	1500	1500	2500	2500	2500	3500	3500
2/3	−100,000	Revenues (savings)	21,000									
		Maintenance cost	0	0	0	1500	1500	2500	2500	2500	3500	3500
3/3	−100,000	Revenues (savings)	28,000									
		Maintenance cost	0	0	0	1500	1500	2500	2500	2500	3500	3500

Table 16.3 Investment alternative – provider B (solar plant)

Efficiency based on location	Investment amount	Cash flows	Year						
			1	2	3	4	5	6	7
1/3	−80,000	Revenues (savings)	16,500						
		Maintenance cost	500	500	1000	1000	1500	1500	1500
2/3	−80,000	Revenues (savings)	22,000						
		Maintenance cost	500	500	1000	1000	1500	1500	1500
3/3	−80,000	Revenues (savings)	29,500						
		Maintenance cost	500	500	1000	1000	1500	1500	1500

Table 16.4 Investment alternative – provider C (wind power plant)

Efficiency based on location	Investment amount	Cash flows	Year									
			1	2	3	4	5	6	7	8	9	10
1/3	−90,000	Revenues (savings)	15,000									
		Maintenance cost	0	0	5000	0	0	5000	0	0	5000	0
2/3	−90,000	Revenues (savings)	19,500									
		Maintenance cost	0	0	5000	0	0	5000	0	0	5000	0
3/3	−90,000	Revenues (savings)	25,000									
		Maintenance cost	0	0	5000	0	0	5000	0	0	5000	0

16.2.6.3 Deliverables

You need to provide us with your final decision supported by your calculations. Then, your calculations will be cross-checked by the game masters.

16.2.7 Spationomy Dragons' Den//Dober Posel//Den D//Die Höhle der Löwen

The Dragons' Den was always the last round, where the main objective for students was to:

1. Present their project in the best light
2. Get the investment from the investors (teachers)
3. Calculate, how much is their business worth, to set the optimal ratio between the percentage they will give to the investor in return for his/her money.

This round was inspired by successful TV series, broadcasted in the United Kingdom under the name "Dragons' Den", in the USA as "Shark Tank", in Czechia as "Den D", in Slovenia as "Dober Posel", and in Germany as "Die Höhle der Löwen", and in more than other 30 countries worldwide. Originally, the TV series format was established in Japan as a "The Tigers of Money". Ordinary people with interesting ideas and business plans presented their products to wealthy investors to get investment for their entrepreneurship.

16.2.7.1 The Rules

Rule 1: The Pitch
Entrepreneurs (Students) must start the meeting by stating their name, the name of the business, the amount of money (up to 1 M EUR) they are pitching for and the percentage of equity they are willing to give away in their company. They must follow this with a pitch of up to 3 min. If it exceeds 3 min, the Dragons (Teachers) can stop entrepreneurs at any point, but they cannot interrupt the initial pitch.

Rule 2: The Questions and Answers
Entrepreneurs DO NOT have to answer all the questions asked, but what they do or do not choose to answer may affect the outcome – for example, if they refuse to reveal net profits. They may ask the Dragons any questions that help them determine whether they are suitable investors for their business.

Rule 3: Opting 'Out'
Also, once a Dragon has declared his or herself 'out', they MUST NOT re-enter negotiation on the deal, and unless there is a compelling reason, they should remain quiet and leave the others to pursue the negotiations.

Rule 4: Investments
The entrepreneur must secure at least the total amount they have asked for at the beginning of the pitch. If a Dragon offers less than the full amount, the entrepreneur must try and make up the total by securing an investment from one or more of the remaining Dragons. Each entrepreneur must leave the Den with at least the full amount they asked for, or they exit empty-handed. The entrepreneur can negotiate more money than was initially requested, as this is usually to redress the sticking point of an entrepreneur giving up more equity than was initially offered.

Rule 5: Multi-Dragon Investments
Each Dragon is working as an individual investor. The Dragons can invest as little or as much of their own money as they want. It is up to the entrepreneur to persuade them to match the required investment or pledge to invest a portion thereof. As above, it is acceptable for the entrepreneur to seek investment from more than one investor to make up the total amount required. A full investment may involve between one and five parties.

Rule 6: Refusing Investments
An entrepreneur can refuse investment from a Dragon if they think they are an unsuitable investor or the deal on the table isn't right for them.

Rule 7: The Deal
The deal agreed on the day is an unwritten agreement that depends on due diligence checks, and relies on the integrity of both investor and entrepreneur to freely enter the transaction and be fully committed to seeing it through. However, the deal is solely between the Dragon and the entrepreneur, and after additional meetings, if an agreement cannot be reached, neither party is legally obliged to complete the deal.

16.2.7.2 Deliverables

Each team tries to negotiate the investment within the most profitable conditions. If the team refuses the investment, it does not affect their overall scoring. This round is evaluated base on a mixture of criteria – e.g. quality of presentation, negotiation style, timing, self-promotion, marketing skills and so on.

16.3 Evaluation of the Simulation Game

This chapter is dedicated to the qualitative assessment of the Spationomy simulation game. The first part is devoted to the internal feedback among staff members and is mainly based on personal experiences with the game concept, and technical and practical findings from playing itself. The second part of this chapter focuses on the evaluation from students which was conducted via an online questionnaire after at least one month of playing the game. We provide a summary of students' answers and feedback, grouping it into four main aspects of the game.

16.3.1 Staff Evaluation

From the process of individual rounds creation, we have to mention that the idea of developing several versatile and self-independent rounds has been very pragmatic. Since all of the Spationomy team members focus on different topics, moreover grouped by their institutions/departmental field of study (i.e. geography, geoinformatics, economy, management, business informatics, quantitative methods), it is a very complex task to coordinate and to develop flawless "chain" of follow-up rounds. Therefore, we applied an approach where each project partner developed their round (or more rounds) that do not need results from the previous round as an input, except the very first initial round and virtual cash balance. Also, every single game round creates some intermediate output/result that is used in a next game round (again, only teams' cash balance is counted every round). Therefore, when developing such complex simulation games requiring interaction and immediate action, the versatility and independence of individual rounds is highly recommended. Besides, it is possible to include a "dramatic" aspect into the game dynamics by drawing rounds by chance.

Another interesting feature of the simulation game, from the staff perspective, was that the students were immediately engaged and keen to play. The reason why playful learning and teaching is perceived to be a more effective method than regular classes is given in Sect. 16.1. But there is another important aspect of explaining students' quick engagement. It is rather simple, coming from the Spationomy project design itself – students play the simulation game at the very end of the project after they went through all the activities that involve cooperation within the team, and supports healthy competition with other teams. Therefore, students feel motivated to play against others and win the game, also due to strong bonds within the team they built during the project.

On the other hand, the simulation game concept represented a new experience for most of the staff members, and many more or less serious problems evolving even during the game itself had to be solved on the spot. With more previous experiences these issues could have been prevented, but fortunately, none of these problems endangered the game. Examples of "hot" issues are listed in the following points which are based on authentic notes and commentaries recorded during the game by the "observing" staff members:

- Can't build a factory in an existing building – not even inside
- Wrong price of bikes
- Cost of pollution – min/max
- 1 round = 1 year
- Bonuses work with at least one location buffer
- Save everything to give us at the end of the game (gdb, pictures, maps, calculations, etc.)
- Flood zones were not 100% real
- Cutting the tree in order to build the factory is at no cost? – should not be
- Do back-ups – computers wipe out every night (in Maribor)
- How to collect and share data – google form + whole project – Dropbox/Google Drive link
- No SW for graphics editing – logos, etc.
- Be EXACT about size!!!
- Some groups did wrong calculations – we should assign points based on that
- Relative criteria of performance measurement!!!!
- Floods according to the DTM
- Zone 4 in floods was missing, so they were not sure, how to intersect without the zone
- Figure out how students can easily deliver the outcomes to us!

All the comments above come from direct observations of the students' playing the game, biases in rules or guides we encounter, as well as from students questions. We now provide a brief commentary on some of the points. For example, the first point about the factory placement – students are smart and creative in the sense of finding gaps in rules, so they asked if they can build their bicycle factory in a quadrangle of an existing block of flats (in the inner part/courtyard of buildings). Although it is a simulation game, we had to avoid such unrealistic actions. We also had to justify several mistakes/uncertainties "on-the-fly", such as the wrong price of bicycles, adjusting values for pollution costs, the timing of rounds, bonuses, figuring out how students will deliver results to game master in the most effective way (via email, USB drives, sharing system etc.). From the technical point of view, we faced difficulty with computers in Maribor (Slovenia) which are set to be rebooted and formatted overnight to be "clean" for another days' lecture. Therefore, every workstation that was used had to be backed-up before leaving the room. Another technical drawback was a software availability – although students were taught in specific software tools by intention during the Spationomy course, they wanted to use the software they are more familiar with. For example, some of the geoinformatics students were more skilled in ArcGIS platform from their regular studies, but due to lack of licenses in Maribor, computers were equipped by open source QGIS. The same happened with a statistical software usage – SPSS platform was substituted for open-source R Project.

In general, from the staff's perspective, the preparatory phase of the simulation game was demanding in the sense of careful design of individual rounds. We had limited experiences to anticipate potential problems if the round tasks are understandable if the data sources are rich enough, and especially the timing of individual rounds could not be estimated. However, in most cases, the time allocation of 90 min was sufficient. Fortunately, it never happened that a team is not able to deliver desired outputs at the end of the round. There was also uncertainty in the time we, as game masters, need to evaluate/score individual rounds. Since we did it after each round, it had to be done quickly, because the final cash balance of the team was needed for next rounds (the cash balance was the only "dependent" interweaving red-line in otherwise versatile and independent rounds). Before the first simulation gameplay, we were unable to estimate the evaluation time, but because of prepared spreadsheets (Fig. 16.3), we managed to provide quick feedback to students (usually within 30 min).

16.3.2 Students Evaluation

After each year, the student evaluation of the simulation game was organised. Students were asked to evaluate their decision-making process within the team, simulation game design; and technical aspects of the game. Furthermore, gender, level of study, and university were collected as optional information. In total, 33 students replied in the survey, 20 students in the first year and 13 in the second year of the project. In general, students did not have much prior experience with simulation games (85% of students).

16.3.2.1 Decision Making

Although it was different in each round, and the answers varied from 53,8% to 84,6%, the students agreed that "Decision making was done collectively (based on majority agreement)". Other options were "The team was indecisive, and we used "trial-and-error" method" and "There was one leader, and we followed his/her instructions (decision made by the leader)". The percentage of the remaining two possible answers changed depending on the structure and complexity of the task. Regardless the decision-making process, majority of the students indicated that they were satisfied with the final decision of their team – on a 5-point Likert scale 72% students marked 4 or 5 in terms of their satisfaction, where one was lowest, and five was highest.

16.3.2.2 Spationomy Balance

The inherent issue of the whole project – how to connect geographical tasks with economic tasks was also present within the simulation game. We aimed to combine tasks in each round, so students feel that they have something to contribute with. Of course, not every round can be designed to involve equally students from two study fields. Therefore we asked them, "How did cooperation between "geo" and "eco" team-members go on (for all rounds)?". The answers surprised us in a positive sense, 69,7% of students stated that "Cooperation and decision making was well-balanced".

16.3.2.3 Using the Knowledge from the Project

In the evaluation form, there were two questions focused on how did and will the students use the knowledge gain through the project cycle. The first question "Did you use knowledge and skills acquired within the previous Spationomy activities for the simulation game?" was answered very positively with 48% answering "Yes" and 52% answering "Some of it", leaving 0% answering "No". The second question focusing of their future use of knowledge gained during the project formulated as: "Did you find the knowledge, skills, and experiences acquired during simulation game useful for your future studies/profession?", 42% of students answered "Yes", 55% answered "Some of it" and 3% answered "Hope so", leaving again 0% for an answer "No". In general, students were positive about the usability of new skills and knowledge gained during the

Spationomy project, which was very motivating feedback for the organisers. Students would also like to have even more links between economic and GIScience activities, as described by one of the students: "Maybe more focus on the interconnection between "geo" and "eco" activities." (male, MVSO).

16.3.2.4 Playfulness of the Simulation Game Rounds

One of the aspects of the simulation game was the playfulness (Poplin 2014) of the process. We have asked students "How playful was each round?" in the evaluation form, and the students had the opportunity to answer again via five-point Likert scale (1 lowest – 5 highest). The average value in the first year was 4.0, in the second year 3.9. So the rounds were considered quite playful in general. Nevertheless, there was a slight difference. While the least playful round (data filtering) had an average value of 3.0, the most playful round (the Dragon's Den) had an average value of 4.5. Students tend to assign more positive values to the tasks, where they interact with other groups or where there is an element of surprise. This is also reflected by one of the students in the evaluation "add an unexpected situation (for example Economic Crisis)" (male, UPOL). When asked, which features did students find the most exciting, in each year the most answers got "decision making – tactics and strategy".

16.4 Summary

This last part of the book was devoted to the Spationomy simulation game concept. First, we provided a comprehensive theoretical background of geographical/business simulation games with a focus on educational purposes in Sect. 16.1. From the literature review, it is clear that gamification of an education process brings added value to the learning and teaching experience to both learners and teachers. If there is a gaming or at least playful feature in a class, it is usually very highly encouraging students attention, engagement, and finally stimulates their learning outcomes. In general, any kind of gaming or playful activity entails more relevant and attractive pedagogy. However, it has to be noted that we mostly refer to "serious" gaming for educational purposes (not gaming solely for personal amusement). Lately, we described the most important features of the Spationomy simulation game concept. We touched the topics of students' teams' preparations, initial simulation game settings, time management of the game, scoring and ranking system and other aspects of the game.

In Sect. 16.2, we shared original simulation game rounds description and task. It is presented exactly in the form that it was given to students, in order to provide an authentic "feeling" about the simulation game rounds. We must mention that oral explanations have always complemented the written game round description. Sometimes, students expressed their uncertainty about the game rounds, so further clarifications from the game round creator were provided. Generally, the game rounds settings and documentation were sufficient enough, so we did not experience any serious issues in playing the simulation game.

Finally, Sect. 16.3 focuses on the evaluation of the Spationomy simulation game. We described a staff reflection on the simulation game. For all staff members, it was the first time experience conducting a simulation game. Therefore, it is more an authentic testimony about the simulation game rather than systematic evaluation. By contrast, the section devoted to students' evaluation is solely based on the formal questionnaire-based evaluation. Staff and students' evaluation together serve as unique feedback on the simulation game concept (and playing the game itself) which definitely helps to develop the game further use.

References

Abt, C. (1969). *Games for learning* (Occasional Paper No. 7). Cambridge, MA: Educational Services Inc..

Ampatzidou, C., & Gugerell, K. (2019). Mapping game mechanics for learning in a serious game for the energy transition. *International Journal of E-Planning Research (IJEPR), 8*(2), 1–23.

Conolly, G. (1981). The use of four selected games in a tertiary geography programme. *Journal of Geography in Higher Education, 5*(2), 121–132.

Crawford, C. (1982). *The art of computer game design*. Berkeley: McGraw-Hill.

Egenfeldt-Nielsen, S., Smith, J. H., & Tosca, S. P. (2008). *Understanding video games: The essential introduction*. New York: Routledge.

Faria, A. J., Hutchinson, D., Wellington, W. J., & Gold, S. (2009). Developments in business gaming: A review of the past 40 years. *Simulation & Gaming, 40*(4), 464–487.

Gagnon, J. H. (1987). Mary M. Birshtein: The mother of Soviet simulation gaming. *Simulations and Games, 18*(1), 3–12.

Gould, P. R. (1963). Man against his environment: A game-theoretic framework. *Annals of the Association of American Geographers, 53*(3), 290–297.

Healey, M., & Jenkins, A. (2000). Kolb's experiential learning theory and its application in geography in higher education. *Journal of Geography, 99*(5), 185–195.

Heath, R. (2017). *Geography games and simulations, geographical association conference workshop 2017*. Available from: http://themapoff.wixsite.com/mapoff/ga-conf-2017-games

Juul, J. (2003). *The game, the player, the world—Looking for a heart of gameness. In Paper presented at the level up—Digital games research conference*. Utrecht.

Kim, M., & Shin, J. (2016). The pedagogical benefits of SimCity in urban geography education. *Journal of Geography, 115*(2), 39–50.

King, R. (1981). To play or not to play: An introduction to games and simulations in geography teaching: A JGHE special feature. *Journal of Geography in Higher Education, 5*(2), 111–112.

List, J., & Bryant, B. (2014, March). Using Minecraft to encourage critical engagement of geography concepts. In *Society for information technology and teacher education international conference* (pp. 2384–2388). Chesapeake: Association for the Advancement of Computing in Education (AACE).

McCormick, J. (1972). Simulation and gaming as a teaching method. *Programmed Learning and Educational Technology, 9*(4), 198–205.

Minović, M., Milovanović, M., & Starcevic, D. (2011, September). Literature review in game-based learning. In *The world summit on knowledge society* (pp. 146–154). Berlin/Heidelberg: Springer.

Nebel, S., Schneider, S., & Rey, G. D. (2016). Mining learning and crafting scientific experiments: A literature review on the use of Minecraft in education and research. *Journal of Educational Technology & Society, 19*(2), 355.

Nesbitt, W. A. (1971). *Simulation games: The social studies classroom* (Vol. 1, No. 1). New York: Crowell.

Poplin, A. (2014). Digital serious game for urban planning: "B3—Design your marketplace!". *Environment and Planning B: Planning and Design [online], 41*(3), 493–511. B.m.: London: Sage.

Raghothama, J., & Meijer, S. A. (2013). A review of gaming simulation in transportation. In *The international simulation and gaming association conference* (pp. 237–244). Cham: Springer.

Salen, K., Tekinbaş, K. S., & Zimmerman, E. (2004). *Rules of play: Game design fundamentals*. Cambridge MA: MIT Press.

Scarlett, M. (2015). Gaming geography: Using Minecraft to teach essential geographic skills. In *Society for information technology and teacher education international conference* (pp. 838–840). Chesapeake: Association for the Advancement of Computing in Education (AACE).

Strum, I. (1969). *Zeiler, R.: Games for school use*. New York: Centre for Educational Research.

Twelker, P. A. (1970). *Some reflections on the innovation of simulation and gaming*. In Paper presented at the first annual conference of the Society for Academic Gaming and Simulation in Education and Training, Berkshire College of Education, Reading.

Walford, R. (1981). Geography games and simulations: Learning through experience. *Journal of Geography in Higher Education, 5*(2), 113–119.

The manufacturer's authorised representative in the EU is Springer Nature Customer Service Centre GmbH, Europaplatz 3, 69115 Heidelberg, Germany. If you have any concerns regarding our products, please contact ProductSafety@springernature.com

Printed and bound by CPI Group (UK) Ltd, Croydon, CR0 4YY
30/09/2026
02225593-0019